Handbook of Spent Hydroprocessing Catalysts – Regeneration, Rejuvenation and Reclamation

Handbook of Spent Hydroprocessing Catalysts – Regeneration, Rejuvenation and Reclamation

Dr. Meena Marafi
Kuwait Institute for Scientific Research
Petroleum Refining Department
Petroleum Research and Studies Centre
P.O. Box 24885, 13109 safat, Kuwait

Dr. Antony Stanislaus
Kuwait Institute for Scientific Research
Petroleum Refining Department
Petroleum Research and Studies Centre
P.O. Box 24885, 13109 safat, Kuwait

Dr. Edward Furimsky
IMAF group
184, Marlborough Avenue,
Ottawa, Ontario, KIN 8G4
Canada

ELSEVIER

AMSTERDAM • BOSTON • HEIDELBERG • LONDON • NEW YORK • OXFORD
PARIS • SAN DIEGO • SAN FRANCISCO • SINGAPORE • SYDNEY • TOKYO

Elsevier
Radarweg 29, PO Box 211, 1000 AE Amsterdam, The Netherlands
Linacre House, Jordan Hill, Oxford OX2 8DP, UK

First edition 2010

Library of Congress Cataloging-in-Publication Data
A catalog record for this book is available from the Library of Congress

British Library Cataloguing in Publication Data
A catalogue record for this book is available from the British Library

ISBN: 978-0-444-53556-6

Printed and Bound in The Great Britain

10 11 12 13 10 9 8 7 6 5 4 3 2 1

Contents

Preface

All indications suggest that at least for next decade the consumption of hydroprocessing catalysts will continue to increase. Consequently, the contribution of catalyst inventory to the overall cost of operation will be increasing. Driving forces behind these increases include more stringent environmental regulations regarding the emissions from the transportation sector. Moreover, increased refining capacity to accommodate heavy and extra heavy crudes, which offset a steady decline in the supply of conventional crude, will also translate into an increased consumption of catalyst and hydrogen. New challenges in petroleum refining will require some improvement of currently used catalysts and/or development of entirely new catalyst formulations. The corresponding spent hydroprocessing catalysts may require additional attention to ensure safety and environmental compliance during all handling stages.

For more than two decades, spent hydroprocessing catalysts have been classified as hazardous solids because of their flammability as well as their ability to release toxic species on exposure to air and in contact with water. The designation "hazardous solid" requires that special procedures have to be applied during all stages of spent catalyst handling, i.e., removal from reactor, temporary storage, packaging, transportation and disposal in landfills. It is the responsibility of the refiner that all these activities are carried out in accordance with the relevant regulations. To ensure environmental compliance, refineries may establish a partnership with companies and/or consortia of companies with expertise in all aspects of spent hydroprocessing catalysts.

All efforts have been made to minimize landfilling of spent catalysts, which is the last option. Today, a large portion of spent hydroprocessing catalysts is recycled back to the operation after being regenerated. Regeneration is carried out by companies who received the necessary certifications from regulatory authorities. The rejuvenation process has been developed in an effort to recycle also the catalysts that are deactivated by metals. For such catalysts, regeneration as the only step may not ensure desirable recovery of catalyst activity. If regenerated/rejuvenated catalysts cannot be used in the original reactor, they can be cascaded to either less or more severe operations. Recently, significant advancements have been made in reprocessing spent catalysts. In fact, performance of the reprocessed catalysts exceeded that of the corresponding fresh catalysts. Apparently, reprocessing gives a new dimension to recycling of spent hydroprocessing catalysts. A number of non-hydroprocessing catalytic applications for spent catalysts may be identified.

Vigorous research activities focusing on the non-catalytic applications of spent hydroprocessing catalysts have been noted. Thus, a high affinity of transition metals, which are part of hydroprocessing catalysts, suggests that after decoking spent catalysts may be used as sorbents for gas clean-up. Attempts to use spent catalysts for water treatment have also been noted. Construction materials (e.g., cement and bricks) as well as specialties, such as abrasives, alloys, ceramics, etc., represent an additional outlet for spent hydroprocessing catalysts.

Spent hydroprocessing catalysts have been attracting attention as potential sources of metals, such as Mo, W, Co, Ni and V. In most cases, the content of these metals is greater than that in ores used for their production. The methods used for the metals reclamation from spent catalysts reached a commercial stage. In fact, after some modifications, the established hydrometallurgical methods used for metals production from various ores and industrial by-products can be also applied to spent catalysts. In recent years, the efforts to improve existing and/or to develop novel metals recovery methods to suit spent hydroprocessing catalysts have been noted. It should be emphasized that the viability of metals recovery from spent catalysts is influenced by demand and prices, which have been exhibiting significant fluctuations.

This book covers all aspects of spent hydroprocessing catalysts in line with the current and anticipated developments in the petroleum refining industry. Thus, more than 700 references cover earlier studies as well as the most recent information. A condensed chapter describes events occurring during hydroprocessing that are responsible for a gradual change of fresh catalysts into spent catalysts, i.e., catalyst structure, feed origin, deactivation, and operating conditions. The properties of spent catalysts requiring attention for environmental and safety reasons are evaluated, taking into consideration existing and anticipated changes in environmental regulations. The objective of the chapters on regeneration and rejuvenation is to give details of those methods that have been used for recovery of activity of spent hydroprocessing catalysts on a commercial level as well as novel methods, which are in various stages of development. The potential of spent catalysts for cascading and reprocessing for subsequent utilization in a petroleum refinery as well as for catalyst preparation for non-petroleum applications have been assessed. Non-catalytic routes for spent catalysts utilization are compared as well. A comprehensive chapter is devoted to metal reclamation, including laboratory studies and commercial processes. A separate chapter evaluates the world market of metals reclaimed from spent hydroprocessing catalysts.

In efforts to improve efficiency of hydroprocessing, novel catalysts are in various stages of development. It is anticipated that these catalysts will be entering the market in the near future. Of particular importance are the spent catalysts containing precious metals and various acidic supports. In this regard, the state-of-art in this area is given in the chapter discussing the catalysts from different stages of dewaxing operations as part of the preparation of transportation fuels and lube base oil.

List of acronyms

API	American Petroleum Institute
ARDS	atmospheric residue desulfurization
AC	activated carbon
CAC	Clean Air Act
CCR	Conradson carbon residue
CERCLA	Comprehensive Environmental Response Compensation and Liability Act
CUS	coordinatively unsaturated site
CWA	Clean Water Act
DAO	deasphalted oil
DOC	dynamic oxygen chemisorption
DTA	differential thermal analysis
EPA	Environmental Protection Agency
EXAFS	extended X-ray absorption fine spectroscopy
FCC	fluid catalytic cracking
FTIR	Fourier transfer infrared
FTS	Fischer-Tropsch synthesis
HDAr	hydrodearomatization
HDAs	hydrodeasphalting
HCR	hydrocracking
HDM	hydrodemetallization
HDN	hydrodenitrogenation
HDNi	hydrodenickelization
HDO	hydrodeoxygenation
HDS	hydrodesulfurization
HDV	hydrodevanadization
HGO	heavy gas oil
HIS	hydroisomerization
HSWA	Hazardous Solid Waste Amendment
HWTF	hazardous waste trust fund

HYD	hydrogenation
KISR	Kuwait Institute for Scientific Research
LM	Langmuir-Hinshelwood
MSDS	material safety data sheet
NAAQS	National Ambient Air Quality Standard
NESHAP	National Emissions Standards for Hazardous Air Pollutants
NPDWS	National Primary Drinking Water Standards
NSDWS	National Secondary Drinking Water Standards
PAH	polyaromatic hydrocarbons
RCRA	Resource Conservation & Recovery Act
RFCC	residue fluid catalytic cracking
SAPO	silica-alumina phosphate
SDWA	Safety Drinking Water Act
SEM	scanning electron microscopy
SMCRA	Surface Mining Control and Reclamation Act
STM	scanning tunneling microscopy
TCLP	toxicity characteristics leaching procedure
TCM	total catalyst management
TEM	transition electron spectroscopy
TGA	thermal gravimetric analysis
THF	tetrahydrofuran
THFIS	tetrahydrofuran insolubles
TIS	toluene insolubles
TPD	temperature programmed desorption
TPO	temperature programmed oxidation
TPP	temperature programmed pyrolysis
TPR	temperature programmed reduction
TSCA	Toxic Substance Control Act
TSDF	treatment storage and disposal facility
TDGA	Transportation of Dangerous Goods Act
VGO	vacuum gas oil
XPS	X-ray photoelectron spectroscopy
XRD	X-ray diffraction spectroscopy

CHAPTER 1

Introduction

Several types of commercial processes for upgrading various petroleum feeds have been developed. They involve either hydrogen addition to the feed or carbon rejection from the feed. A number of carbon rejecting processes (e.g., visbreaking, delayed-, fluid-, and flexi-coking) and asphaltenes and metals separation processes (e.g., deasphalting) have been used on a commercial scale for several decades [1,2]. The hydrogen addition processes require the presence of an active catalyst. For catalytic processes, it is more difficult to upgrade vacuum residues (VR) than atmospheric residues (AR), whereas far fewer problems have been experienced with catalytic upgrading of vacuum gas oil (VGO), heavy gas oil (HGO), and deasphalted oil (DAO). Decades of refinery experience confirmed that the atmospheric distillates can be upgraded without any difficulties. The difficulty and/or severity of upgrading increases with increasing content of contaminants (e.g., metals, resins, asphaltenes, sulfur, and nitrogen) in the feed. The increase in severity results in the increased consumption of hydrogen and catalyst. Compared with thermal processes, hydroprocessing operations are more flexible, giving higher yields of liquid fractions. However, the costs of high-pressure equipment, catalyst inventory and H_2 required for hydroprocessing have to be offset by the increased yields and quality of liquid products.

In an extreme case (e.g., extra heavy feeds), carbon rejection is the route of choice compared with the hydrogen addition. In this regard, there is little information suggesting that petroleum feeds containing more than 300 ppm of metals can be economically upgraded via catalytic route on a commercial scale, in spite of the fact that some hydrogen addition processes have been designed to handle heavy feeds containing as much as 700 ppm of the metals and more than 20 wt.% of asphaltenes [3–6].

The composition of distillate feeds obtained from conventional crude via distillation may differ from that of a similar boiling range distillates produced using carbon rejecting processes. For the latter, the constituents of primary interest (e.g., S- and N-containing compounds) are of a more refractory nature. Therefore, more severe hydroprocessing conditions are required to achieve a desirable level of hydrodesulfurization (HDS) and hydrodenitrogenation (HDN). Also, these feeds have a higher content of Conradson carbon residue (CCR) forming precursors. Therefore, a higher consumption of hydrogen and catalyst may be anticipated. General trends in crude oil supply indicate on a growing volume of heavy crude. The upgrading of the corresponding heavy feeds, such as atmospheric and/or vacuum residues via non-catalytic processes, will increase the volume of such distillates relative to that of the conventional distillates. Then, even in the case of distillate feeds, a continuous increase in the catalyst consumption and the generation of spent catalysts associated with it can be anticipated.

The catalysts used in the refining processes deactivate with time on stream [8–12]. When the activity of catalyst declines below the acceptable level, it must be replaced with either fresh or regenerated catalyst. However, there is a limit on the number of regeneration cycles. Moreover, it is not always economically attractive to conduct regeneration of spent catalysts [7,13,14]. Thus, after several cycles of regeneration and reuse, the catalyst activity recovery may decrease below acceptable levels. Therefore, further regeneration may not be economically attractive. Then, other options for the spent catalysts utilization have to be considered before they are discarded as solid wastes [14,15].

The market demand for hydroprocessing catalysts was estimated to increase with an annual growth rate of 4.4% [6]. Currently, the market for fresh hydroprocessing catalysts approaches 120,000 tons per year. About half of this amount has been used for hydroprocessing of distillates to produce clean fuels, whereas the other half has been used for residue upgrading [7]. The demand for the hydrocracking catalysts, which is currently about 10,000 tons annually, is also expected to grow at a rate of more than 5% per year. Consequently, the production of spent catalysts will be steadily increasing. Therefore, the actual quantity of spent catalysts discharged from different processing units depends largely on the amount of fresh catalysts used and quality of feeds. This depends on the amount of the processed feed per weight unit of catalyst and the amount of deposits formed during the operation. Therefore, the amount of spent catalysts is generally greater than that of the fresh catalyst by the amount of deposits. For example, spent catalysts from distillate upgrading units contain typically 10–20% coke and 7–15% sulfur together with some hydrocarbons carry-overs. Both organic and inorganic forms of sulfur are present. In the case of residue hydroprocessing operations, metals, such as V and Ni, are present in the feed deposit on the catalyst together with coke. If dispersed solids are present in the feed, they deposit on the front of catalyst fixed bed. The spent catalysts discarded from these units usually contain 7–20% V + Ni, 15–25% coke, 7–15% sulfur, and 5–10% residual oil together with active metals (Mo and Co or Ni) and Al_2O_3 originally present in the catalyst. However, the amount of deposit on catalyst may be decreased on the refinery site by applying de-oiling and drying procedures before unloading spent catalyst from reactor. According to the estimate made by Dufresne [7], the total quantity of spent hydroprocessing catalysts generated worldwide is in the range of 150,000 to 170,000 tons per year. Therefore, with anticipated 5% annual increase in catalyst consumption, the generation of spent hydroprocessing catalysts may exceed 200,000 tons annually within a few years.

Besides hydroprocessing, fluid catalytic cracking (FCC) and reforming units may be another source of solid spent catalysts on refinery site. The feeds for these processes must be subjected to hydroprocessing to minimize catalyst poisoning by nitrogen bases and metals. Because of the hazardous nature, the procedures applied for handling of spent hydroprocessing catalysts may differ from those used for the other types of catalysts. For example, particle size of spent FCC catalysts is much smaller than that of spent hydroprocessing catalysts. Therefore, all precautions have to be taken during the handling of the former catalysts.

The reasons for significant increase in the production of spent hydroprocessing catalysts in recent years may be summarized as follows:

(1) A rapid growth in the distillates hydrotreating capacity to meet the increasing demand for ultra-low sulfur transportation fuels.

(2) Reduced cycle times due to higher severity operations in diesel hydrotreating units to meet stringent fuels specifications.

(3) A steady increase in the processing of heavier feedstocks having high sulfur, resins, asphaltenes, and metal contents to distillate by hydrogen addition technology.

(4) Rapid deactivation and unavailability of reactivation process for residue hydroprocessing catalysts.

Disposal of spent catalysts requires compliance with stringent environmental regulations. Spent hydroprocessing catalysts have been classified as hazardous wastes by the environmental protection agency (EPA) in the USA. The EPA added spent hydrotreating catalyst (K171) and spent hydrorefining catalyst (K172) to its hazardous waste list in August 1998 [16] because of their self-heating behavior and toxic chemicals content. In regulatory documents, these spent catalysts are referred to as K171 and K172 spent catalysts, respectively. Spent hydrocracking catalyst was added to the list in 1999 [17,18]. Metals, such as Co, Ni, and V, that are present in spent hydroprocessing catalysts from dual operations (e.g., simultaneous hydrotreating and hydrocracking) are included in the list of potentially hazardous wastes published by the Environment Canada. These metals can be leached by water after disposal and, as such, pollute the environment. Trace metals, such as As, Cr, Cd, Hg, Se, and Pb, may also be present. Besides the formation of leachates, the spent hydroprocessing catalysts, when in contact with water, can liberate toxic gases. The formation of the highly toxic HCN gas from the coke deposited on hydroprocessing catalysts that contains a substantial amount of nitrogen has been reported [19,20]. Spent catalysts also fall under the controlling terms of the Basel Convention and Organization for Economic Cooperation and Development (OECD) rules. According to these rules, spent catalysts cannot be exported to third world countries. Furthermore, the producing works and/or generators have a legal obligation to ensure that their spent catalysts are properly disposed of or safely recycled. The USA pioneered the principle that waste remains forever the generator's responsibility and this is being adopted throughout the rest of the developed world. It should be noted that the 20 years liability in effect before has been replaced by unlimited liability.

Because of the stringent environmental regulations regarding all phases of handling and disposal of spent hydroprocessing catalysts, research on the development of processes for recycling and reutilization of these solid wastes has been receiving considerable attention. Earlier studies on the environmental, disposal, and utilization aspects of spent refinery catalysts have been reviewed by Furimsky [15], Trimm [14], and Clifford [21]. It should be noted that these studies focused on different aspects of spent catalysts each and, therefore, they are complementary. A review by Marafi and Stanislaus [22], which was part I in the series of two, covered spent catalyst waste minimization methods, and utilization to produce useful materials (excluding metal recovery). The information available in the literature on spent hydroprocessing catalyst waste reduction at source by using improved, more active and more stable catalysts, regeneration, rejuvenation and reuse of deactivated catalysts in many cycles, and reusing in other processes were reviewed in detail. Available methods for the preparation of active new catalysts and the valuable products, such as fused alumina, synthetic aggregates, anorthite glass-ceramics, refractory cement, etc., from spent hydroprocessing catalysts are also reviewed in this paper focusing on recent developments. Another review by these authors [23] focused on the progress in research on metal recovery from spent hydroprocessing catalysts and treatment methods for safe disposal as well as on the recent developments in commercial processes for metal recovery from spent catalysts.

The focus of this book is on all aspects of spent hydroprocessing catalysts starting with their generation on refinery site. All commercial activities and/or options, such as transportation, regeneration, rejuvenation, metal reclamation, reprocessing and production of novel materials, storage, and disposal are being evaluated in details. Special attention is paid to the environmental and safety issues, including developments in regulatory affairs. Cursory accounts of the events, which are responsible for the change of fresh catalysts into spent catalysts, i.e., deactivations are presented as well. This is deemed to be desirable because the subsequent treatment of spent catalysts is influenced by the extent of deactivation. In the course of this project, the wealth of information on all aspects of spent hydroprocessing catalysts in the technical and scientific literature has been noted. Growing interests in this topic has been indicated by vigorous research activities in this field, in recent years. After thoroughly evaluating all developments in petroleum refining and relevant environmental and regulatory affairs, it was concluded that it is now a right time to have all this information consolidated in a book. In addition, as the result of more than two decades of involvement, the authors gathered and have access to an only of its kind information from petroleum refineries and in house research. Some of this information was not yet communicated. It is anticipated that the book will serve as a benchmark for developing strategic plans in petroleum refineries. Regulatory authorities will benefit from various environmental and safety issues frequently discussed in the book. The book may serve as the model for undertaking similar projects in other sectors of chemical and petrochemical industry where catalysts may be used in non-petroleum refining applications.

CHAPTER 2

Developments in Petroleum Refining

Decreasing supply of conventional crude on the world market, offset by the gradual increase in the supply of medium heavy and heavy crude, resulted in the significant change in the structure of refinery and refining strategy. In addition, petroleum refineries must respond to a continuous change in the demand of transportation fuel in favor of increased demand for diesel fuel relative to that of gasoline, as it is shown in Fig. 2.1 [24]. Moreover, the feed imbalance (Fig. 2.2) further complicates the situation in petroleum refining. Simultaneously with theses changes, fuel specifications have been continuously evolving, i.e., becoming more stringent. This is illustrated in Fig. 2.3 using the continuous changes in specifications in Europe [25]. These changes require a redesign of refining units and/or development of entirely new refining concepts. Some of these requirements could only be met with an increased consumption of hydrogen and catalysts. In any case, significant additional costs had to be incurred by petroleum refining industry during the change from conventional refinery to the advanced refinery capable of processing more difficult crude and at the same time to comply with all environmental regulations.

It has been realized that the integration of refining operations with non-refining (e.g., utility, incineration, etc.) improved the overall refining efficiency. Moreover, it provided a viable mean for emissions control, particularly in the large size refineries. Indeed, general trends around the world indicate the preference of large size refineries compared with small or medium size refineries. In some situations, the operation of a small refinery was discontinued, whereas in other, a small refinery has been gradually revamped and upgraded to a modern type of refinery [26].

Table 2.1 [27] shows the properties of several well-known crudes. The differences in properties influence the extent of refining. A conventional refinery, requiring little modification, may be suitable for processing light crudes such as Ekofisk and Arab Light. The higher content of metals in Kuwait export and Arab Heavy crudes would require a redesign of catalytic reactors in the case that further processing of distillation residues is considered. Advanced refining schemes have to be capable of processing heavy crudes such as Maya, Boscan and Cold Lake crudes. For such heavy feeds, the choice of the primary upgrading step, i.e., hydrogen addition versus carbon rejection may be critical.

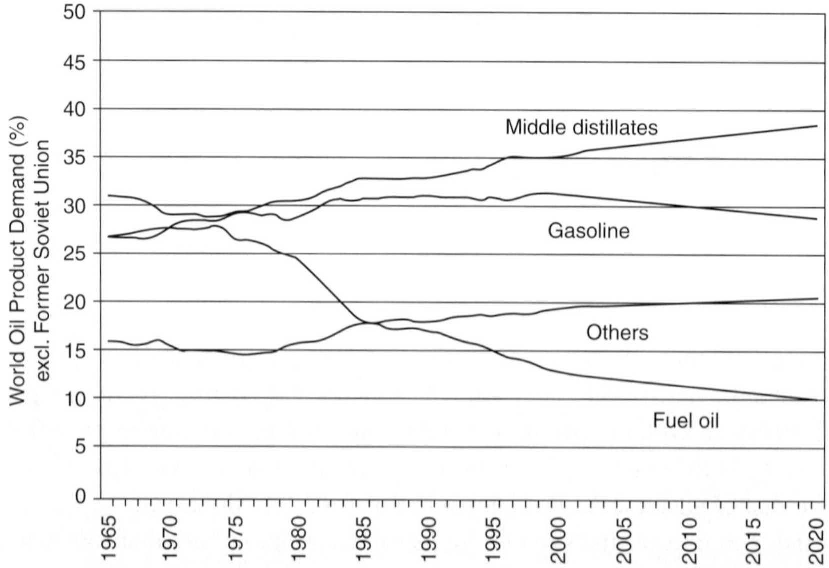

Figure 2.1: Trends in world demand for transportation fuels [From ref. 24. Reprinted with permission].

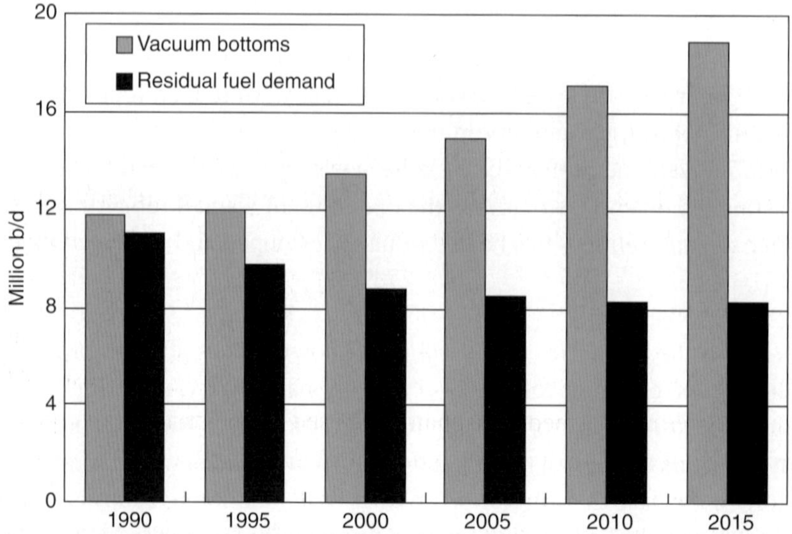

Figure 2.2: Trends in vacuum bottoms supply and demand for residual fuels [From ref. 24. Reprinted with permission].

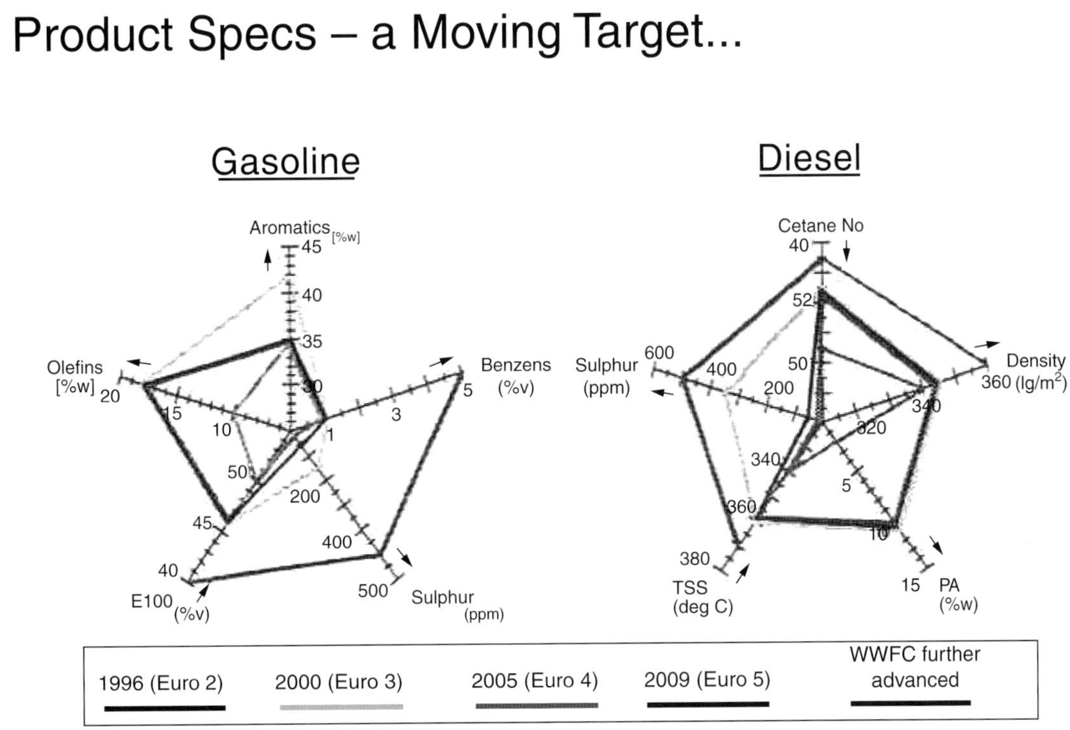

Figure 2.3: Trends in fuel specifications in European Union [From ref. 25. Reprinted with permission].

2.1 Conventional Refinery

Simple refining schemes such as shown in Fig. 2.4 were adequate for processing light-sweet crude. They involved atmospheric distillation to obtain fuel fractions. If desirable, vacuum distillation of atmospheric residue yielded vacuum gas oil (VGO), which could be either the additional source of fuels or used for production of lubricating oils. In early stages of refining, the hydrocarbon gases from distillation were simply flared, in spite of their high heating value. Today, in most of the refineries, the hydrocarbon gases are efficiently utilized on site (e.g., in preheaters). In some site-specific situations, it was advantageous to use atmospheric residue as fuel to produce steam and electricity. In other situations, vacuum distillation yielding additional liquids such as VGO and vacuum residue were added to the overall refining scheme. Vacuum residue, sometimes termed as asphalt, was used as construction material. In the case that market for asphalt dried out, a delayed coker was added to produce additional liquids from the asphalt. The liquids from coking had to be stabilized via a catalytic step. The coke from delayed coking was suitable for production of various carbon products such as carbon electrodes, activated carbon, etc. The surplus of delayed coke was utilized on refinery site to generate steam and electricity via combustion.

Table 2.1: Yields of atmospheric (345 °C) and vacuum residues (565 °C) as volume percent of crude [From ref. 27. Reprinted with permission].

Crude	Arab Light	Arab Heavy	Maya	Boscan	Cold Lake	North Sea Ekofisk	Kuwait export
Yield							
345 °C	44.6	53.8	56.4	82.9	83.1	52.6	45.9
565 °C	14.8	23.2	31.2		50.0	18.0	21.8
Density (kg/L)	0.86	0.89	0.93	1.04	1.00	0.88	0.89
Sulfur (wt.%)	1.8	2.9	3.8	5.2	4.9	0.4	4.1
Nitrogen (wt.%)	0.1	0.2	0.3	0.5	0.6	0.2	0.4
Vanadium (ppm)							
In crude	18	50	273	1220	160	4	55
In 345 °C	40	93	484	1470	190	8	120
In 565 °C	120	215	870		320	22	250
Nickel (ppm)							
In crude	4	16	50	120	80	2	20
In 345 °C	9	30	90	145	96	4	43
In 565 °C	27	70	161		160	11	90
CCR (wt.%)	3	7	15	20	19	4	11

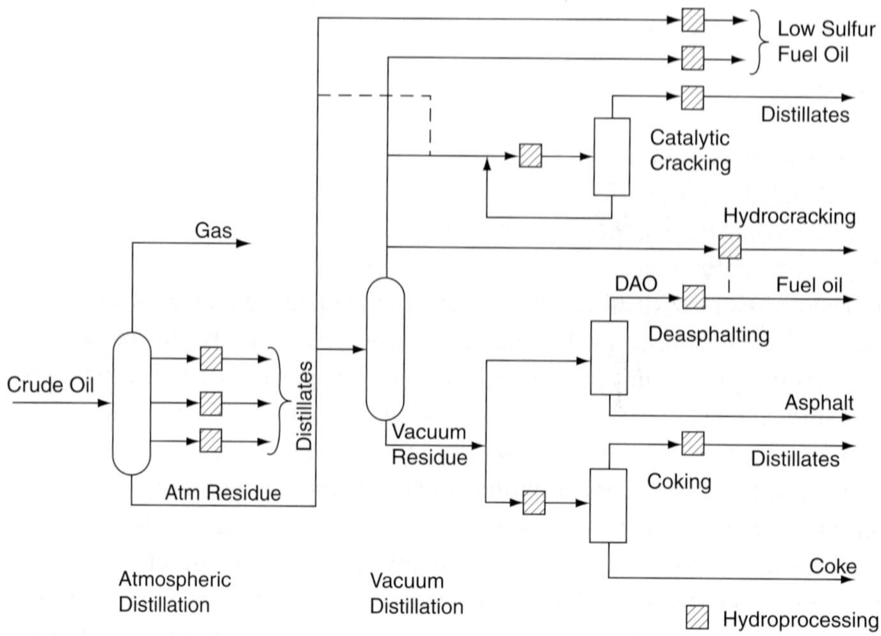

Figure 2.4: Simplified flowsheet of petroleum refinery.

Deasphalting of atmospheric and vacuum residues is another option for producing additional liquids. However, deasphalted oil (DAO) requires additional hydroprocessing to be suitable for further utilization, e.g., as the feed for fluid catalytic cracking (FCC), lube base oil production, etc. Compared with asphalt (vacuum residue), the asphalt from deasphalting has limited applications because of undesirable mechanical properties.

An extreme case would be a conventional refinery processing sweet crude, i.e., Ekofisk and Arab Light crudes (Table 2.1) [27]. However, even for such crude, commercial fuels could not be produced without some catalytic treatment even during the period when lax environmental regulations were in effect. Thus, it is unlikely that straight run distillates could be used directly as fuels. For example, for gasoline, a reforming step would be necessary to attain desirable level of octane number unless significant amount of an additive (e.g., tetraethyl lead) were added. Because of the noble metals containing catalysts employed, the feed for reforming has to be subjected to hydroprocessing to remove sulfur and nitrogen. Otherwise, the life of reforming catalyst would be affected. To ensure stability, the reformate may require additional hydroprocessing to remove olefins.

In the case of diesel fuel and aviation fuels, desirable cold flow properties (e.g., pour point, cloud point and freezing point) have to be attained. The values of these parameters specified by performance standards can be attained by removing straight chain paraffins from the feed. For this purpose, a catalyst selectively enhancing the hydrocracking (HCR) and hydroisomerization (HIS) of n-paraffins is necessary. Again, another hydroprocessing step may be required to ensure stability of the produced fuel. There is an option to conduct all these steps in the same reactor vessel employing several layers of different catalysts. For example, a front layer acidic catalyst would perform HCR and HIS functions, whereas the end-layer the hydrogenation (HYD) function. In another arrangement, several sections with different catalysts in the same reactor vessel may be employed.

Assuming the following range of the severity of hydroprocessing conditions, i.e., low < moderate < high < extra high, conventional refineries were dominated by catalytic hydroprocessing units operating under low severity and moderate severity conditions. The former were suitable for upgrading atmospheric distillates, whereas a moderate severity would be necessary for the conversion of VGO and DAO to fuels. The upgrading units comprised either stationary fixed-bed reactors or multilayer beds in the same reactor vessel. Such systems fulfilled all requirements during the period when environmental regulations just began to evolve. All spent hydroprocessing catalysts generated by conventional refineries were regenerable. In fact, an acceptable level of activity recovery could be achieved after several utilization–regeneration cycles before other utilization options for spent hydroprocessing catalysts would have to be considered.

2.2 Revamped Conventional Refinery

Figures 2.1 and 2.2 [24] indicate the increased consumption of middle distillates relative to that of gasoline and fuel oil as well as the growing imbalance between the availability of vacuum residues and residual fuel demand, respectively. At the same time, the emissions specifications for diesel oil, the main product among middle distillates, have been becoming more stringent [28]. For refiners, these developments mean a higher consumption of H_2 and catalyst inventory in the case that existing upgrading units are employed. This translates into a significant increase in operating costs. The costs can be offset by a more efficient upgrading of middle distillates to produce diesel fuels. In this regard, improvements in the performance of hydroprocessing can by achieved by developing more active catalysts. However, refining experience has shown that revamping of existing reactors and other units can improve the efficiency of H_2 and catalyst utilization as well.

Recent developments in petroleum refining industry indicate on significant efforts aiming at improvement of middle distillates upgrading achieved by revamping of catalytic reactors [29]. For example, installation of Shell internals resulted in the improved flow distribution in the reactor as evidenced by a lower temperature profile in catalyst bed. With these changes, the resistance to catalyst fouling was increased. Key features of reactor internals developed by Albermarle include a highly efficient mixing of process streams and almost an ideal liquid distribution [30]. Similarly, the UltraMix internals developed by the Universal Oil Products (UOP) prevent channeling, maldistribution, hot spots and bypassing. Bypassing is a phenomenon in which a portion of process streams passes through the unit without significant reaction [31]. Thus, in production of the ultra low sulfur diesel fuel, only 0.1% feed bypassing could jeopardize the ability to meet product specification for sulfur. Patel et al. [32] reported that the tray levelness must be carefully considered to avoid bypassing. This is shown in Fig. 2.5 comparing three different distributors. It is evident that among three trays tested, only vapor lift distributor ensured uniform flow profile. It was further observed that perforated plate or sieve tray were less suitable because perforations could become easily plugged. Therefore, bypassing phenomenon must receive adequate attention, particularly when sulfur specifications of diesel fuel are the target. This is confirmed in Fig. 2.6 [32] correlating the amount of feed bypassing and the content of sulfur in products. In this regard, a number of patents describing various designs of distributors should be noted [33–37].

The efficiency of refining operation may be increased by maximizing the utilization of reactor volume. This can be achieved by specially designed reactor internals as confirmed by Swain et al. [38]. This is demonstrated on the example shown in Fig. 2.7. Thus, by combination of three catalyst beds in the pre-revamped reactor into one, bed capacity of the original reactor shell was increased by about 30%. This was accomplished by installing a high dispersion tray in the revamped reactor. The features of several other dispersion trays, described in the study

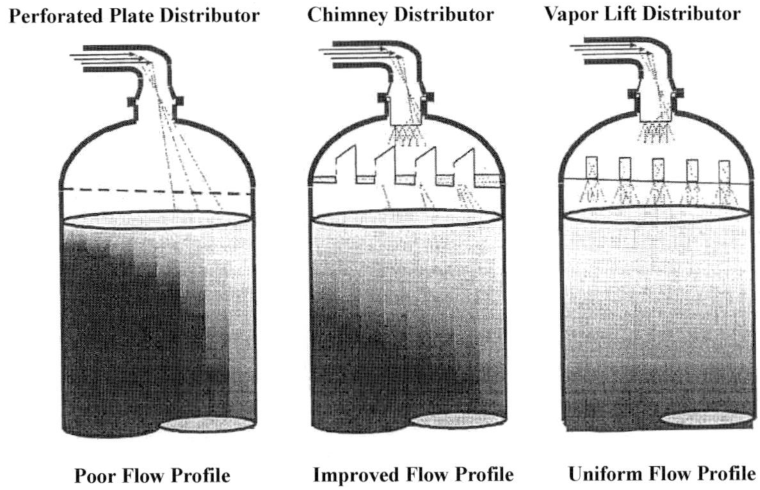

Figure 2.5: Effect of tray levelness on liquids flow [From ref. 32. Reprinted with permission].

of Swain et al. [38], should be noted. It is emphasized that these improvements were achieved for distillate feeds rather than for residual feeds.

Gragnani [39] reported that an additional operating time might be gained by employing the inter-stage treatment of gaseous streams. Figure 2.8 shows schematics of the conventional two-stage system and the same system but with the incorporated gas treatment. The advantages of the latter are clearly evident from Fig. 2.9 and they can be ascribed to the removal of gaseous hydrocarbons and H_2S in particular. The build-up of the latter and a gradual decrease in H_2 partial pressure will ultimately increase the H_2S/H_2 ratio. It was

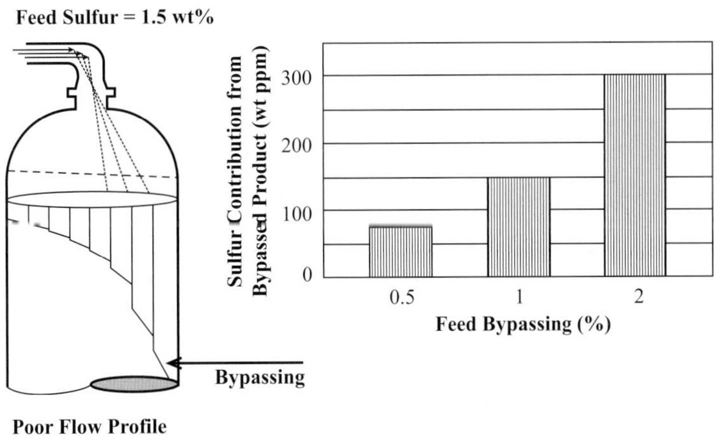

Figure 2.6: Effect of bypassing on sulfur in products [From ref. 32. Reprinted with permission].

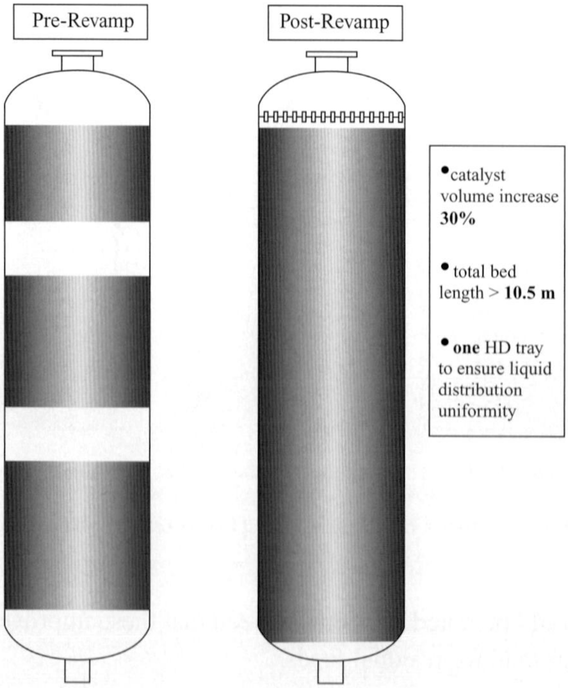

Figure 2.7: Effect of revamping on catalyst volume in fixed bed reactor [From ref. 38. Reprinted with permission].

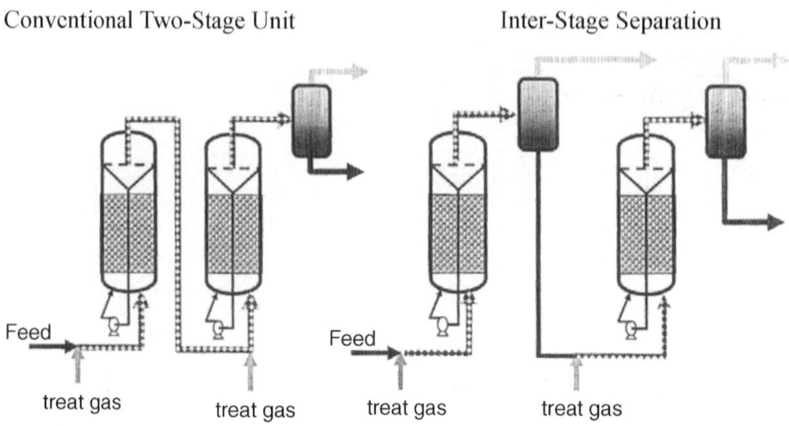

Figure 2.8: Two-stage unit with and without inter-stage treatment of gas [From ref. 39. Reprinted with permission].

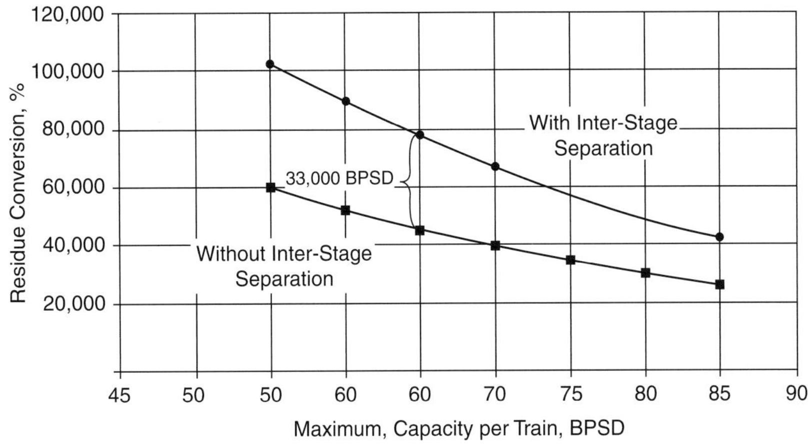

Figure 2.9: Effect of inter-stage separation (Fig. 2.8) on residue conversion [From ref. 39. Reprinted with permission].

confirmed that at high H_2S/H_2 ratios, some catalytic sites were inhibited due to the excessive adsorption of H_2S [40]. It is believed that this phenomenon is responsible for the difference in system performance indicated in Fig. 2.9 [39].

It is evident that any form of revamping has a direct effect on catalyst performance, i.e., overall utilization, activity and stability. Consequently, catalyst consumption, as well as the amount of generated spent catalyst, is decreased. However, there might be limits beyond which any form of revamping may not bring necessary benefits. In such circumstances, more advanced processes have to be employed. The decision between revamping and employing a novel process becomes more critical with increasing content of contaminants in the feed.

2.3 Advanced Refineries

Table 2.1 [27] indicates that for light crudes such as North Sea Ekofisk and Arab Light, the yields of distillates (atmospheric and vacuum) exceed 80%. For little heavier crude, such as Arab Heavy and Kuwait export, the yields are between 75 and 80%. On the other extreme, only about 50% and less than 40% of the distillates can be obtained from Cold Lake and Boscan crude, respectively, unless the residues (e.g., 50 and 64%, respectively) are subjected to additional upgrading. This comparison shows that no conventional refinery would be ready to undertake such tasks without significant modifications and/or revamping of existing units combined with the entirely new systems added to the refining operation. Also, an integration of refining operation with non-refining processes may be necessary to improve viability of the former.

Besides refining issues associated with decreasing quality of crude, significant modifications in refining strategy were necessary in response to stringent environmental regulations. For example, as it was indicated above, the ultra deep HDS and even deep HDS could not be achieved using conventional refining methods. Using the unmodified reactors, a deep HDS could only be achieved with the significantly lowered feed/catalyst ratio and increased H_2 consumption. It became soon evident that HDS can be enhanced by modifying the reactor design to improve hydrodynamics in the fixed-bed of catalyst. Of course, new types of catalysts with the enhanced activity and selectivity have been developed and/or are in various stages of development. The improved properties of the catalysts ensured several utilization-regeneration cycles before other options have to be found.

Distillation residues have been receiving much attention as the source of additional liquids. In this regard, new upgrading concepts involved multistage catalytic systems comprising several fixed-bed reactors in a series. Compared with the distillate feeds, for residues, the feed/catalyst ratio significantly decreases. The regenerability of catalysts in such systems increased from the first reactor contacting the feed towards the last reactor contacting a much upgraded feed. Moving-bed and ebullated-bed reactors were introduced with the aim to accommodate more problematic feeds, i.e., those with content of metals approaching 300 ppm of V + Ni. The regenerability of spent catalysts from such systems is rather low however; other methods of the catalyst reactivation (e.g., rejuvenation) have been developed. It should be noted that the management of spent catalysts has been an integral part of all advanced refineries.

Non-catalytic, carbon-rejecting processes were introduced to deal with the most problematic feeds, e.g., those containing more than 300 ppm. The primary liquids from coking processes require hydroprocessing steps to attain specifications of commercial fuels. Generally, the content of sulfur, nitrogen and aromatics in such liquids is much greater than that in the conventional distillates of a similar boiling range. Therefore, a successful upgrading of such liquids may not be accomplished with fixed-bed reactors used for hydroprocessing conventional distillates. In this regard, modifications may include multireactor systems and/or a single fixed-bed comprising several layers of different catalysts. Systems comprising several sections with different catalyst in the same reactor vessel is another alternative. The presence of more refractory compounds (e.g., S-, N-containing heterorings and aromatics) in coking distillates requires more severe hydroprocessing conditions compared with conventional distillates. This can only be accomplished by incorporating advanced catalytic reactors into refining schemes, as it is shown in Fig. 2.10 [41]. Otherwise, a coking process or deasphalting process has to be employed. Nevertheless, spent catalysts from upgrading coking liquids should still exhibit good regenerability, although the number of utilization-regeneration cycles may be lower than that for spent catalysts from upgrading distillates having a similar boiling range, but of a conventional origin.

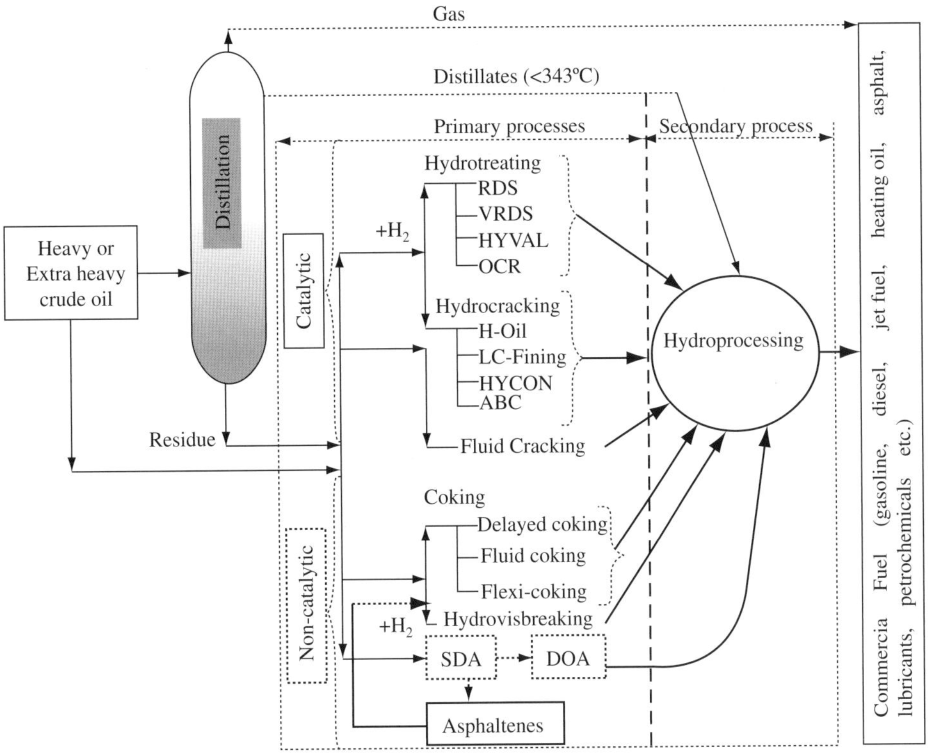

Figure 2.10: Advanced refinery for upgrading heavy and extra heavy feeds [From ref. 41. Reprinted with permission].

The transition from a conventional refinery (Fig. 2.4) to a more advance refinery resulted in the increased consumption of electricity, steam and hydrogen per unit of processed crude and/or commercial fuel produced. The flowsheet shown in Fig. 2.11 [42,43] indicates that the advanced refinery must be capable of utilizing refinery residues to the extinction. This may involve the production of additional liquids as well as that of electricity, steam and hydrogen required for the operation of refinery. In the first step, the conversion of residues (e.g., coke) is accomplished via gasification. The integration of gasification with petroleum refining significantly improved the control of solid and gaseous emissions [26,42,43]. No other technology besides gasification can offer such benefits. The generation of by-products such as coke, asphalt and elemental sulfur as well as waste materials such as caustics, sludge and contaminated water also increases compared with conventional refining. Today, the established methods for utilization, handling and disposal of these by-products are available.

It has been recognized that the overall costs of handling the by-products and wastes per unit of processed crude decrease with increasing size of refinery. The trends around the world indicate on the integration of refineries with gasification technology allowing the conversion of coke

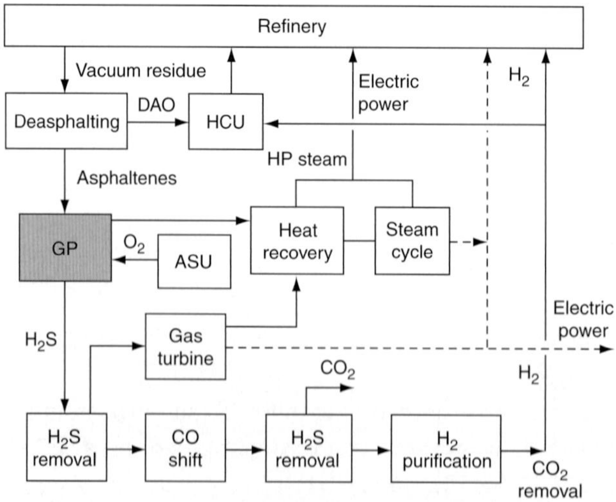

Figure 2.11: Flow sheet including units for gasification (GP) of residues to produce hydrogen, steam and electricity [From ref. 24. Reprinted with permission].

and asphalt to electricity, steam and hydrogen. At the same time, sludge and some contaminated water can be coprocessed with coke and/or asphalt and as such being incinerated. This concept is particularly suitable for large upgrading complexes producing synthetic crude. Such complexes are situated on or near the sites of heavy crude reservoirs and tar sands deposits, frequently occurring in remote locations. However, in the case of spent catalysts, the cost and/or means of transportation to a regenerating company or a metal reclaimer may be a factor to be considered.

CHAPTER 3

Hydroprocessing of Petroleum

As the most advanced hydrogen addition method, hydroprocessing has been an essential process for conversion of various petroleum fractions and residues to commercial fuels and other products. Depending on the origin of the feed, wide ranges of operating conditions have been employed in commercial hydroprocessing units. To suit processing requirements, refiner may choose from among reactors employing different types of beds, e.g., fixed bed, moving bed, ebullated bed and slurry bed. An efficient hydrogen transfer to reactant molecules cannot be facilitated without the presence of an active catalyst. The design of catalytic reactors, particularly their internals, may have a pronounced effect on the operation. A high performance of hydroprocessing units requires an optimal matching of the type of feed with that of catalyst and reactor. This ensures that the rate of catalyst deactivation is kept at a minimum.

3.1 Feeds for Hydroprocessing

Most, if not all, spent catalysts generated during hydroprocessing of the feeds derived from petroleum require special management procedures. One may also include in this category of the feeds synthetic crude obtained from heavy oils and tar sands. Non-petroleum feeds include those derived from bio crude, oil shale and coal-derived liquids. There is little known about the generation of spent catalysts during hydroprocessing of these materials on a commercial scale. There has been decades of commercial experience in the production of synthetic crude via Fischer-Tropsch (FT) synthesis. The upgrading of this crude to commercial products is conducted under conditions approaching those applied during hydroprocessing of petroleum feeds, although some fundamental differences in operating parameters should be noted. The procedures used for handling spent catalysts from this source are similar as well. Nevertheless, the following discussion is focusing primarily on the feeds of petroleum origin. Because of the extensive information on various aspects of hydroprocessing feeds readily available in the literature, only a general and brief account of their properties is given. A significant difference between the operating conditions applied during hydroprocessing of the metals free feeds and those containing metals and asphaltenes should be noted.

The properties of several distillate feeds and an atmospheric residue are shown in Table 3.1 [44]. It was indicated earlier that the severity of hydroprocessing depends on the properties of the feed and it always increases from light feeds towards heavy feeds. For example, for feeds in Table 3.1, the severity will increase in the following order: kerosine > gas oil > atmospheric

Table 3.1: Properties of some distillate feeds and atmospheric residue [From ref. 44. Reprinted with permission].

	Kerosene	Gas oil	Atmospheric residue
Density	0.7952	0.8967	0.978
Distillation (360+ °C)			
IBP	89	232	
50	202	363	
90	262	424	
FBP	291	440	
Sulfur (wt.%)	0.45	2.29	4.2
Nitrogen (ppm)	200	800	2450
Asphaltenes (wt.%)	0	0	~4
CCR (wt.%)	0	0	~12
Vanadium (wt.%)	0	0	67
Nickel (wt.%)	0	0	20

CCR: Conradson carbon residue; IBP: Initial boiling point; FBP: Final boiling point.

residue. The consumption of catalyst increases with increasing severity as well. Consequently, the amount of feed processed per unit weight of catalyst will decrease. Figure 3.1 shows the correlation between hydrogen consumption and hydrogen pressure. It is evident that much more hydrogen is consumed during hydrocracking (HCR) than during hydrotreating (e.g., deasphalted oil-HC [DAO-HC] vs DAO-HT).

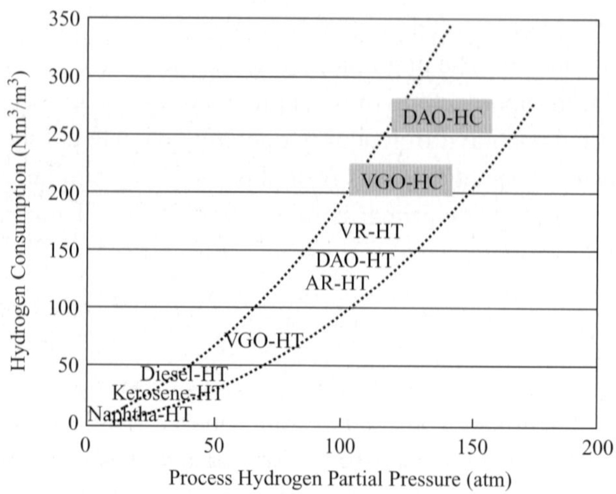

Figure 3.1: Effect of H_2 pressure and feed origin on hydrogen consumption.

3.1.1 Light Feeds

According to the flowsheet of conventional refinery shown in Fig. 2.4, atmospheric distillates, coking distillates, and vacuum gas oil (VGO) are among the fractions requiring hydroprocessing. Coking distillates are usually fractionated to naphtha and heavy gas oil (HGO) fractions. Figure 2.4 shows that VGO is subjected to hydroprocessing to obtain fuels. But, VGO can also be the feed for fluid catalytic cracking (FCC), particularly if gasoline is a preferred product. These feeds are free of contaminants, such as metals, resins, and asphaltenes. However, depending on the origin of crude, the composition of the virgin fractions having a similar boiling range may exhibit a great variability. Because of a higher temperature employed, a higher content of olefinic and aromatic structures is expected for coking and FCC distillates. From the processing point of view, the content of sulfur and nitrogen is of primary interest. These heteroatoms are in the form of heterocyclic rings. The stability and/or refractory nature of these rings increase with increasing molecular weight. The distribution of hydrocarbon groups must be also adjusted to meet the specifications of commercial products. For example, the content of n-paraffins must be low to ensure desirable cold flow properties of fuels and lubricants. The content of aromatics in diesel fraction must be kept below specified limits as well.

The primary products from FT synthesis contain no metals and little of sulfur and nitrogen. Heteroatom containing compounds is dominated by oxygenates and small concentrations of water, which is the product of FT synthesis. Most of the oxygenates have aliphatic structures. Hydrocarbon groups of the FT synthetic crude are dominated by straight chain paraffins and olefins. Small quantities of aromatic and naphthenic structures can also be present. The principal objective of the upgrading FT products is the hydroisomerization (HIS) of n-paraffins and n-olefins to i-paraffins. However, these reactions are affected by the presence of oxygenates. Therefore, hydroprocessing, as the primary step during upgrading of the FT products, must be conducted to extend the life of HIS catalysts, which usually contain noble metals.

3.1.2 Medium Heavy Feeds

For the purpose of this book, a medium heavy crude is characterized as one having less than 100 ppm of V + Ni and less than 10 wt.% of Conradson carbon residue (CCR). According to Table 2.1 [27], the atmospheric residues (340+ °C) derived from Arab Light crude and North Sea Ekofisk crude containing 49 and 12 ppm of V + Ni, respectively. For the latter crude, even vacuum residue (VR) (33 ppm of V + Ni) is classified as a medium heavy feed. Therefore, with respect to metal content, the VR (565+ °C) derived from North Sea Ekofisk crude represents a rather unique case of VR. Some DAO can also be classified as medium heavy feed. The medium severity hydroprocessing conditions, such as encountered using two or more fixed bed reactors in series, would be necessary to upgrade these residues, respectively.

Figure 2.4 includes a deasphalting of residues to produce DAO. The content of metals and asphaltenes in DAO depends on the type of deasphalting solvent and on the origin of the feed from which the DAO was derived. Thus, among several DAOs, it is not unusual to have one which contains more metals than residues, particularly when the former was obtained from heavy crudes (e.g., Boscan, Maya, Orinoco, Zuata, etc.). For example, the DAO studied by Reyes et al. [45] contained ~230 ppm of metals. However, one report suggests that the amount of metals in the DAO obtained from the Boscan crude by hexane deasphalting approached 510 and 60 ppm of V and Ni, respectively [46]. For such feeds, the deposition of metals is expected to be the predominant mode of catalyst deactivation from the early stages on stream, particularly when the content of asphaltenes in the DAO was much lower than that in a VR containing similar amount of metals. In some situations, it was more beneficial to use the blend of VGO with DAO, particularly when both were derived from a conventional crude [47]. Subsequently, the blend may be hydroprocessed to obtain the feed either for FCC or dewaxing. Correspondingly, the severity employed during the hydroprocessing of VGO/DAO or DAO and catalyst deactivation associated with it would be somewhere between that used during hydroprocessing of VGO and atmospheric residue.

3.1.2.1 Heavy and Extra Heavy Feeds

For the purpose of this book, a heavy feed and extra heavy feeds are defined as those having the total metal content between 100 and 200 ppm and 200 to 300 ppm, respectively. Based on this classification, the atmospheric residues (345+ °C) obtained from Kuwait export and Arab Heavy crudes contain 163 and 123 ppm of metals, respectively, whereas the VR (565+ °C) derived from the Arab Light contained 147 ppm of metals. Both the atmospheric and VRs derived from other crudes (except Arab Light and Ekofisk) in Table 2.1 [27] contain more than 300 ppm of metals and, as such, are classified as extra heavy feeds.

Decades of commercial experience confirmed that both heavy crudes and extra heavy crudes can be upgraded using hydroprocessing method. However, an optimal match of the properties of the feeds with type of catalyst and reactor systems becomes much more critical than that for medium or medium heavy feeds. The ultra heavy feeds, i.e., those containing more than 300 ppm of metals, can also be processed catalytically, however, not without a significant catalyst inventory and excessive hydrogen consumption [43]. Potential for the hydroprocessing of such feeds improves when the catalytic step is preceded by pretreatment, such as deasphalting. Otherwise, carbon-rejecting processes (delayed coking, fluid/flexi coking, etc.) must be employed for the primary upgrading step of extra heavy feeds. Most of the spent catalysts from the upgrading of extra heavy feeds are non-regenerable, however, in some cases; a desirable level of catalytic activity may be still recovered when the regeneration process is combined with rejuvenation.

3.2 Hydroprocessing Reactions

The presence of large molecules indicates a significant complexity of the reactions occurring during hydroprocessing of heavy feeds compared with light feeds. Because of the increasing involvement of asphaltenic molecules, the complexity increases from distillates, through VGO/HGO towards VRs and topped heavy crude. In every case, the primary objective is the conversion of large molecules to those present in distillates. This may be accomplished via HCR of resins and asphaltenes simultaneously with the conversion of porphyrin structures. Therefore, for heavy asphaltenic feeds, a high rate of the hydrodeasphaltization (HDAs) is required to achieve a desirable rate of hydrodemetallization (HDM). Thus, the desirable level of the removal of metals cannot be achieved before most of asphaltenes are depolymerized to smaller entities.

In the case of VGO, HGO, and DAO feeds, a high level of hydrogenation (HYD) of aromatics, i.e., a high rate of hydrodearomatization (HDAr), must be achieved, when the feed preparation for FCC is the objective. Moreover, for such feeds, even traces of metals and asphaltenes as well as nitrogen have to be removed to prevent poisoning of FCC catalyst, unless a more advanced process, i.e., residue FCC (RFCC) process, is used. In the case of catalytic dewaxing of these feeds, catalyst must posses an adequate HCR activity and selectivity to ensure a high yield of middle distillates and lube base oil fractions. For dewaxing catalysts, the HIS of n-paraffins to isoparaffins becomes an important catalytic functionality to ensure low freezing point and pour point of the final products. In addition, to be suitable for preparation of lubricants, lube base oil must exhibit good viscosity behavior. For this purpose, aromatic structures must be converted to naphthenic compounds. For most of the VGO, HGO, and DAO, desirable properties of the products (e.g., lube base oil and diesel oil) cannot be attained in one stage.

To various degrees, hydrodesulfurization (HDS), hydrodenitrogenation (HDN), and HDO reactions occur simultaneously with HYD, HCR, HDM, HDAs, and HIS. The mutual effects of these reactions are rather complex. Kinetic measurements can be used to quantify the progress of these reactions. The kinetic and mechanistic aspects indicate that the mutual effects of hydroprocessing reactions are rather complex even for light feeds. To a certain extent, these effects may be influenced and/or controlled by the properties of catalysts as well as by the experimental conditions.

The extensive information on various aspects of the mechanism of hydroprocessing reactions has been published in the literature. Several authoritative reviews were devoted to specific reactions, i.e., HDS [1,2,27,48], HDN [49,50], HDO [51], and HYD [52]. Focus has been on both model compounds and real feeds. Usually, the objective of hydroprocessing of the conventional feeds boiling below 350 °C has been the removal of heteroatoms and HYD of aromatics to meet specifications of the conventional fuels. The distillate fractions (e.g., naphtha) derived from heavy feeds by carbon-rejecting processes may contain olefins, which have to be removed to ensure stability of the final products. Again, the mechanism of reactions occurring during the hydroprocessing of distillate feeds is well documented [49–53] compared with that for heavy feeds, particularly those containing resins, asphaltenes, and metals.

The reactions occurring during hydroprocessing of the feeds boiling below 350 °C are common with those for the feeds boiling above 350 °C, such as VGO and HGO. However, for the latter, HCR and HIS reactions may be an important part of the overall mechanism, particularly if the production of middle distillates for transportation fuels and lube base stock are the objectives of hydroprocessing. In this case, a high level of dewaxing and HDAr may be necessary to meet specifications of the final products.

For asphaltenes and metals containing feeds, HCR, HDAs, and HDM are the most important reactions, as it is documented later in the book. In multistage systems, hydroprocessing will be dominated by different reactions in different stages. The HDM and HDAs are always the main reactions occurring in the first stage. While these reactions may be still important, the conversion of resins may become important in the second stage and stages following after until the overall hydroprocessing is governed by HYD, HDS, HDN, and HDO reactions in the final stage. For atmospheric and VRs, the reactions occurring during the final stage resemble those occurring during hydroprocessing of DAO and VGO. However, the extent of these reactions in different stages depends also on the origin of heavy feed and the type of catalyst. Therefore, the selection of catalysts for every stage requires attention.

It is generally known that the structural changes of hydrocarbons increase with the increasing acidity of catalysts. This supports the involvement of the HIS and HCR reactions. To a certain extent, such reactions proceed via a carbocation mechanism. Because the thermal scission of the C–C bond to form free radicals begins above 600 K, the latter may be formed under typical hydroprocessing conditions. Therefore, both carbocations and free radicals may be part of the overall mechanism of hydroprocessing.

3.3 Hydroprocessing Catalysts

These are extensive information on all aspects of hydroprocessing catalysts. This information has been periodically reviewed by several authors [15,27,44–54]. For the purpose of this book, a brief and general account of the chemical composition and physical properties of hydroprocessing catalysts will only be given. The Mo(W)-containing supported catalysts, promoted either by Co or Ni, have been used for hydroprocessing for decades. The γ-Al_2O_3 has been the predominant support. In recent years, other supports, e.g., silica-alumina, zeolites, TiO_2, etc., have been gradually introduced with the aim of improving catalyst performance. The enhancement in the rate of HCR and HIS reactions was the reason for using more acidic supports. The operating (sulfided) form of the catalysts contains the slabs of the

$Mo(W)S_2$. The distribution of the slabs on the support, i.e., from a monolayer to clusters, depends on the method used for the loading of active metals, conditions applied during sulfiding, operating conditions, properties of supports, etc.

3.3.1 Structure and Chemical Composition

The unsupported $Mo(W)S_2$ catalysts exhibit a hexagonal coordination. Apparently, the same coordination is retained in the supported catalysts. Under hydroprocessing conditions, the corner and edge sulfur ions in $Mo(W)S_2$ can be readily removed. This results in the formation of the coordinatively unsaturated sites (CUS) and/or sulfur ion vacancies, which have the Lewis acid character. The double and even multiple vacancies can be formed. Because of the Lewis acid character, CUS can adsorb molecules with the unpaired electrons (e.g., N-bases) present in the feed. They are also the sites for hydrogen activation. In this case, H_2 may be homolytically and heterolytically split to yield the Mo-H and S-H moieties, respectively [55].

Catalytic functionality of a catalyst could not be established without its ability to activate hydrogen. The active hydrogen is subsequently transferred to the reactant molecules adsorbed on or near CUS. Part of the active hydrogen can be spilt over on the support and to a certain extent protect slabs of the active phase from deactivation by coke deposits. In the course of operation, size of the latter (on the bare support) is progressively increasing [56,57]. In this regard, the protective role of surface hydrogen may be enhanced by optimizing the method of catalyst preparation and presulfiding.

The promoters, such as Co and Ni, decorate $Mo(W)S_2$ crystals at the edges and corners sites of the crystals. In the presence of promoters, CUS are considerably more active than those on the metal sulfide alone. Consequently, the rate of hydrogen activation is enhanced. The H_2S/H_2 ratio is the critical parameter for maintaining the optimal number of CUS. It has been confirmed that above 673 K, the -SH moieties on the catalyst surface possess the Bronsted acid character [53]. The presence of the Bronsted acid sites is desirable for achieving a high rate of HDN. Otherwise, other hydroprocessing reactions would be inhibited because of the prolonged adsorption of the N-compounds on CUS. Besides preventing other reactants from being adsorbed on active sites, the N-containing species on CUS may slow down hydrogen activation process. These adverse effects are the main reason for the catalyst poisoning by N-bases [49,55]. In addition, the formation of coke and metal (predominantly V and Ni) deposits on CUS will diminish the availability of active site.

During industrial operations, the oxidic form of catalysts is converted to the sulfided form, unless the catalyst sulfidation was conducted before the operation. Practical experience favors the catalyst presulfiding prior to contact with feed. The structure of such catalysts is rather complex. In this regard, published information is dominated by results on the evaluation of either fresh sulfided catalysts or spent catalysts under significantly different conditions than those encountered during industrial operations [55]. Thus, little information is available on the form of catalyst during the steady state operation, i.e., under in situ conditions. Inevitably, under hydroprocessing conditions (e.g., 600–700 K and 5–15 MPa of H_2), some properties of catalysts, i.e., interaction of active phase with support, lattice vibrations, interaction of promoting metal with base metal of active phase, etc., will differ from those observed under conditions employed during catalyst characterization. Therefore, it is desirable that a testing protocol, which could closely simulate practical situation, is developed, although this would appear to be rather challenging task.

3.3.1.1 Co(Ni)-Mo(W)-S Phase

Several research groups have been involved in determining the structure of hydroprocessing catalysts. The contributions of Topsoe et al. [53] to the understanding of these issues should be noted. In the case of the CoMo/Al_2O_3 catalyst, several species could be detected on the γ-Al_2O_3 surface. Thus, presence of the species, such as MoS_2, Co_9S_8, and Co/Al_2O_3, was clearly confirmed. Moreover, the Mossbauer emission spectroscopy provided clear evidence for the presence of the phase in which Co was associated with MoS_2, i.e., Co-Mo-S phase. Similar structures were also found in the NiMo/Al_2O_3, CoW/Al_2O_3, and NiW/Al_2O_3 catalysts, e.g., Ni-Mo-S, Co-W-S, and Ni-W-S, respectively. In this phase, enhanced concentrations of Co and/or Ni promoters at the edge planes of MoS_2 crystals have been confirmed. The occurrence of these promoters in the same plane as that of Mo ruled out the intercalation of the former between the layers of MoS_2. In the Co-Mo-S phase, the Mo–S bond is weaker than in the unpromoted MoS_2. Then, the CUS required for hydroprocessing reactions can be facilitated more readily. Temperature and the H_2S/H_2 ratio are among the important operating parameters for controlling the CUS concentration.

The structure of the Co-Mo-S phase is temperature-dependent [53,58,59]. Thus, the type I phase formed at lower temperatures was still chemically bound with the support, as it was evidenced by the presence of the Al-O-Mo entities. This phase was favored at low Mo loading on the γ-Al_2O_3. The occurrence of this phase was an indication of the incomplete sulfiding. The sulfiding at higher temperatures facilitated the transformation of the type I phase into type II phase. Consequently, the Al-O-Mo entities were not present indicating a diminished interaction of the active phase with the γ-Al_2O_3 support. The existence of the type II phase was further confirmed in the unsupported Co/MoS_2 system as well as in the CoMo catalyst supported on carbon [58], suggesting that type I phase requires the presence of oxygen on the support to facilitate the interaction with the active phase. Because of a lesser interaction with the support, the structure of type II phase is dominated by the multiple stacks of slabs compared with more or less monolayer-like distribution occurring in type I phase. Generally, the former phase exhibits a higher catalytic activity. This suggests that the active sites are present at the edges and corners of the Mo(W)S crystallites. The proportion of such sites in the type II phase is much greater than in the type I phase because the latter may still be attached to γ-Al_2O_3 via Mo–O bonds.

The study on the effect of support on the structure of active phase conducted by Bouwens et al. [59] revealed that type II phase on carbon supports resembled type I phase on SiO_2 and γ-Al_2O_3 supports, i.e., in the former case, type II phase approached a monolayer-like form. This was consistent with the significant dispersion of active metals on some carbon supports. In this regard, the presence of surface defects on carbons may play an important role. For example, much more efficient dispersion of active metals should be achieved on activated carbon (AC) compared with that on pristine graphite [55]. For both NiMo/AC and NiMo/Al_2O_3 catalysts, only two forms of metal sulfides were detected [60]. One was type II form, such as Ni-Mo-S, and the other Ni_3S_2. The latter was detected after the Ni/Mo ratio exceeded 0.48 and 0.56 for the NiMo/AC and NiMo/Al_2O_3 catalysts, respectively.

The evolution of the Co-Mo-S phase in the AC supported catalysts appeared to be H_2 pressure-dependent, as it was observed by Dugulan et al. [61]. These authors reported that the Mossbauer spectra of the CoMo/AC catalyst sulfided at 573 K under high H_2 pressure (e.g., 4 MPa) differed from those obtained at atmospheric pressure. Under high H_2 pressure, the stability of the Co sulfide species as part of the Co-Mo-S phase was affected compared with the CoMo/Al_2O_3 catalyst. This suggests that under high H_2 pressure conditions, properties of the Co-Mo-S phase on carbon supports may differ from those on the γ-Al_2O_3 support.

3.3.1.2 Brim Sites Model

Further insight into the structure, morphology and activity of MoS_2, Co-Mo-S, and Ni-Mo-S phases were obtained by Topsoe et al. [62,63] using a combination of novel experimental and theoretical methods like STM, DFT, and HAAD-STEM. The STM method showed the atom-resolved images of the catalytically active edges of MoS_2, Co-Mo-S, and Ni-Mo-S nanoclusters. The edge was found to exhibit a special electronic edge state identified as brim sites. Detailed analysis using DFT revealed that the brim sites have metallic character. It was postulated that because of metallic character, brim sites may bind sulfur-containing molecules, and when hydrogen is available at the neighboring edge sites in the form of SH groups, hydrogen transfer and HYD reactions can take place. The brim sites are thus catalytically active for HYD reactions. But, the brim sites are not CUS. It was generally accepted for a long time that CUS were the key sites involving in both HYD and hydrogenolysis reactions. It was believed that MoS_2 or Co-Mo-S structure with higher (> 2) sulfur vacancies at the corners are primarily responsible for HYD by Π adsorption and that hydrogenolysis site could be edge site with lower (1 or 2) sulfur vacancies [64,65]. The new “brim site” model, proposed by Topsoe et al. [62,63], is consistent with many inhibition steric and poisoning effects, which have been difficult to interpret using “vacancy” model. DFT calculations have helped to gain detailed insight into the HDS of thiophene under industrial conditions. Thus, it was suggested that the HYD reactions take place on brim sites, whereas the direct sulfur removal can take place at both edges. The mechanism involving brim sites in HYD allows the understanding of many observations, which were difficult to explain using previous models. Since brim sites are fully coordinated sites, they do not adsorb H_2S. This explains lack of inhibition of HYD reactions by H_2S. The brim site model also explains the lack of steric hindrance of alkyl substituents in the HYD pathway of molecules, such as 4,6-DMBT. The brim sites are very open sites and, therefore, they allow the adsorption of the refractory sterically hindered molecules, which need to be removed in the ultra low sulfur diesel (ULSD) production. The brim sites and their neighboring protons can interact strongly with basic N-containing molecules. This interaction is stronger than the interaction with simple aromatic compounds like benzene [66]. In this way, the observed strong inhibition of the HYD pathway by basic N-compounds may be explained.

It should be noted that the introduction of "brim" sites model represents a highlight of hydroprocessing catalysis in recent years, although most of the observations were made for a simple molecule, such as thiophene. However, because of unique approach, the authors [62,63] were able to describe the most early stages and intimate state of the reactions of thiophene. This has never been achieved before. The information on the reactions of more complex molecules is desirable to enhance the validity of the "brim" sites model. Nevertheless, it appears almost certain that during hydroprocessing, several types of active phase may facilitate catalytic reactions occurring either in parallel or consecutively.

3.3.1.3 Co-Mo-C(S) Phase

The presence of carbon on catalyst and conditions encountered during hydroprocessing support the presence Co(Ni)-Mo(W)-S phase. The same was supported by the study of Wen et al. [67] who showed that formation of the $Mo_{27}S_xC_y$ cluster was thermodynamically favorable. In this case, the edge sulfur atom on MoS_2 could be readily replaced by carbon atom. Similarly, Chianelli and Berhault [68] suggested that carbon could play an important role in stabilizing active phase. They proposed that the excess of sulfur on the surface of MoS_2 could be replaced by carbon to give stoichiometric MoS_xC_y phase. The clusters with three different S/C (i.e., 1.83, 1.68, and 8.27) were proposed [69]. According to Kasztelan [70], the replacement of sulfur with carbon on the edge of MoS_2 can be accommodated crystallographically. Therefore, the Co(Ni)-Mo(W)-S-C phase may be part of the overall hydroprocessing catalysis, particularly for the carbon supported catalysts. In this regard, the recent article published by Kibsgaard et al. [71] should be noted. These authors used STM spectroscopy to study the MoS_2 nanoclusters supported on graphite. A limited dispersion of MoS_2 clusters was achieved on pure graphite. However, a high dispersion was observed after introduction of small density defects. It is speculated that some form of bonding with the surface, presumable involving Mo–C bonds, was responsible for the increased dispersion. During operation, a modifying effect of carbon from coke on catalytically active phase cannot be ruled out. This indicates on the coexistence of the Co(Ni)-Mo(W)-S-C phase and Co(Ni)-Mo(W)-S phase and potentially other phases (e.g., brim sites). Therefore, because of the availability of carbon, the former phase may be present and participate during hydroprocessing reactions, even for the catalysts supported on γ-Al_2O_3 and other supports.

3.3.1.4 Effect of Support

The γ-Al_2O_3 is the support that has been most frequently used for the preparation of hydroprocessing catalysts. Depending on the conditions applied during preparation, γ-Al_2O_3, varying widely in surface properties, such as surface area, pore volume and pore size, can be prepared. In addition, desirable mechanical properties of γ-Al_2O_3 can be attained during preparation. After loading active metals, the surface properties and mechanical strength of catalyst are determined by those of γ-Al_2O_3. Generally, for the preparation of catalysts for hydroprocessing light feeds, the γ-Al_2O_3, possessing a high surface area and porosity predominantly in a mesoporous region, is suitable, whereas for heavy feeds, a macroporous, low surface area γ-Al_2O_3 is used. A more detailed account of the effects of surface properties of catalysts during hydroprocessing is given later in the book.

It has been generally known that supports other than γ-Al_2O_3 can have a pronounced effect on the activity and selectivity of hydroprocessing catalysts [72]. Attempts have been made to modify catalytic functionalities of the catalysts used for hydroprocessing of heavy feeds by replacing γ-Al_2O_3 with different supports. For example, a suitable acidity of the catalyst for achieving a desirable conversion of the large hydrocarbon molecules to light fractions can be maintained with the aid of support. General trends suggest that acidity has been a target parameter in designing the catalysts used for hydroprocessing of VGO, HGO, and DAO, whereas porosity for that of residues. This is not to say that for the former feeds as well as for residues, porosity and acidity, respectively can be ignored. Supports, such as carbon, SiO_2–Al_2O_3, zeolites, ZrO_2, and various mixed oxides, have been studied using a wide range of feeds [73,74]. The detailed review of the carbon supported hydroprocessing catalysts in relation to those supported on conventional supports, i.e., γ-Al_2O_3 was also published [75]. The recent information indicates on a growing interest in TiO_2 as the support either alone or in the combination with Al_2O_3 and SiO_2 [76,77]. However, the γ-Al_2O_3 modified with a small amount of alkali metals, such as Na and Li, as well as alkali earth metals, such as Ca and Mg, was also tested as the support for catalysts used during hydroprocessing of heavy feeds [78–80].

Differences in catalytic activities due to changes in support arise mainly from variations in catalytic acidity and metal-support interactions. Abotsi and Scaroni [81] showed that the acidity of carbon supports is markedly lower than that of the most frequently used γ-Al_2O_3 support. This was further confirmed by the NH_3 temperature programmed desorption (TPD) results of an AC, γ-Al_2O_3, and corresponding FeMo catalysts [82]. These results showed that the NH_3 adsorption on AC was negligible compared with that on γ-Al_2O_3. The addition of metals to AC enhanced the NH_3 adsorption. It is obvious that in the case of AC, the created acidity was associated with active metals. As expected, the acidity of the FeMo/Al_2O_3 catalyst was greater than that of the γ-Al_2O_3 support. It has been shown that the acidity control became critical for achieving a high level of HDS (deep and/or ultra HDS) of distillates [75,83]. This was confirmed by a much higher HDS activity of the CoMo catalysts supported on carbon support compared with that of the corresponding catalysts supported on γ-Al_2O_3 [75]. The latter catalysts were more sensitive to poisoning by N-containing bases present in distillates.

Support interactions also play a key role in the dispersion and morphology of the active phases (e.g., Co-Mo-S and Ni-Mo-S) [73,84]. Studies have shown that strong interactions between the molybdate ions and support lead to the formation of low-active type I Co-Mo-S structures, which are incompletely sulfided, and have some remaining Mo-O-Al linkages [53]. The application of high-resolution electron microscopy has provided valuable information on the degree of stacking in MoS_2 and Co-Mo-S structures prepared with different supports [85,86]. Very weak support interaction resulted in the formation of multistacking of type II Co-Mo-S phase. The degree of stacking can be controlled by carefully controlling support properties. Formation of small stable single slabs MoS_2 crystallites on alumina support have been observed. Such slabs will have a high MoS_2 edge concentration and dispersion and, as such, can accommodate more Co and Ni atoms to form higher activity single slab type II Co-Mo-S and Ni-Mo-S structures.

3.3.2 Physical Properties

The chemical composition of catalysts may not be so important unless suitable surface properties have been established. This is desirable for maintaining a long life of catalyst during the operation. Besides surface properties, the optimal size and shape of particles have to be chosen to achieve optimal performance of catalyst. Furthermore, the catalyst utilization usually increases with the decreasing size of catalyst particles. The influence of porosity as well as that of the size and shape of catalyst particles is evident even for relatively light feeds, such as AGO, VGO, and HGO [71]. Of course, for the asphaltenes and metals containing feeds, the design and selection of the catalysts have become a much more challenging task.

Among the surface properties, pore volume and pore size distribution as well as the mean pore diameter of the catalyst are much more important than surface area when heavy feeds are considered. At the same time, for light feeds, surface area may be a reasonable indication of the catalyst suitability. A high surface area and moderate pore volume catalysts are very active for HDS because of the efficient dispersion of active metals in the pores. However, in the case of heavy feeds, these pores become gradually unavailable because they are deactivated by pore mouth plugging. On the other hand, the catalysts with a small surface area and a large pore volume are less active because of the lower concentration of active sites. However, they are more resistant to deactivation by pore mouth plugging and their metal storage capacity is greater, therefore such catalysts may be suitable for HDM and HDAs. Apparently, the relation between surface properties and catalyst activity is more complex as it is indicated by numerous

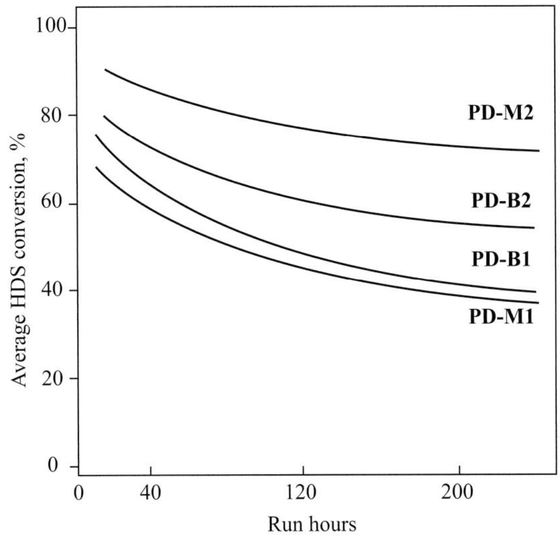

Figure 3.2: Effect of catalyst type (Table 3.2) on average hydrodesulfurization (HDS) activity [From ref. 87. Reprinted with permission].

studies in the literature. For example, the change in HDS conversion with time on stream shown in Fig. 3.2 for catalysts in Table 3.2 clearly confirms a significant effect of catalyst type [87]. The results in Fig. 3.2 were obtained using the Kuwait VR containing about 120 ppm of metals in trickle bed reactor (713 K and 12 MPa). The highest activity of the PD-M2 catalyst confirmed that an optimal combination of mesoporosity, surface area and pore volume has to

Table 3.2: Properties of catalysts [From ref. 87. Reprinted with permission].

Property	Catalyst			
	PD-M1	PD-M2	PD-B1	PD-B2
MoO_3 (wt.%)	13.2	11.9	11.6	13.2
NiO (wt.%)	4.0	2.8	2.5	4.0
Surface area (m^2/g)	85	228	136	312
Pore volume (mL/g)	0.60	0.53	0.73	0.76
Mesopore distribution (nm %)				
3–10	4	35	7	55
10–25	11	60.5	34	8
25–50	27	1.5	19	8
Macropore distribution (nm %)				
50–100	15	0	6	6
100–300	43	0	16	21
>300	0	0	18	2

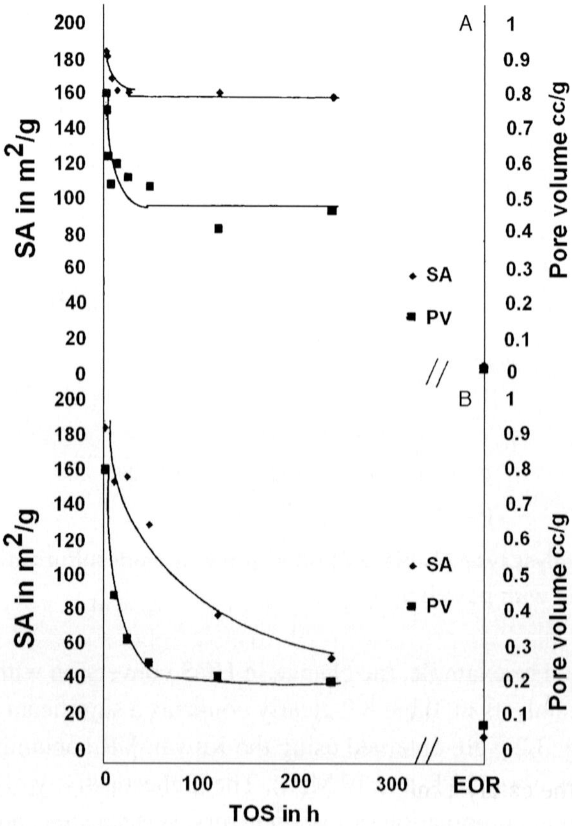

Figure 3.3: Effect of feed origin on loss of surface area and porosity of catalysts. A. Kuwait atmospheric residues (AR). B. Boscan feed [From ref. 89. Reprinted with permission].

be established to achieve a high catalyst performance. This would suggest that the feed was not heavy enough to observe the importance of macroporosity.

The above discussion suggests that there is an optimal combination of the surface area and porosity giving the highest catalyst activity [88]. The optimum may be different for different feeds and catalysts. This is evident from the results in Fig. 3.3 [89] showing the effect of feed origin on the loss of porosity and surface area of catalysts. Naturally, one would expect such effects when the properties of the relatively light Kuwait residue are compared with the Boscan feed. However, the optimal combination of surface area and pore size distribution was also crucial for achieving a high activity during the HDS of several gas oil of variable boiling range [77]. Another example of the effect of the feed origin is shown in Fig. 3.4 [90]. In this case, for HGO, the steady catalyst performance was maintained for an extended period, whereas a continuous catalyst deactivation was observed during hydroprocessing of the atmospheric residue. For the latter, the catalyst was deactivated both by coke and metal deposits.

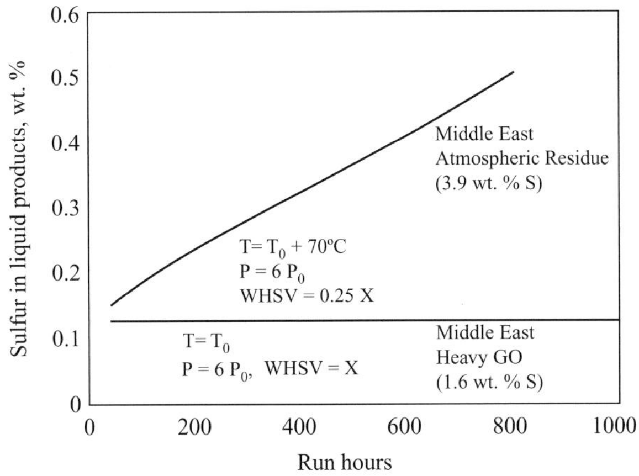

Figure 3.4: Effect of feed origin on hydrodesulfurization (HDS) activity ($CoMo/Al_2O_3$) [From ref. 90. Reprinted with permission].

It is again emphasized that an optimal pore size and volume distribution are critical for hydroprocessing of the high metal content feeds, particularly those derived from heavy crudes. This results from the large molecular diameter of the V- and Ni-containing porphyrin molecules, i.e., for microporous catalysts, the diameter may exceed that of pores. For small pore diameters, most of the metals will deposit on the external surface of the catalyst particles and the diffusion into the catalyst interior becomes the rate-limiting factor. It is therefore expected that the tolerance of catalyst to metals will increase with the increasing pore diameter as it is shown in Fig. 3.5 [78]. At the same time, the catalyst activity will decrease. At a certain

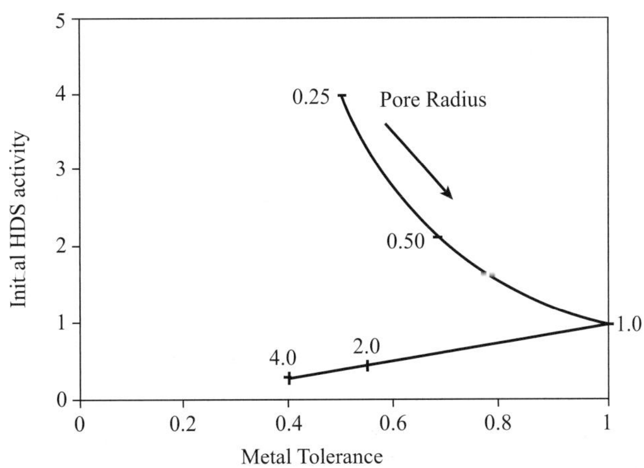

Figure 3.5: Effect of pore radius on metal tolerance and hydrodesulfurization (HDS) activity [From ref. 78. Reprinted with permission].

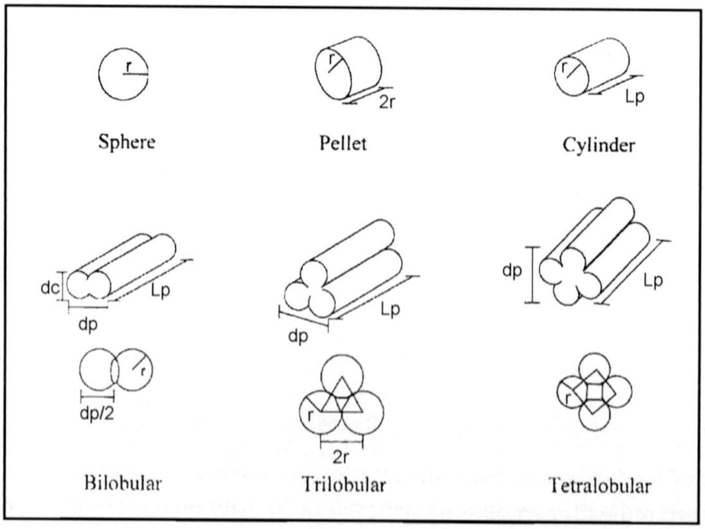

Figure 3.6: Typical shapes of commercial hydroprocessing catalysts [From ref. 75. Reprinted with permission].

pore radius, the tolerance to metals abruptly decreases, whereas the decrease in activity was less pronounced.

In an effort to enhance the overall catalyst utilization and to improve the reactor performance, various shapes and sizes of catalyst particles have been developed. Typical shapes of particles are shown in Fig. 3.6 [75]. In the case of fixed bed reactors, the development of pressure drops can be diminished by selecting an optimal shape of particles. In this regard, the method of catalyst loading, i.e., dense versus sock, is also important as it is evidenced by Fig. 3.7 [91]. The shape and size of particles as the method of loading may play an important role if the in situ regeneration of spent catalyst bed is considered.

3.3.3 Improved Hydroprocessing Catalysts

Catalysts used in petroleum refining processes deactivate with time during the operation. The rate of deactivation and catalyst life depends primarily on the catalyst structure, operating severity and feedstock quality. For a given feedstock, the operating severity can be reduced and the life of the catalyst can be extended if more active and stable catalysts are used in the process. Therefore, the development of better catalysts is one of the alternatives to minimize the utilization of fresh catalysts and generation of spent catalysts. In this regard, numerous attempts to improve catalyst performance reported in the scientific literature have been noted. These improvements together with better catalyst loading procedures and improved feed distribution in reactors have increased run-lengths significantly and reduced spent catalyst waste generation.

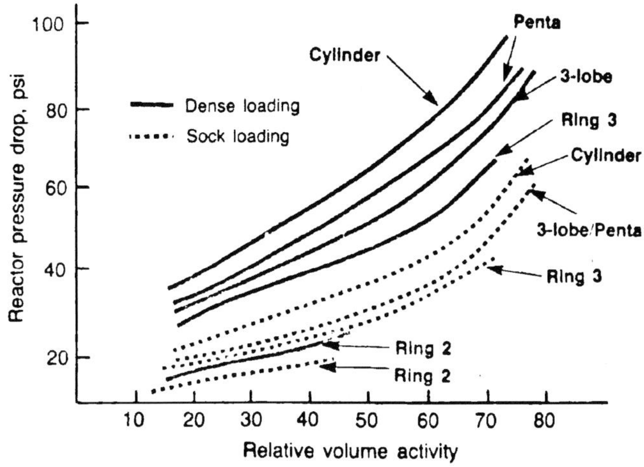

Figure 3.7: Effect of particle shape and relative volume activity on reactor pressure drop [From ref. 91. Reprinted with permission].

Remarkable improvements have been made in recent years in the performance of catalysts used in distillate and residual oil hydroprocessing units [27,92–95]. Development of improved hydrotreating catalysts has been possible through a clear understanding of the key properties, namely, nature of the active sites and their structure, and the textural characteristics of supports, more specifically pore size, that have significant influence on the catalysts performance [53,96–102]. The scientific basis for the high activity of the new generation hydrotreating catalysts have been presented in many reviews and in some recent papers [53,102–108]. Haldor Topsoe has introduced a number of catalysts, such as TK-573, TK-574, TK-911, and TK-915, which not only significantly improved HDS activity, but also tackled density and aromatics reduction.

Recently, Topsoe et al. [108,109] have developed a new catalyst preparation technology, giving highly active hydroprocessing catalysts. This new proprietary BRIM technology not only optimizes the brim site HYD functionally, but also increases the type II activity sites for direct desulfurization [109]. The first two commercial catalysts based on the brim technology were Topsoe's TK û 558 BRIM (CoMo) and TK û 559 BRIM (NiMo) for FCC pretreatment service. This was followed by a new series of high performance TK û 576 BRIM (CoMo), TK û 575 BRIM (NiMo) and TK û 605 BRIM catalysts for ultra low sulfur diesel production and for hydrocracker feed pretreatment. Akzo Nobel (now Albemarle) came up with the STARS (KF 757, KF 767, KF 848, etc.) catalysts series [110,111], which almost doubled the HDS activity. More recently, the company started to market a new catalyst, the NEBULA, which is considered a breakthrough in hydrotreating catalysis [93]. The new catalyst is almost four times as active as the conventional $CoMo/Al_2O_3$ catalyst as used for the hydroprocessing of gas oil [113]. This is indicated in Fig. 3.8 [114] by a significantly decreased weight average

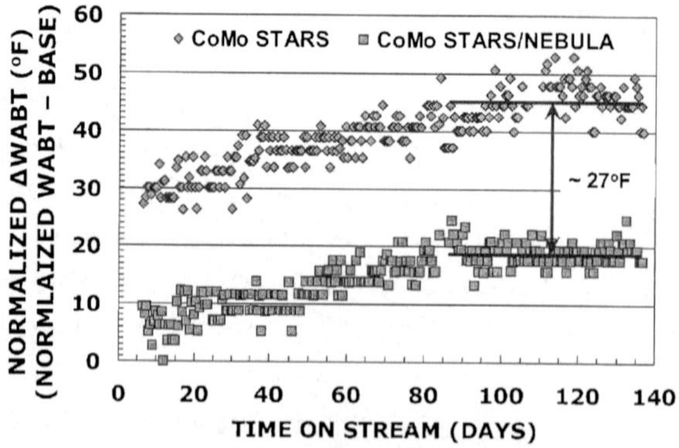

Figure 3.8: Normalized above base temperature versus time on stream [From ref. 114. Reprinted with permission].

base temperature increase required to maintain a similar conversion. AXENS has also introduced a series of catalysts some of which have superior HDS activities over their conventional middle distillate HDS catalysts [115]. This is shown in Fig. 3.9. Thus, conventional HDS catalyst was initially more active, but with time on stream, the stability of the improved catalysts was quite evident. Kuwait Catalyst Company has also introduced two new catalysts, i.e., HOP-414 and HOP-467 catalysts, which as a combination can achieve the target low sulfur levels using existing diesel hydrotreating facilities [116]. Criterion has introduced the CENTINEL Ascent and CENTINEL Gold series of catalyst that are designed to meet the ultra low sulfur specifications for diesel [117,118]. Finally, ART has developed the sulfur minimization by ART (SMART) catalyst system with a remarkably high activity than predecessor hydrotreating catalysts [119].

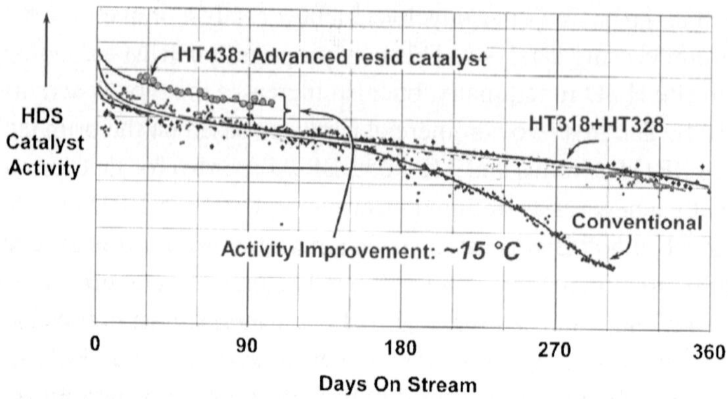

Figure 3.9: Effect of catalyst type on hydrodesulfurization (HDS) activity [From ref. 115. Reprinted with permission].

The improved high performance of HDM, HDM/HDS, and HDN catalysts has also been marketed by the above-mentioned catalyst companies for residual oil hydroprocessing. The HDM catalysts are designed to maximize metals (V and Ni) removal from the residual oil feed. They have large pore volume with balanced amounts of wide pores and mesopores to enhance the diffusion of the metals-containing large reactant molecules into the active surface within the catalyst pores and to allow for even more deposition of the removed metals within the pores. New generation HDM catalysts have a high capacity for storage of the metals removed from the feed, while retaining high activity and stability for metals removal. They are used in the front-end reactors, and in effect, they protect the valuable HDS and HDN catalysts that follow in the second, third and, possibly, fourth reactors, from deactivation by metals contamination. HDM/HDS catalysts that are used in the middle reactors (second and, possibly, third) are designed with two functions. First, they remove some of the remaining metals, not picked up by the front-end demetallization catalysts, and secondly, they have significant activity for HDS. A third type of catalyst with very high surface area and a sharp narrow pore distribution is usually placed in the last reactor. This catalyst is known as the tail-end catalyst. It possesses the highest HYD activity. The major responsibilities of this catalyst are HDN, hydroconversion and HYD in addition to HDS. New generation tail-end catalysts have higher stability which is essential for the increased length of cycle at severe operations. By using a combination of these improved catalysts in multiple reactor residue hydrotreating units, the on-stream efficiency of the catalyst system has been increased considerably.

In addition to the development of highly active and more stable new generation hydrotreating catalysts, improvements in the feed distribution in reactors by using better trays, better catalysts loading and process revamps and optimization have been made in recent years to improve hydrotreating reactor performance. These improvements have increased run-lengths significantly and, thereby, reduced spent catalyst waste generation.

3.4 Hydroprocessing Reactors and Processes

The detailed reviews of the commercial and emerging processes used for hydroprocessing of petroleum feeds were published elsewhere [11,120,121]. Simplified schematics of the conventional and advanced refineries shown in Figs. 2.4 and 2.10, respectively indicated the presence of several catalytic units operating in a hydroprocessing mode on the site of the petroleum refinery. It is evident that during the transition from the conventional refinery to advanced refinery, the number of catalytic reactors has been further increased. Properties of the feeds and those of the anticipated products after hydroprocessing of the former determine the selection of catalysts and the extent of the process modifications. Thus, for heavier feeds, revamping or modifications of reactors may not be sufficient, therefore, several reactors operating in series may be needed to achieve a desirable conversion and quality of anticipated products.

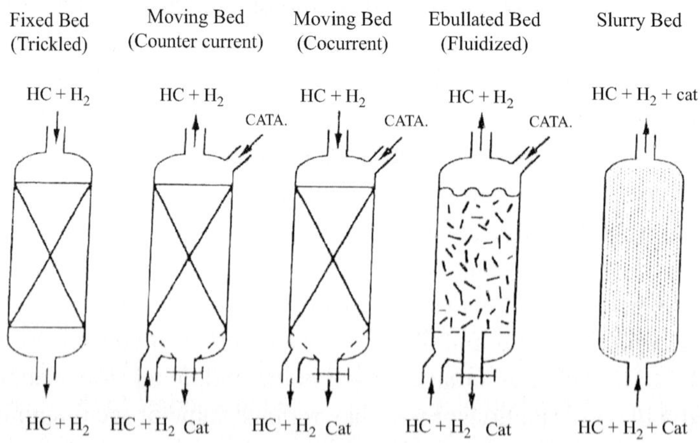

Figure 3.10: Simplified features of catalytic reactors for upgrading heavy feeds [From ref. 120. Reprinted with permission].

Simplified schematics of the catalytic reactors, which have been used commercially, are shown in Fig. 3.10, whereas the typical operating ranges of these reactors are summarized in Table 3.3 [120]. The features of these reactors indicate the importance of the proper selection of catalysts, the size and shape of the catalyst particles in particular, to ensure an efficient and continuous operation. In addition, the properties of feeds must be taken into consideration to achieve the optimal matching of catalysts with reactor.

Table 3.3: Operating conditions during hydroprocessing of heavy feeds in different reactors [From ref. 120. Reprinted with permission].

	Fixed-bed	Swing fixed-bed	Moving-bed	Ebullated-bed	Slurry bed
V + Ni maximum (ppm)	120	500	700	> 700	> 700
Pressure (MPa)	10–20	10–20	10–20	10–20	10–30
Temperature (K)	655–693	655–693	655–693	655–713	693–753
LHSV (h^{-1})	0.1–0.5	0.1–0.5	0.1–0.5	0.2–1	0.2–1
Maximum conversion to 550 °C	50–70	60–70	60–70	70–80	80–95
Cycle length (month)	6–12	12	CO	CO	CO
Catalyst particles size (mm)	~1.2 × 3	~1.2 × 3	~1.2 × 3	~0.8 × 3	~0.002
RCC	1	1	0.55–0.70	1.4–2	

CO: continuous operation; RCC: relative catalyst consumption for the same feed for one year; LHSV: Liquid hourly space velocity.

There has been decades of experience in the operation of fixed bed reactors, though for the hydroprocessing of light feeds. Progressively, fixed bed reactors have been modified to achieve the steady and prolonged operation using heavier feeds. The degree of modification increased with the increasing amount of asphaltenes and metals in the feed. Therefore, it is believed that for atmospheric distillates derived from conventional crudes, desirable conversions could be achieved with a single fixed bed and/or a fixed bed comprising several layers of different catalysts. Fixed bed reactors consisting of several sections in the same vessel may also be suitable.

For metals- and asphaltenes-containing feeds, frequent shutdowns of the operation and catalyst replacement could not be avoided using single fixed bed reactors. This problem can be alleviated by using several fixed bed reactors connected in a series. In this case, the primary function of the first reactor, termed as "guard reactors", is to remove most of the metals with the aim to extend catalyst life in the downstream reactors. A high HDM activity and/or metal storage capacity is the requirement for the catalyst to be used in the guard reactor.

In some cases, a "guard chamber" is placed upstream of the guard reactor, which operates mainly in the HDM mode. The function of the former is the removal of inorganic solids dispersed in heavy feeds. Therefore, guard reactor is filled primarily with a catalyst possessing a high metal storage capacity. At the same time, guard chamber is filled with the lower value solid materials (e.g., clays, minerals, alumina, etc.) with the aim to filter off the inorganic solids dispersed in heavy feed. Some removal of the V and Ni from heavy feed may be achieved in the case that the guard material includes the γ-Al_2O_3 of a suitable porosity. Most likely, part of these solids was formed during the non-catalytic reactions of V and Ni porphyrins with H_2 and H_2S rather than via catalytic reactions. The number of reactors downstream of the guard reactor increases with increasing content of metals and asphaltenes in the feed. Because of the different properties of the feed (product from the preceding reactor), each reactor may require a different type of catalyst. Again, this depends on the origin of the feed and anticipated slate of the products. Therefore, a special attention must be paid during catalyst selection to achieve a synchronized operation of a multistage catalytic system.

To avoid frequent shutdowns due to catalyst replacement, more advanced hydroprocessing reactors, which have provision for either continuous or periodic addition and withdrawal of catalyst during the operation, had to be developed. Figure 3.10 [120] shows that one type of the advanced catalytic reactor employs an expanded and/or ebullated bed of catalyst, whereas the other type employs moving beds. In the latter case, the catalyst is added at the top and progressively moves towards the bottom for a periodic withdrawal co-currently with liquid streams. In ebullated bed reactors, the slurry of catalyst in gas oil is continuously added at the top and spent catalyst withdrawn at the bottom of the reactor. An ebullated bed reactor can be operated without any difficulties even in the presence of inorganic solids dispersed in heavy feed. Thus, difficulties associated with the development of pressure drops, channelling, etc. encountered in fixed bed reactors are not present in the ebullated bed reactors.

Attempts have been made to further advance the existing or develop new catalytic systems for hydroprocessing of heavy feeds. In this regard, the focus has been on the countercurrent reactors compared with co-current reactors, which have been used predominantly on a commercial scale. The former reactors employ a co-current flow of the liquid and gaseous streams [122]. In countercurrent reactors, a structured catalytic bed in which catalyst particles are enclosed within a packed system is being used. Various features of catalytic reactors for hydroprocessing are in different stages of development. It should be noted that for the purpose of this book, only reactors which are part of commercial processes, i.e., those which generate spent catalysts, are being discussed.

With the aim of decreasing the cost of catalyst inventory, once through, low-cost catalytically active solids have been receiving attention. This included throw-away by-products from metallurgical and aluminium industries and fly ash from combustion of petroleum coke and coal as well as naturally occurring clays and minerals containing catalytically active metals, such as iron. In this case, a pulverized form of these solids, slurried with a heavy feed, is being introduced into the reactor operating under more severe conditions than typically employed during the hydroprocessing of the topped heavy crudes and VRs. The suitability of this approach for hydroprocessing of heavy feeds containing more than 300 ppm of metals (V + Ni) has been demonstrated on a commercial scale [43]. Definitely, in a pulverized form under otherwise similar conditions, conventional hydroprocessing catalysts would exhibit a much higher activity than the throw-away solids. However, for such a system, an economic method for the recovery of metals for reuse has not yet been developed. In this case, metals would have to be isolated from the VR obtained after distillation of the products unless the residue was further converted to liquid products and petroleum coke in a coking process. If such option was chosen, the catalyst metals together with the metals contained in the heavy feed would end up in the ash providing that the petroleum coke was utilized via a combustion and/or gasification technology. It is noted that the catalysts, such as used in slurry bed reactors, are not covered by this review.

3.4.1 Fixed Bed Reactors Systems

Several decades of experience in the operation of fixed bed reactors using conventional feeds containing neither metals nor asphaltenes were the basis for their adaptation and/or modification to suit hydroprocessing of more complex feeds. Many years of the experience confirmed that it is easy and simple to operate fixed bed reactors for atmospheric distillates as well as for VGO and HGO. Fixed bed reactors can be operated in the upflow and downflow mode [123]. The latter, so-called trickle bed mode, has been used predominantly. However, the upflow reactors ensure better catalyst wetting at low and high mass velocities for both the cylindrical and shaped catalyst particles regardless of the catalyst loading procedure. In trickle bed reactors, the catalyst wetting can be improved by choosing the loading procedure which ensures a minimal horizontal orientation of particles in the reactor. There is a lesser probability of malfunctioning of trickle bed reactors caused by channeling than that in the upflow reactors.

The fixed bed can comprise either a single stationary bed (Fig. 3.10) of the same catalyst of the same particle size and shape or layers of different catalysts. The layers may consist of the catalyst having the same chemical composition, but different size and shape of particles as well as the different pore size and pore volume distribution. For example, the layers may include the HDM catalyst at the reactor inlet, on the top of a HCR catalyst, followed by the HDS/HDN catalyst near the reactor outlet. The choice of catalysts and number of layers depends on the origin of heavy feed as well as on the anticipated quality of the final products.

There are some advantages of the fixed bed systems consisting of several sections in the same vessel with an empty space between the sections (Fig. 3.11). The sections may contain the same or a different catalyst each. In any case, with this arrangement, the make up H_2 can be introduced between the sections to quench the heat released by exothermic reactions. Also, some systems have a provision for scrubbing ammonia and H_2S from the gaseous effluent from the first section before it enters into the next section. This enables control of the H_2S/H_2 ratio, which is critical for a high conversion of HDN reactions [27,49]. Otherwise, the excessive poisoning of catalysts by N-bases would affect the operation. Indeed, it has been

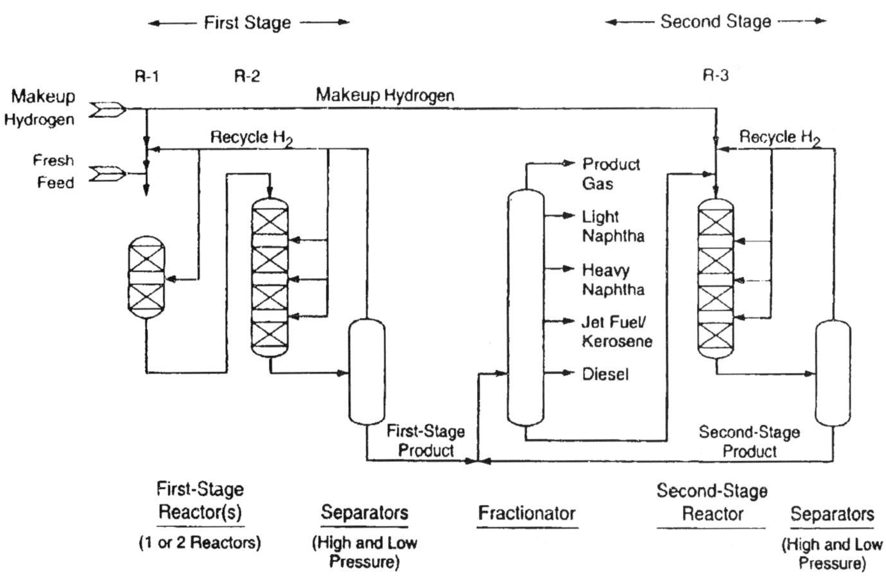

Figure 3.11: Modification of Unicracking process for dewaxing petroleum feed [From ref. 130. Reprinted with permission].

generally observed that the coke build-up in fixed bed reactors increased from the inlet towards the outlet, whereas metal deposition usually exhibits opposite trend. The H_2S/H_2 ratio increased in the same direction [27]. Consequently, the variable structure of spent catalyst between the inlet and outlet of the reactor should be expected.

It has been observed that the performance of fixed bed reactors depends on the method of catalyst loading, i.e., either dense loading or sock loading [124]. In the latter case, many catalyst particles will reach the loading surface together, having little time to attain a favorable resting position. Then, particles lay against one another, bridge and maintain random pattern. In this case, large voids are created to hold particles. The bridges may collapse if some forces are exerted on such fixed bed. For example, this may be caused by pressure drop, which may develop during the operation. When catalyst is loaded slowly, particles can settle into place before being inferred by other particles. This prevents bridging and creation of the oversized voids. The bed will have a higher density and shrinkage will be prevented. The advantages of the dense loading compared with the sock loading include the increase in the relative volume activity and decrease in the start of run temperature [124]. An increased start of run pressure drop is a negative effect of dense loading. One may anticipate that more problems are expected with the dense loaded beds when an in situ regeneration of spent catalyst is considered. In fact, it is unlikely that an in situ regeneration of such beds at the end of operation can be performed without significant problems.

An optimal combination of the bed void and activity per reactor volume giving the acceptable pressure drops has to be determined to ensure a steady performance of the fixed bed reactors. In this regard, the shape and size of the catalyst particles are important [125,126]. This is clearly shown in Table 3.4 [90]. There is a limit on the maximum pressure drop at which fixed bed can be operated. This depends on the type of the feed as well on the size and shape of catalyst particles. Thus, for light feeds, the particle shape and size may be chosen for dense loading to obtain maximum activity per reactor volume. However, for the high asphaltenes and metal

Table 3.4: Effect of particle size and shape on hydrodesulfurization (HDS) activity [From ref. 90. Reprinted with permission].

Shape	Dimensions (mm)	V_p/S_p (mm)	Activity
Cylinder	0.83 OD × 3.7 length	0.189	9.7
Cylinder	1.2 OD × 5.0 length	0.268	7.9
Cylinder	1.55 OD × 5.0 length	0.345	5.7
Ring	1.62 OD × 0.64 ID × 4.8 length	0.233	8.7
Ellipse	1.9 OD × 1.0 ID × 5.0 length	0.262	8.4
3-lob	1.0 OD × 5.0 length	0.295	8.2
Crushed	0.25–0.45	~ 0.04	14.0

OD: outside diameter; ID: inside diameter.

feeds, a small particle size may be needed to achieve a desirable level of catalyst utilization. Then, the shape of the catalyst particles must be chosen to obtain the fixed bed with a sufficient level of voidage. For example, this may be achieved by sock loading of the ring and lobe particles giving 35 and 10% higher voidage, respectively, compared with the cylinders [90].

Refinery experience indicates that the heavy feeds containing less than 120 ppm of V + Ni can be successfully hydroprocessed using several fixed bed reactors in a series [127]. Under optimized conditions, a high activity and the relatively low metal tolerance catalyst may be suitable for heavy feeds containing less than 25 ppm. A dual catalyst system may be required for feeds containing between 25 and 50 ppm of metals. In this case, the first stage catalyst should possess a high metal tolerance, whereas the second stage, a high catalyst activity for HDN and HDS. For heavy feeds containing between 50 and 100 ppm of metals, at least a three-stage system employing fixed bed reactors may be necessary. In this case, the catalyst in the first reactor should possess a high HDM activity and a high metal storage capacity to ensure the long life of catalysts in the subsequent reactors. It is believed that heavy feeds containing more than 150 ppm of metals can still be hydroprocessed in fixed bed reactor systems providing that some modifications were undertaken. This may include the use of two guard reactors, one in operation and the other on stand-by. Such guard reactors are part of the HYVAHL process [120]. The sizing of these guard reactors, i.e., the total metal storage capacity would need to be matched with the content of metals in the heavy feed. An uninterrupted operation could be ensured by switching to the guard reactor with the fresh catalyst as soon as the total metal storage capacity of the reactor on stream was approached [120]. The addition of another reactor downstream may also be considered as an option. However, such a step may drive costs of the operation to an unacceptable level.

Commercial processes employing fixed bed reactors have similar features, although they are licensed by different process developers. The number of stages and/or reactors included in the process is determined by the content of asphaltenes and metals in heavy feeds, the projected daily throughput of the heavy feed and the anticipated quality of liquid products. It is unlikely that for heavier feeds, a desirable level of hydroprocessing can be achieved in one stage. Thus, even VGO may require a graded system, e.g., either multilayer bed or multisections reactor, particularly when the objective is to produce the feed for FCC or to increase the yield of middle distillates in the products.

Entirely different configurations of the fixed bed reactors and systems may be necessary when the lube base oil is the targeted product. In this case, catalytic dewaxing reactor may be part of the overall hydroprocessing of VGO and DAO followed by hydrofinishing step performed under milder conditions as usually applied during hydroprocessing. It should be noted that the catalyst formulations required for dewaxing and hydrofinishing may differ from those of the conventional hydroprocessing catalysts.

3.4.1.1 Unibon Process

Typically, this process has been used downstream of the deasphalting unit. It may also be used for hydroprocessing of either VGOs and HGOs or the blend of VGO with DAO. Depending on the feed, the process can be used as a single stage or two-stage configuration. For example, the commercial configuration of the Unibon process is using DAO as the feed consisted of two single fixed reactors: one operating predominantly in HDM mode (guard reactor) and the other in the HDS mode [128,129]. The DAO feed contained about 27 ppm of V + Ni and less than 1% of asphaltenes. The blend of VGO and DAO can also be used. To suit refinery requirements, different configurations of the Unibon process, i.e., BOC Unibon, RCD Unibon, etc., have been licensed [130]. For example, the Unicracking, the residue desulfurization (RDS) version of the Unibon process, shown in Fig. 3.11, was designed primarily for the HDS of ARs and VRs derived from the conventional crudes [131]. In Fig. 3.12, besides guard reactor and two HDS reactors, all necessary downstream and upstream units are shown as well. Most of these units are common for other similar commercial systems employing fixed bed reactors. During dewaxing of VGO and DAO to produce lube base oil, HIS and HCR are important functionalities besides other hydroprocessing reactions. This can only be achieved using several types of catalysts. For this purpose, the modified Unibon process, such as Unicracking process (Fig. 3.11), comprising two reactor vessels with several sections in each, has been used.

A number of other commercial processes employing fixed bed reactors have been licensed. For example, the asphaltenic bottom conversion (ABC) process developed in Japan has similar features as the Unibon process [130]. A modified version of this process includes the recycling of asphalt from the deasphalting unit to the HDM reactor for further processing, i.e., recycle to

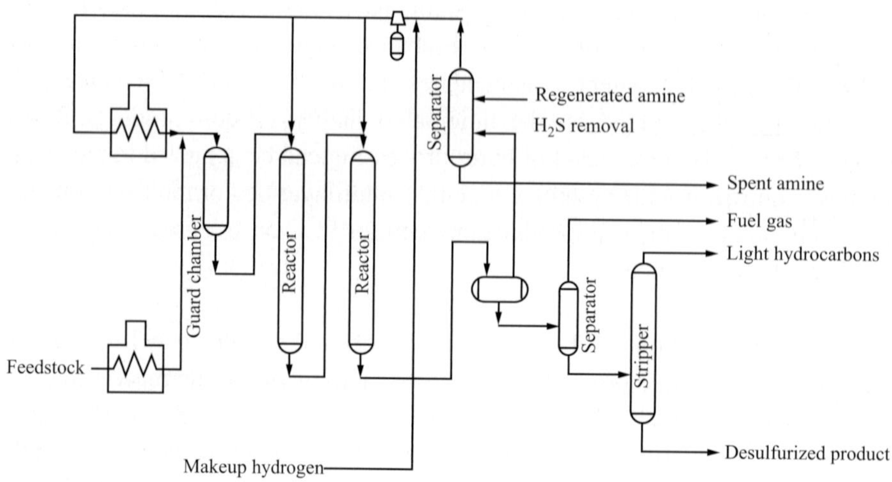

Figure 3.12: Simplified flowsheet of Unibon process [From ref. 130. Reprinted with permission].

the extinction. Apparently, almost complete conversion of the atmospheric residues (AR) could be achieved. The fixed bed reactors, which are part of the Gulf RDS process [132], consist of several sections in one reactor vessel, similarly as it is in the Chevron RDS (VRDS) process. Using these processes, a high level of HDS could be achieved with a proper catalyst selection. The Chevron RDS process has also been used downstream of deasphalting unit for the upgrading of DAO [133]. With the proprietary catalyst designed for this process, a high level of HDS and a low H_2 consumption could be achieved. The EXXON Residfining process consists of a guard reactor and the catalytic reactor comprising several sections [134]. This process was designed for the HDS of the atmospheric residues obtained from conventional crudes with the aim to produce fuel oils meeting all commercial specifications.

3.4.1.2 Atmospheric Residue Desulfurization (ARDS) and HYVAHL Processes

The new processes employing fixed bed reactors (Fig. 3.13) comprising various combinations of reactors and catalysts were developed in response to new developments in refining industry. A brief description of the HYVAHL process and ARDS process is given as an illustration of the efforts to modify fixed bed reactors for hydroprocessing of the asphaltenes and metals containing heavy feeds. The ARDS process was developed by Unocal for hydroprocessing of atmospheric residues. Simplified schematic of this process is shown in Fig. 3.13. Apparently, this is an extension of the Unibon process to accommodate more problematic feeds. There are many years of experience in the commercial operation of this process using Kuwait AR, typically containing about 85 ppm of V + Ni and about 12 wt.% of CCR [135]. In this case, the process consists of two trains each having design capacity of 33,000 barrels per day. Each train comprises one guard reactor and three main reactors with a common fractionation section

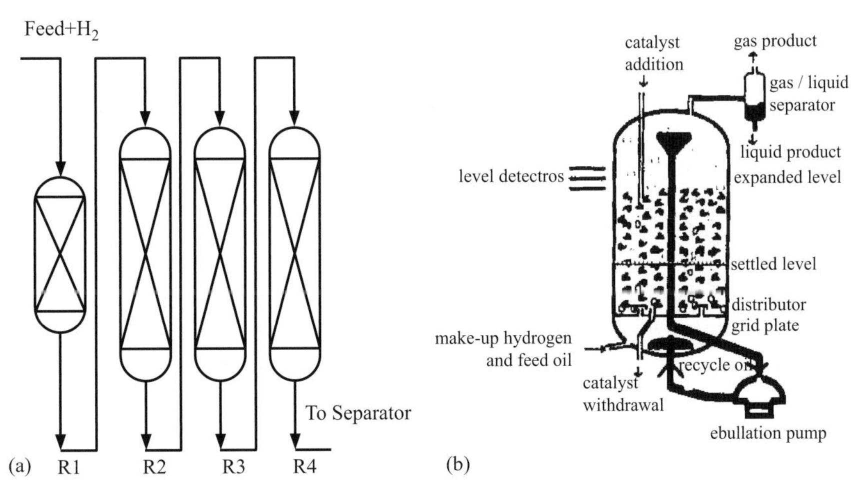

Figure 3.13: Simplified schematics of (a) atmospheric residue desulfurization (ARDS) process and (b) ebullated bed reactor.

attached. The guard reactor contains about 7% of the total catalyst inventory and its main function is HDM of the feed. It is however believed that this amount depends on the content of metals in the feed. Other three reactors contain 31% of the catalyst inventory each. All three reactors employ a graded bed consisting of either the same catalyst but of different particle sizes and shapes or catalysts of a different composition. The purpose of using the graded bed is to diminish the reactor pressure drop particularly in the front of catalyst system, which is contacted with only partially converted and/or unconverted feed. Because the guard reactor only removes a portion of metals, the catalysts in the downstream reactor must possess an adequate HDM activity. Thus, a relatively large amount of metals was still present in the spent catalysts from all three main reactors [136]. However, this problem may be alleviated by an optimal selection of catalyst for the guard reactor and the subsequent reactor. The modified Unicracking/HDS process comprises five reactors in a series [137]. It has similar features as ARDS process. In this case, the first reactor was in fact a guard reactor containing a high metal storage capacity HDM catalyst. With this arrangement, heavy feeds containing as much as 150 ppm of V + Ni were successfully hydroprocessed.

The HYVAHL process was developed and licensed by the French Institute of Petroleum [138]. This process was successfully tested for hydroprocessing of various heavy feeds, i.e., DAOs, atmospheric and VRs. The process consists of the guard reactor placed upstream of the two HDM reactors. The guard reactor is sized and optimized to achieve a satisfactory length of the cycle. To protect catalyst in the HDS section, two more HDM reactors are placed downstream from the guard reactor. This version of HYVAHL process, known as the swing reactor concept, ensured a continuous operation of the process approaching one year using heavy feeds, the metal content of which was in the range 500 ppm of V + Ni. In this case, the process included two guard reactors that were switchable during the operation. With this concept, the replacement of catalyst in the guard reactor does not require shutdown of the operation [139]. The guard reactor and two HDM reactors represent about 40% of the total catalyst volume. Of course, the exact amount of catalyst required for guard reactor depends on the amount of metals in the feed and metal storage capacity of the catalyst.

3.4.2 Moving Bed and Ebullated Bed Reactors

It has been evident that for fixed bed reactors, the difficulties in handling heavy feeds could be overcome either by frequent catalyst replacements or by adding more reactors in the series. At a certain point, both these options become economically unattractive. Also, it is not easy to maintain synchronized operation of so many fixed bed reactors in a series. Because of these problems, reactor design and catalyst development has reached entirely new levels. In this regard, attention has been focussing on the development of a process enabling catalyst replacement on stream without interrupting the operation. The bed of catalyst moving vertically through the reactor was one option which had been explored. Several moving bed catalytic reactors reached a commercial scale. Among those, the best known are bunker reactor and quick catalyst replacement reactor (QCR). It should be noted that moving bed reactors require special equipment and procedures for safe transfer of catalyst into and out of the high-pressure and high-temperature vessels and reactors. This may include several high-pressure vessels upstream and downstream of the reactor. With respect to the generation of spent hydroprocessing catalysts, the processes employing moving bed reactors are unimportant.

The first process employing ebullated bed reactor (Fig. 3.13) was known as the H-Oil process developed jointly by the City Services with Hydrocarbon Research Institute (HRI). The HRI was joined by Texaco and later by Institut français du pétrole (IFP) to license H-Oil process, whereas City Services jointly with Lummus and Amoco have been licensing similar process known as LC-Fining. The changing crude supply patterns are reflected by a gradual increase in the demand for ebullated bed reactors as it is supported by Fig. 3.14 [24].

The ebullated bed reactors were designed to handle the most problematic feeds such as VRs and toped heavy crudes having high contents of metals, asphaltenes, sediments as well as dispersed clays and minerals. The flexibility of the operation of the ebullated bed reactors was successfully demonstrated during coprocessing using the mixtures of VRs with coals as well as VRs and plastics. Table 3.3 [120] shows some operating parameters which confirm that the ebullated bed reactors are suitable for hydroprocessing heavy feeds containing more than 700 ppm of metals. This, however, cannot be achieved without significant catalyst inventory. Because of the catalyst being in a continuous motion, particle size less than 1 mm can be used without any difficulties. This ensures a high level of catalyst utilization. However, for such thin particles, mechanical strength requires an attention to prevent their breaking in the reactor, as it

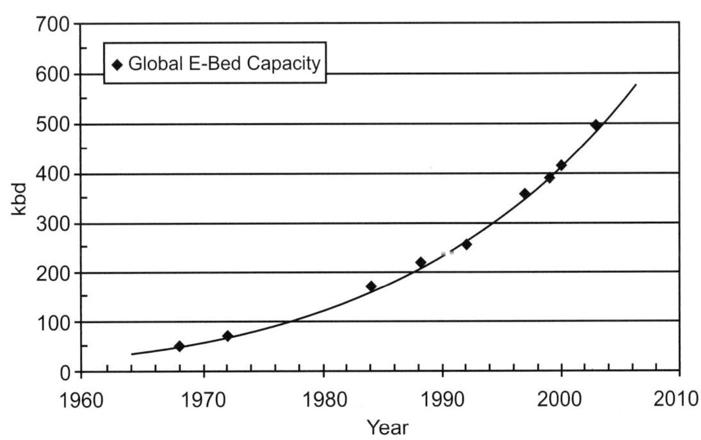

Figure 3.14: Trends in demand for ebullated-bed reactors [From ref. 24. Reprinted with permission].

was observed by Al-Dalama and Stanislaus [12]. To be cost competitive, this process must produce enough additional liquid products compared with the non-catalytic options, i.e., deasphalting and coking, to compensate for the costs of catalyst inventory and excessive hydrogen consumption. Also, the additional high-pressure vessels and equipment upstream and downstream of the reactor are necessary to ensure safety of the operation similarly as it was noted for moving bed reactors. This adds to the capital cost of the processes employing ebullated bed reactors compared with the fixed-bed reactors.

The most important features of the ebullated bed reactors include their capability to either periodically or continuously add/withdraw catalyst without interrupting the operation. The bed design ensures an ample free space between particles allowing entrained solids to pass through the bed without accumulation and plugging as well as without increasing pressure drop. Under such conditions, the catalyst particles having a diameter smaller than 1 mm (e.g., 1/32 in extrudates) can be utilized. This results in the considerable increase in reaction rate because of the significantly diminished diffusion limitations. Moreover, under such conditions, the catalyst utilization is significantly enhanced. Depending on the operating strategy of the refinery, the process can operate either in a high conversion mode or in a low conversion mode [27].

The information on the LC-Fining and H-Oil reactors is quite extensive [129–131]. It is again noted that these reactors have similar features. In ebullated bed reactor (Fig. 3.13), the heavy feed and H_2 enter at the bottom and move upwards through the distributor plate at a sufficient velocity to expand the catalyst above the grit into a state of random and turbulent motion. The expanded bed is maintained about 35% above the settled level of catalyst. This can be achieved by controlling the speed of the recycle oil pump. In this regard, the operation is monitored using the density detectors. The suction of the recycle pump is supplied from near the top of the reactor. The recycle pan is used for disengaging the gas before recycling the liquid. The advanced design of the ebullated bed reactor used in the H-Oil process incorporates an improved internal recycle cup enabling a complete separation of gas from the recycled liquid. With this modification, the throughput of heavy feed was increased. On a commercial scale, usually three ebullated bed reactors are used in the series (Fig. 3.15). The first reactor serves as a guard reactor, the primary function of which is HDM. The main functions of the second and third reactors are HDS, HDN and HCR. In some situations, the ebullated bed reactor can be used as the guard reactor upstream of the fixed bed reactors. However, in the case of a large amount of inorganic solids in heavy feed, part of these solids may not be trapped in the ebullated bed reactor. Such solids may then be carried out with liquid streams to the subsequent fixed bed reactor.

Figure 3.16 [132] shows the simplified diagram of the catalyst handling system consisting of three sections, i.e., fresh catalyst handling, the daily addition/withdrawal of catalyst to and from reactors and spent catalyst handling system. The fresh HDM catalyst is carried as slurry

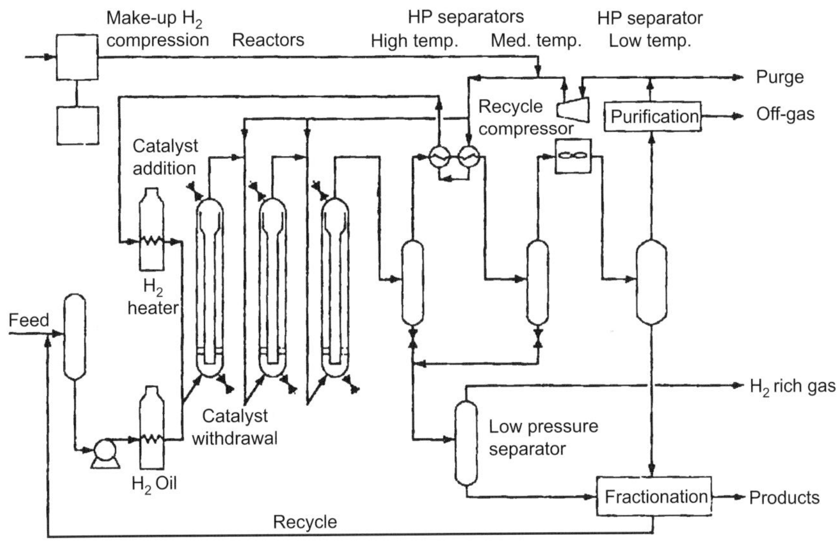

Figure 3.15: Process employing ebullated-bed reactors.

from the high-pressure vessel to the first reactor. The equilibrium catalyst is withdrawn from the third reactor and transported as slurry to the second reactor. The spent catalysts are withdrawn from the first and second reactors to the transfer vessel. It is then washed, cooled and transferred to the spent catalyst inventory vessel. Further utilization of spent catalysts from

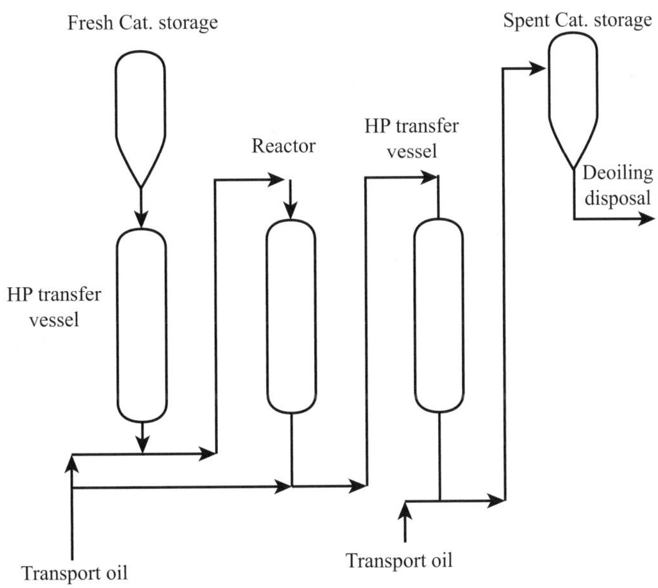

Figure 3.16: Catalyst handling system for ebullated-bed reactors [From ref. 132. Reprinted with permission].

the ebullated bed reactors depends on the level of deactivation, particularly on the amount of deposited metals such as V and Ni.

3.4.3 Comparison of Hydroprocessing Reactors

Fixed bed reactors have always been chosen for hydroprocessing distillate feeds. There is a wide range of modifications to fixed bed reactors to suit different feeds, available commercially. If properly designed and loaded with a suitable catalyst, any fixed bed reactor can be used for hydroprocessing light feeds. Moreover, an optimal selection of conditions such as temperature, H_2 pressure, feed rate, etc. can ensure an efficient and steady operation of fixed bed reactors. Morel et al. [120] estimated ranges of the yields and of the properties of the products from hydroprocessing of the Safania VR in different types of reactors. The properties of the VR are shown in Table 3.5, whereas those of the products together with their yields in Table 3.6 [120]. With respect to the content of contaminants (e.g., sulfur, nitrogen and CCR) in products, fixed/moving bed reactors were the most efficient followed by ebullated bed reactor. Because of the higher temperature employed, the latter reactor gave the larger yields of naphtha and gas oil. In the slurry bed reactors employing throw-away solids, the conversion to liquid products have exceeded 80%. This resulted from the temperatures which were higher than those typically used during conventional hydroprocessing. The residence time was usually longer as well.

The quality of products (Table 3.6) [120] from different reactors reflects the difference in operating conditions. The lower quality for ebullated bed reactor compared with fixed/moving bed reactor is attributed to a higher temperature used in the former. This may be offset by a lower yield of VR in the products from ebullated bed reactor. The lowest quality products are obtained in slurry bed reactors, most likely, because of the highest temperature used compared with the other reactors. This suggests that a significant hydroprocessing of the liquid products from the slurry bed reactors would be required to achieve specifications of the commercial fuels. Moreover, feasibility of the slurry bed reactors may be affected by availability of the catalytically active solids. Thus, the plant processing 10,000 tons per day of heavy feed, requiring about 0.5 wt.% of catalyst to achieve acceptable conversion, would consume about

Table 3.5: Properties of Safania vacuum residue [From ref. 120. Reprinted with permission].

Specific gravity (kg/L)	1.035
Sulfur (wt.%)	5.28
Nitrogen (ppm)	4600
CCR (wt.%)	23.0
Asphaltenes (heptane)	11.5
V + Ni (ppm)	203

CCR: Conradson carbon residue.

Table 3.6: Yields and properties of products from different reactors [From ref. 120. Reprinted with permission].

	Fixed/moving	Ebullated	Slurry
Naphtha			
Yield/feed (wt.%)	1-5	5-15	10-15
Density (kg/L)	0.71-0.74	0.71-0.72	0.72
Sulfur (wt.%)	<0.01	0.01-0.2	0.06
Nitrogen (ppm)	<20	50-100	200
Gas oil			
Yield/feed (wt.%)	10-25	20-30	40-45
Density (kg/L)	0.850-0.875	0.840-0.860	0.866
Sulfur (wt.%)	<0.1	0.1-0.5	0.7
Nitrogen (ppm)	300-1200	>500	~1800
Vacuum gas oil			
Yield/feed (wt.%)	20-35	25-35	20-25
Density (kg/L)	0.925-0.935	0.925-0.970	1.010
Sulfur (wt.%)	0.25-0.50	0.5-2.0	2.2
Nitrogen (ppm)	1500-2500	1600-4000	4300
Vacuum residue			
Yield/feed (wt.%)	30-60	15-35	10-20
Density (kg/L)	0.990-1.030	1.035-1.100	1.160
Sulfur (wt.%)	0.7-1.5	1-3	2.7
Nitrogen (ppm)	3000-4000	>3300	11000
Asphaltenes (heptane)	5-10	>20	26

50 tons per day of the catalytically active solid. Therefore, the integration with an industrial process (e.g., aluminium production) generating low-cost solids would enhance the viability of slurry bed reactors

The safety aspects of hydroprocessing operations deserve attention. Decades of the experience using heavy feeds varying widely in properties shows that it is quite easy and safe to operate fixed bed reactors. The additional high-pressure equipment upstream and downstream of the moving bed and ebullated bed reactors adds to the complexity of the operation. More severe conditions, i.e., higher temperatures and pressures, than in fixed bed reactors indicate that ebullated bed reactors may require special materials for the construction of equipment, similarly as it is for slurry bed reactors. However, the simple features, i.e., no need for internals, suggest that the design of the slurry bed reactor may be less challenging compared with that of the moving and ebullated bed reactors, although the selection of material for the construction of the former may be more demanding.

CHAPTER 4

Catalyst Deactivation

Deactivation is evidenced by decline in the rate of hydroprocessing reactions with time on stream. In refinery practice, the activity decline is offset by increasing temperature, as it is shown in Fig. 4.1. According to Fig. 4.1, three regions of the activity decline are generally observed. The efforts have been made to develop more active and stable catalysts with the aim to minimize the loss of activity. Then, the catalyst consumption and generation of spent catalysts would be decreased as well. The extent of deactivation depends on several parameters, e.g., properties of feeds, operating conditions, structure of catalysts, etc. For example, Fig. 4.2 [143] shows the effect of temperature and H_2 pressure on carbon and vanadium deposited on catalyst as well as on surface area. The increase in V deposition with increasing temperature may be attributed to enhance hydrodemetallization (HDM) via non-catalytic route [27]. However, the H_2 pressure must be maintained to ensure an optimal H_2S/H_2 ratio. Thus, importance of the H_2S/H_2 ratio for controlling the coke deposition on catalyst during hydroprocessing operations has been confirmed [27,49,53,96]. Obviously, a significant difference between the catalyst deactivation during hydroprocessing of heavy feeds and that of light feeds can be anticipated.

For atmospheric distillates, the catalyst deactivation is dominated by the poisoning involving N-compounds and coke deposition. The N-bases, which are always present in every petroleum feed, contribute to the catalyst deactivation by preferentially adsorbing on active sites and as such slow down the hydrogen activation process [49,55,56,144]. General trends indicate that during the operation, nitrogen accumulates in coke as indicated by its increasing content with time on stream as coke becomes more refractory. It has been established that the relative contribution of N-bases to the overall loss of catalyst activity increased from residues towards vacuum gas oil (VGO)/heavy gas oil (HGO) feeds and atmospheric distillates. In other words, poisoning effect increases with decreasing molecular weight of N-compounds in the feed. Furthermore, it was indicated that the conversion of N-compounds to hydrocarbon products might be influenced by the H_2S/H_2 ratio [10]. There may be the optimal H_2S/H_2 ratio for which the conversion (e.g., hydrodesulfurization [HDS] and hydrodenitrogenation [HDN]) reached a maximum [40]. The optimal ratio may exhibit a continuous change with time on stream because of the catalyst deactivation. This further contributes to the complexity of deactivation mechanism. This ratio may vary from feed to feed and from catalyst to catalyst. In

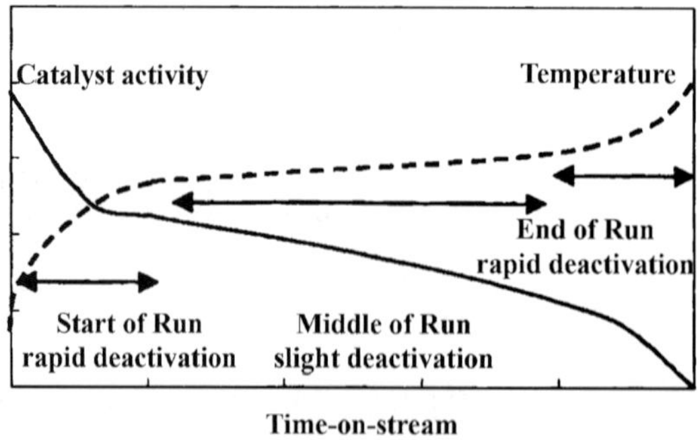

Figure 4.1: Temperature increase with time on stream to offset catalyst deactivation.

the fixed-bed reactor, this ratio may change between the inlet and outlet of the reactor. The situation becomes even more complex in multilayer and/or multistage catalytic systems.

For the asphaltenes and metals containing feeds, the catalyst deactivation by coke and metals occur simultaneously. For HDM catalysts, some reports suggest that more than 50% of catalyst

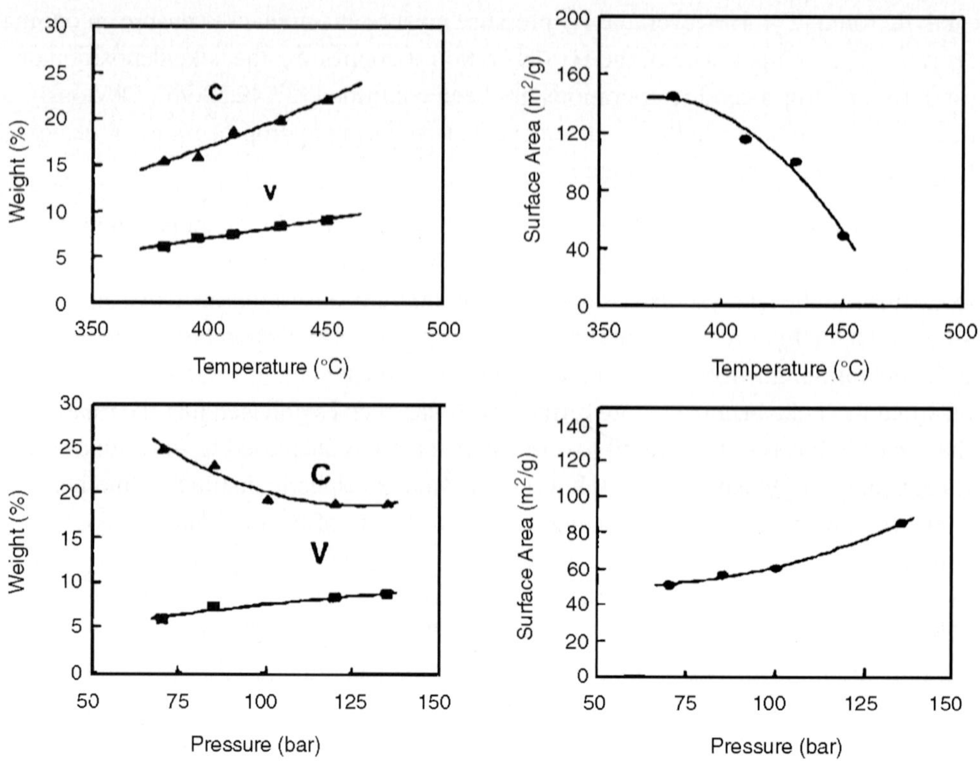

Figure 4.2: Effect of temperature and hydrogen pressure on deposition of carbon and vanadium on catalyst as well as on catalyst surface area [From ref. 143. Reprinted with permission].

deactivation is caused by coke [80], whereas the overwhelming information confirmed metals as the main cause of deactivation [10,145–147]. However, these statements and/or information tend to oversimplify the actual events, particularly in the case of fixed-bed reactors. Evidently, for a high metal content feed, front of catalyst bed will be deactivated by metal deposits. The contribution of metals to deactivation will then decrease and that of coke increase towards the outlet of the catalyst bed. However, the front zone of the metals deactivated catalyst bed will gradually move towards the outlet until the entire bed is deactivated. Before this point, HDM reactions occurring near the end of catalyst bed are affected by the coke deposited during the initial stages of the operation. It is therefore obvious that the relative contribution of coke and metals to deactivation will vary between the inlet and outlet of catalyst bed.

The above discussion suggests that relative contribution of coke and metals to catalyst deactivation will also vary with time on stream. Thus, during very early contacts of catalyst with a heavy feed (start of run region in Fig. 4.1), the coke deposition may dominate catalyst deactivation. At this point, little contribution of metals to the overall loss of catalyst activity may be evident. General trends (Fig. 4.3) show that coke deposition reaches a steady state, while the contribution of metals increases almost linearly with time on stream.

For the graded systems comprising either several layers of the different catalysts in one fixed-bed or several fixed-bed reactors containing different catalyst each connected in a series, e.g., HDM, HDM/HDS and HDS in the first, second and third reactor, respectively, the contribution of metals to deactivation will decrease from the first towards the third reactor [12,148]. At the same time, the contribution of coke to deactivation will increase. It should be noted that N-compounds in the feed are gradually converted to the hydrogenated N-containing intermediates. The basicity of the latter is greater than that of the N-compounds originally present in the feed [144]. This indicates an increased contribution to catalyst poisoning with

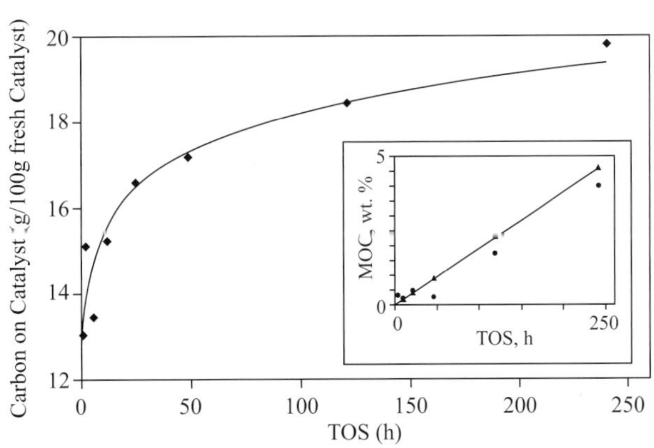

Figure 4.3: Deposition of carbon and metals on Mo/Al_2O_3 catalyst in atmospheric residue desulfurization (ARDS) process versus time on stream.

time on stream before the N-intermediates are completely converted to hydrocarbons. The N-intermediates may be at least partly responsible for the increased coke formation in the downstream reactors.

The study of Al-Nasser et al. [149] gave detailed accounts of the selection of the optimal catalyst bed combinations for the graded system comprising four fixed-bed reactors, which are part of the commercial atmospheric residue desulfurization (ARDS) process. Thus, every stage required a different catalyst. In addition, a control of the H_2S/H_2 ratio between the stages may be necessary to ensure that the concentration of H_2S is not in the inhibition region. To certain extent, this may be achieved by withdrawing a portion of gaseous products between the reactors and replacing them by a make up H_2 to ensure that the optimal H_2S/H_2 ratio is maintained. An option involving scrubbing H_2S from the gaseous effluent of the reactor before entering the subsequent reactor may be less practical.

The study of Rana and Ancheyta [150] indicated the complex deactivation patterns, which resulted from different experimental conditions. In this case, the bench scale downflow reactor and the upflow microreactor were used to study the Maya heavy crude and 50/50 blend on the Maya crude and diesel oil, respectively. The former reactor could accommodate 10 times more catalyst (e.g., 100 mL/85 g). The experiments in microreactor could not be conducted without blending the crude. Moreover, the experimental conditions were different, i.e., 653 K, 5.4 MPa and 10 L/h of H_2 in microreactor, compared with 673K, 7.0 MPa and 100 L/H of H_2 in bench scale reactor. The summary of these results (after 120 h on stream) using the $CoMo/Al_2O_3$ and $NiMo/Al_2O_3$ catalysts is shown in Table 4.1 [150]. The experimental conditions had the most pronounced effect on hydrodeasphaltization (HDAs). Thus, for the $CoMo/Al_2O_3$ catalyst, significant decrease in the HDAs conversion in the bench scale unit compared with the microreactor was observed, whereas the opposite trend was observed for the $NiMo/Al_2O_3{\cdot}TiO_2$ catalyst. At the same time, for both catalysts, the loss of the HDM, activity was more evident in the bench scale unit than in the microreactor. These results may be used to illustrate how the observations and conclusions reached during the catalyst deactivation studies can be influenced by experimental conditions.

Table 4.1: Catalyst activities in microreactor (MR) and bench scale reactor (BS) [From ref. 150. Reprinted with permission].

Catalyst	Conversion (%)		
	HDS	HDM	HDAs
$CoMo/Al_2O_3$-MR	56.6	50.0	40.0
$CoMo/Al_2O_3$-BS	56.6	31.2	14.9
$NiMo/Al_2O_3$-MR	37.8	35.4	34.0
$NiMo/Al_2O_3$-BS	35.6	15.7	45.7

HDM: hydrodemetallization; HDS: hydrodesulfurization; HDAs: hydrodeasphalting.

4.1 Deactivation Due to Structural Change of Catalyst

During the hydroprocessing of light feeds, the operation lasting several years may be anticipated. A prolonged exposure to operating temperatures may result in recrystallization of catalyst causing a change in porosity. In addition, an increase of size of the MoS_2/WS_2 crystallites in normal direction may occur. This would be an indication of the conversion of the type I active phase to a more active type II active phase. At the same time, the growth in a lateral direction would have an opposite effect. In this regard, the studies of Yokoyama et al. [151] showed that the lateral growth of MoS_2 crystallites in the $CoMo/Al_2O_3$ catalyst was partly responsible for the loss of activity during the hydroprocessing of VGO at about 660 K and 5.9 MPa,

Because the catalytically active site comprises coordinatively unsaturated sites (CUS), it is essential that their stability during the operation be maintained for a long period of time on stream. This requires the suitable H_2S/H_2 ratio to ensure a desirable size of CUS comprising a sulfur vacancy and SH groups at its proximity [55]. It was established that the number of vacancies decreased with increasing H_2S/H_2 ratio. On the other hand, at low H_2S/H_2 ratios, the catalyst over reduction may occur. Loss of sulfur from the active phase during the reaction has been reported to be the main cause of initial deactivation of hydroprocessing catalysts [152]. Such situation favors the adsorption of N-bases as well as deposition of coke and disfavors hydrogen activation.

In an effort to simulate deactivation, Tanaka et al. [153] conducted the accelerated aging in the pilot plant at a higher temperature than that used in commercial units. In the former case, the activity loss due to the lateral crystal growth was more pronounced than that observed in the commercial unit operating at lower temperatures for much longer time on stream than that used during the accelerating aging experiments. This agreed with the observations made by Gamez et al. [154] who studied the spent $CoMo/Al_2O_3$ catalyst used for hydroprocessing of the mixture of atmospheric gas oil (AGO) and VGO. Thus, only a minor change in morphology of the MoS_2 crystallites was observed after 12 months on stream in a commercial unit operating at lower temperature than that used by Tanaka et al. [153] during the accelerating aging. These observations suggest that temperature may be the main parameter influencing the catalyst recrystallization. It is therefore apparent that the catalyst deactivation patterns may not be properly identified by accelerating aging.

Promoter segregation from the mixed Co-Mo-S and Ni-Mo-S phases has been reported in several studies [152,155]. Eijsbouts et al. [156,157] demonstrated that catalyst deactivation by MoS_2 sintering and segregation of promoters (Ni- or Co-sulfides) was quite extensive in hydroprocessing units operating at high temperatures.

Usman et al. [158] studied the thermal stability of Co-Mo-S structure by chemical vapor deposition (CVD) technique. The results revealed that the CoMoS structure was thermally stable to the treatment at 623 K, slightly unstable at 673 K and remarkably destroyed above 773 K. It was suggested that at high temperatures, part of the Co atoms is detached from the Co-Mo-S structure, leaving some MoS_2 edge sites vacant, while the detached Co atoms form catalytically inactive Co-sulfide clusters. The catalyst support influenced the thermal stability of Co-Mo-S structure; Co-Mo-S structure supported on Al_2O_3 was thermally more stable than that supported on either boron modified Al_2O_3 or SiO_2. Quantitative calculations suggested that at 873 K about 30% of the Co-Mo-S structure supported on Al_2O_3 was decomposed after a 2-h treatment in 10% H_2S/H_2 stream, in contrast to about 50% of that supported on boron-modified Al_2O_3 and SiO_2.

Recently, Guichard et al. [159] investigated the stability of Co-Mo-S and Ni-Mo-S phases in working state using a variety of techniques such as X-ray photoelectron spectroscopy (XPS), TEM, energy dispersive X-ray (EDX), density functional theory (DFT), calculations and catalytic measurements. They concluded that part of Co and Ni were destabilized and segregated from the edges of the Co-Mo-S and NoMoS crystallites under the reaction conditions (high temperature and highly reductive environment).

The chemical composition of the original active phase on catalyst may be gradually modified by the metals deposited from heavy feeds during the operation. The effect of deposited V and Ni on the catalyst activity is rather complex. Moreover, the deactivating patterns will change with progressive deposition with time on stream. For example, the deposits had beneficial effects on HDM reaction as it was demonstrated by a gradual increase in catalyst activity up to maximum attained between 15 and 20 wt.% of the deposited metals [154,160]. Then, the HDM activity began to decline with further increase in the metal deposition. Almost certainly, the activity decline resulted from the change in pore size distribution, which affected the diffusion of reactant molecules into the catalyst pores. Therefore, for an active HDM catalyst, porosity may be at least as important parameter as is its chemical composition. Thus, industrial experience showed that about 5 wt.% of MoO_3 in the absence of promoter on the γ-Al_2O_3 support possessing suitable porosity resulted in the active catalyst for HDAs and HDM. Such catalysts have been used industrially. Other catalyst functionalities, e.g., HDS and HDN, were influenced by the metal deposits differently than HDM. This resulted from the transformation of the Co(Ni)MoS phase into the VMoS phase, which was less active than the former [161]. Moreover, it was reported that the unsupported V_2S_3-sulfide alone exhibited some activity for hydrogenation (HYD) and HDS [162–164]. However, this was only demonstrated for model compounds rather than for heavy feeds.

4.2 Deactivation by Coke and Nitrogen Bases

For distillate feeds, coke deposition and poisoning by N-bases are the main causes of catalyst activity decline. To various extends, the deactivation by coke and N-compounds occur in parallel. For N-compounds, deactivation results from their strong adsorption on the catalytic sites. This slows down the activation of hydrogen, the availability of which is crucial for hydroprocessing reactions to occur as well as for slowing down coke formation [55]. Moreover, the prolonged adsorption of N-compounds diminishes the access of other reactant molecules to catalytic sites. Therefore, at least part of the coke is formed as a consequence of the catalyst poisoning by N-compounds. The extensive information on catalyst deactivation by coke and N-compounds was reviewed in details and published elsewhere [49]. It should be noted that a deep or even ultradeep HDS could not be accomplished without minimizing poisoning effect of N-compounds. There may be a difference between the poisoning effects of the N-compounds present in VGO compared with that in HGO. This is supported by the presence of the fractions boiling below 350 °C in the latter, whereas such fraction is not present in VGO. Thus, depending on the preparation of HGO, this may represent as much as 30% of the HGO fraction. It was indicated earlier that the poisoning effect of N-compounds increased with decreasing boiling range of the fractions [10,49,165]. For example, the rate of the HDN of quinoline, which was added to the 616–666 K, 706–756 K and 797+ K fractions, increased in the same order [166]. In practical situation, e.g., between the inlet and outlet of the fixed-bed reactor, the inhibiting effect of N-compounds may exhibit a maximum before most of the N-compounds were converted to hydrocarbons [115]. Similarly, in a multistage system, the inhibiting effect of N-compounds will increase from the first stage and reach a maximum in one of the downstream reactors.

The poisoning effect of N-compounds on catalyst activity was clearly demonstrated in the study published by Kaernbach et al. [167] on HDS of the distillate feed derived from the Russian crude. In this case, N-compounds were separated from the feed by ion exchange chromatography prior to the experiments performed at 633 K and 7 MPa in the continuous fixed-bed reactor. As expected, the HDS conversion was much greater in the absence of N-compounds. Similarly, the HDS activities increased by almost 60% after the N-compounds were removed from the feed by adsorption with silica-alumina [168]. The poisoning by N-compounds decreased with increasing temperature because of their diminished adsorption on catalytic sites. The adverse effect of N-compounds in the feed on catalyst activity was also confirmed by Massoth et al. [169,170].

The catalyst samples taken after 12 months on stream from the different depths of the single fixed-bed used for hydroprocessing of a VGO (633 to 673 K; 8 MPa) had different coke deposition patterns [171]. The amount of coke increased with the increasing depth of the bed. The graphitic nature of coke increased towards the end of the bed as well. The predominantly amorphous structure of coke on the inlet and graphitic structure on the outlet of catalyst bed observed by Koizumi et al. [172] is in agreement with the results of Anemia et al. [171]. It was proposed that the increasing temperature towards the end of fixed-bed (because of the increased rate of exothermic reactions) was the main contributor to the difference in coke structure. Almost certainly, the increased rate of poisoning by N-compounds was an important contributor as well. Thus, the HYD of N-heterorings, occurring near the front of fixed-bed,

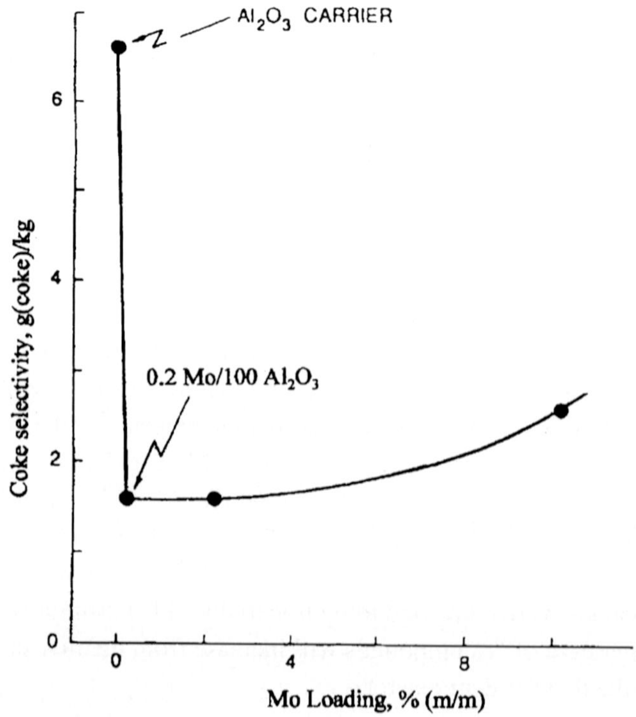

Figure 4.4: Effect of Mo loading on coke selectivity (vacuum gas oil [VGO], 3 MPa) [From ref. 174. Reprinted with permission].

resulted in the formation of N-intermediates possessing a higher basicity than the corresponding-containing reactants in the feed [144]. An ultimate result of this was the diminished availability of the active surface hydrogen [55].

In the studies of de Jong et al. [173–175] on hydroprocessing of a VGO, the coke formation was observed to be dependent on the catalyst structure. Thus, according to Fig. 4.4, the addition of a small amount of Mo to γ-Al_2O_3 resulted in the significant decrease in coke formation [173,174]. The coke build-up increased with the increasing amount of Mo, i.e., with the increasing catalyst activity. Therefore, the formation of this coke, termed as the "chemical reaction coke", was associated with hydroprocessing reactions. The influence of catalyst structure on the coke formation was further demonstrated in the study on aging of the $CoMo/Al_2O_3$ and Mo/Al_2O_3 catalysts [175]. The aging was conducted at 723 K and 3 MPa in the fixed-bed reactor using VGO. For both catalysts, the amount of deposited coke was similar. However, the former catalyst was more deactivated because of the much greater coverage by coke, i.e., about 90 and 50% for the $CoMo/Al_2O_3$ and Mo/Al_2O_3 catalysts, respectively. For the latter catalyst, the islands of coke were present, whereas for $CoMo/Al_2O_3$ the coke was more evenly distributed. The detailed spectroscopic evaluations of the spent catalysts from the hydroprocessing of a VGO conducted by van Dorn et al. [176–178] provided the information on morphology of the coke deposited on the catalyst surface. They concluded that most of the coke was located far from the active phase in the form of the irregularly shaped structures covering the entire surface.

Some information [179] showed that for the model feeds such as carbazole and alkylated carbazoles, as well as polyaromatic hydrocarbons (PAHs), the initial coke deposition (less than 5 wt.%) occurred predominantly on the Al_2O_3 support. Consequently, little deactivation was observed during this initial period. However, when a VGO was used under similar conditions, the catalyst poisoning by N-compounds and PAH present in the feed was observed few minutes after the start of the run [180]. Similarly, the initial poisoning by N-compounds was also observed during the hydroprocessing of the Maya heavy crude, although to a lesser extent [181]. While using Kuwait atmospheric residue (AR), Matsushita et al. [182] concluded that the coke formed during the early stages of the operation deposited on the support rather than on active phase. At the same time, N-compounds in the feed adsorbed near and/or on the active sulfides phase and as such contributed to the initial catalyst deactivation. Therefore, the catalyst deactivating patterns observed for model compounds [179] were rather different than those observed for the real feeds [180–182].

Figure 4.5 shows the decrease in the H/C ratio of coke with time on stream [183]. At the same time, the N content of coke increased linearly. This suggests that during hydroprocessing of the diluted vacuum residue derived from the Chinese heavy crude, the accumulation of carbon and nitrogen in the coke deposited on catalyst increased, whereas that of hydrogen decreased with the increasing time on stream. An ultimate result of this change was the increase in the content of graphite-like structure of the coke on catalyst. The experiments were performed at 683 K and 8 MPa in an autoclave. During these experiments, no attempt was made to decouple the contribution of metals to deactivation from that by the coke and N-compounds. The duration of these experiments (e.g., maximum 5 h) suggests that these observations reflect the events

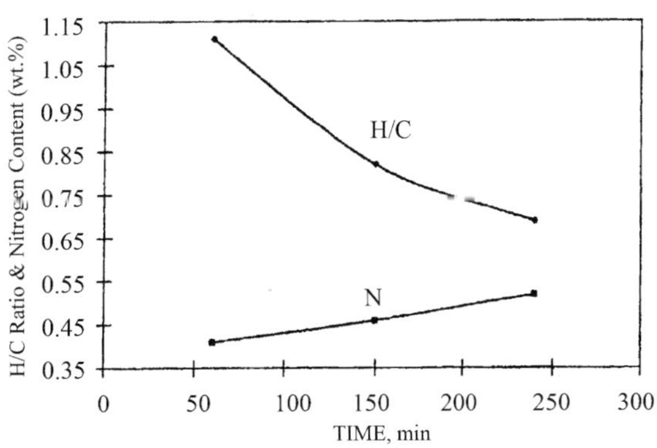

Figure 4.5: Effect of time on stream on H/C ratio and nitrogen content of coke on catalyst [From ref. 183. Reprinted with permission].

occurring during the initial stages of operation [183]. Contrary to these observations, Callejas et al. [181] reported a decrease in the nitrogen content in coke with time on stream for the Maya heavy crude during the early stages of the experiment. This suggests that this coke was rather "young", still possessing some reactivity. But, the "young" coke, which is more soluble, had more deactivating effect on HDS activity than less soluble coke [184]. Also, in area of active phase, coke deposits were thinner than on the bare support [185]. For extrudates used in hydroprocessing of several gas oils, typical M-shape profiles of coke were observed [186]. This suggests that coke was deposited by sequential deactivation mechanism. It is then evident that the observed trends in coke formation and its structure depend on the origin of crude, type of catalyst and operating conditions. The time on stream at which the coke evaluation was conducted is important as well. In the case of residues, metals deposited on the catalyst surface during the operation may modify deactivating pattern by coke compared with distillate feeds. For residues, the contribution of N-compounds to the overall deactivation will increase with the progress of hydroprocessing, i.e., in the fixed-bed reactor, from the inlet towards the outlet.

4.3 Combined Effect of Coke and Metals on Deactivation

Multireactor systems have to be used for residues upgrading. For example, the ARDS process discussed in Chapter 3 has been used for the hydroprocessing atmospheric residue derived from Kuwait crude. The extent of deposition of metals and coke in four reactors, which is a part of the process, is shown in Fig. 4.6 [12]. In this case, the feed enters reactor 1 and products exit reactor 4. As expected, the deposition of metals and associated deactivation decreased from reactor 1 towards reactor 4, whereas reversed trend was observed for coke deposition. The loss of pore volume and surface area exhibited similar trends as the coke deposition (Fig. 4.7). This suggests that coke had a more detrimental effect on the pore volume and surface area than metals. Similar set of catalysts as shown in Figs. 4.6 and 4.7 [12,187]

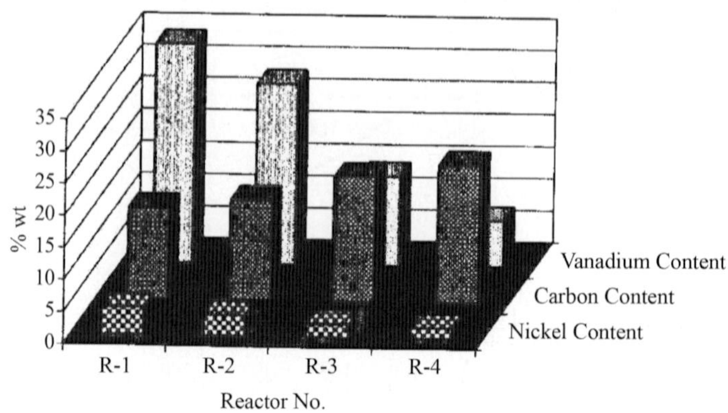

Figure 4.6: Content of vanadium, nickel and carbon in spent catalysts from atmospheric residue desulfurization (ARDS) reactors [From ref. 12. Reprinted with permission].

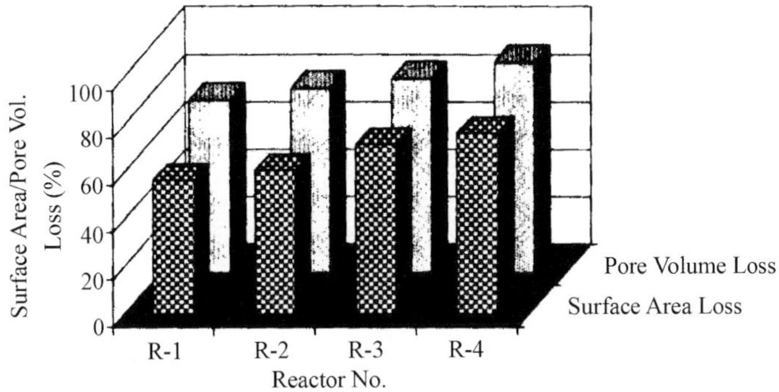

Figure 4.7: Loss of surface area and pore volume in spent catalysts from atmospheric residue desulfurization (ARDS) reactors [From ref. 12. Reprinted with permission].

were used to estimate level of the surface area recovery on the oxidative regeneration [12]. For catalysts taken from reactors 1 and 2, the recovery was rather low, suggesting a permanent deactivation by metals. A significantly higher level of the surface area recovery was achieved for the catalysts taken from reactors 3 and 4. For these catalysts, coke deposition was the primary cause of catalyst deactivation.

Figure 4.8 identifies major factors causing the catalyst deactivation during hydroprocessing of the asphaltenes and metals containing feeds in the three-stage ebullated-bed reactors process [188]. The results were obtained using the heavy feed containing about 400 ppm of V + Ni. For every stage, the last point on the curve was recorded after 110 days on stream. Therefore, for the stage 3 catalyst, deactivation was caused mainly by coke deposition, whereas for stage 1 catalyst, the deposition of metals and restricted diffusion were the predominant modes of deactivation with the contribution of the latter increasing with time on stream until it became the main cause of the loss of activity. Furthermore, the relative contribution of these factors depends on the properties of heavy feeds. Deactivation patterns observed in stage 3 may

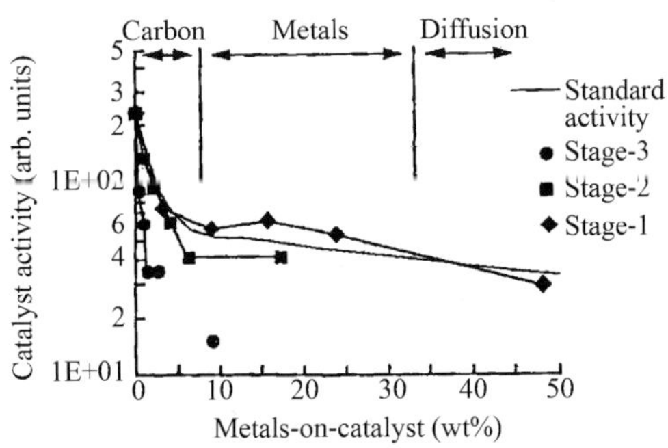

Figure 4.8: Major factors causing catalyst deactivation versus metals on catalyst [From ref. 188. Reprinted with permission].

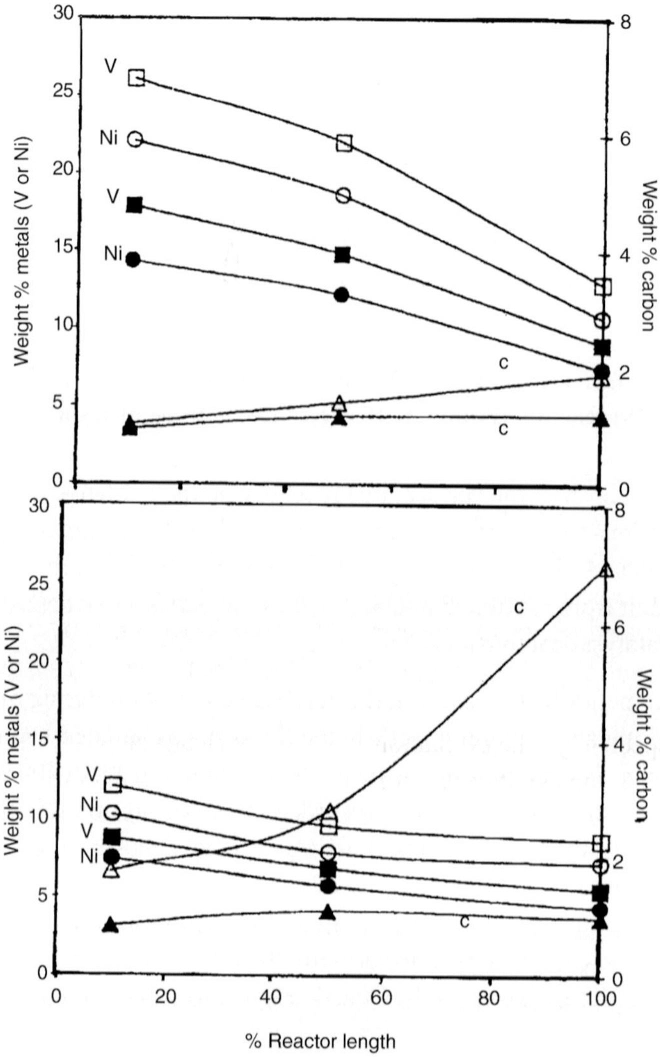

Figure 4.9: Deposition of metals and coke along the length of 1st and 2nd reactors for catalysts in Table 7.1; open symbols catalyst A, full symbols catalyst D [From ref. 189. Reprinted with permission].

approach those observed during hydroprocessing of VGO. Then, different types of catalysts may be required for every stage.

Figure 4.9 [189,190] shows the deposition of metals and coke in two trickle-bed reactors connected in a series, used for hydroprocessing of the deasphalted oil (DAO) containing 27 ppm of V + Ni and less than 1 wt.% of asphaltenes. This DAO was obtained by deasphalting of the vacuum residue derived from conventional crude. The properties of catalysts A and D used for this study are shown in Table 4.2 [190]. The catalysts (in baskets) were placed in the

Table 4.2: Properties of catalysts [From ref. 189. Reprinted with permission].

	Catalyst A	Catalyst D
Mo (wt.%)	8.1	8.0
Ni (wt.%)	2.3	2.0
Particle size (mm)	1.0	2.1
Surface area (m^2/g)	320	146
Average pore diameter (A)	126	233
Surface area (m^2/g) in pore size range (A)		
0–0	32	4
60–00	80	9
100+	208	133

central axis of the two fixed-bed reactors, which were part of a commercial unit. In the first reactor, baskets were placed at the top and middle of the bed, whereas in the second reactor in the middle and bottom of the bed. The objective of the commercial run was to produce feed for the subsequent fluid catalytic cracking (FCC) [189,190]. The system operated at the total pressure of about 10 MPa. To compensate for deactivation, the temperature between the start-up and shutdown was increased from 603 to 628 K and from 646 to 658 K in the first and second reactor, respectively. The evaluation of the catalyst was performed after 241 days on stream. For catalyst A, the significant increase in the coke formation towards the end of the second reactor should be noted compare with a little change for catalyst D. It is suggested that in the former case, poisoning of the catalyst by N-bases was the main cause of the catalyst deactivation. Catalyst A exhibited a greater metal storage capacity than catalyst D in spite of the larger average pore diameter of the latter. Most likely, smaller particle size of catalyst A than that of catalyst D ensured more efficient catalyst utilization. Moreover, the surface area of the former catalyst, in the 60–100 A pore range, was nine times greater than that of the catalyst D.

It is expected that the amount of coke deposited initially is governed by the content of asphaltenes and resins in the feed. This is supported by the results in Fig. 4.10 [191]. In this case, the feeds with different content of resins and asphaltenes were obtained by solvent deasphalting of the two vacuum residues and one atmospheric residue derived from different crudes each. A close examination of the scatter of data in Fig. 4.10 indicates that the amount of deposited coke was influenced by the origin of the asphaltenes and resins. To certain extent, the observations made by Morales and Solari [192] complement the results in Fig. 4.10. These authors used several heavy feeds and established the correlation between the content of asphaltenes in the heavy feed and its HDS, HDM and Conradson carbon residue (CCR) conversions. Thus, the conversions decreased with the increasing content of asphaltenes but they leveled off when about 20 wt.% of asphaltenes in the feed was approached. However, it is

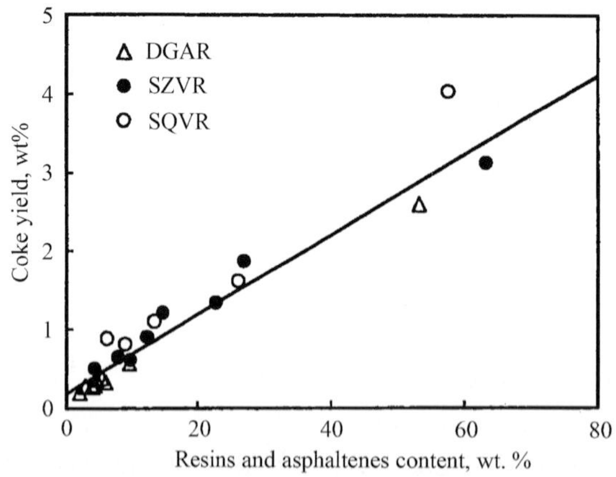

Figure 4.10: Coke on catalyst as function of the content of resins and asphaltenes in fractions from Dagang AR (DGAR), Saudi light VR (SQVR) and Saudi medium VR (SZVR) at 673 K and 8.5 MPa of H_2 over $NiMo/Al_2O_3$ [From ref. 191. Reprinted with permission].

unlikely that these observations can be generally applied to all heavy feeds because the chemical structure of asphaltenes may be another parameter influencing coke deposition. Thus, for heavy feeds having a similar content of asphaltenes, but of different chemical structure, the coking propensity increased with the increasing aromaticity of asphaltenes.

It is believed that during very early stages of the operation, there is little effect of metals on coke formation. On the other hand, the coke formed initially can have a pronounced effect on the rate of the metal deposit formation because of the partial pore plugging by coke. Moreover, this part of the support on which metals could deposit was already occupied by coke. It is therefore critical that the rate of coke formation is kept at minimum to ensure a high HDM activity of catalysts. In this regard, the results in Fig. 4.4 [174,175] can have important implications on the design and preparation of the HDM catalysts, although they were obtained for a VGO feed. Thus, the coke formation may be kept at a minimum by selecting an optimal composition of catalyst. At the optimal composition, formation of the "chemical coke" associated with hydroprocessing reactions is slow, thus ensuring a high HDM activity due to the diminished interference by coke. However, this was not confirmed in the study involving the Kuwait atmospheric residue (90 ppm of V + Ni; 3.6 wt.% asphaltenes) conducted by Marafi et al. [193] who compared the Mo/Al_2O_3 (3 wt.% Mo) with $NiMo/Al_2O_3$ (8 wt.% Mo and 2 wt.% Ni) catalysts having pore volume of 0.7 and 0.5 mL/g, respectively. Typically, the catalysts were used for HDM and HDS, respectively. Between 633 and 693 K and at 12 MPa, consistently more coke was deposited on the HDM catalyst. As expected, the H/C ratio of coke on the HDM catalyst was much lower than that on the HDS catalyst because of the higher HYD activity of the latter. The contradictory results reported in the literature underline complexity of the simultaneous deactivation of catalyst by coke and metals, particularly during the initial stages. This may be attributed to the differences in experimental conditions. This is evidenced by the different deactivation patterns for different feeds and different catalysts observed initially. In this case, the method used for catalyst presulfiding may be an important factor for controlling the initial coke deposition.

Undoubtedly, during the hydroprocessing of heavy feeds containing metals, the structure of coke on catalyst will be progressively influenced by the metals deposited from the feed. This was indeed observed by Zeuthen et al. [194,195]. In this case, the coke formed in proximity of the deposited V was more refractory, i.e., it had lower H/C ratio than the coke in the interior of pores. This suggests that in the course of the experiment, the V enhanced dehydrogenation of coke. Then, different forms of coke may be present on catalyst surface. The influence of metals on properties of coke was reported by Galiasso Tailleur and Caprioli [196]. They observed that, initially, coke filled pores before depositing on the exterior of the catalyst particles. However, because of its permeability, the liquid phase could reach catalyst surface. The permeability of the coke was gradually decreasing before catalyst was completely deactivated. The permeability decrease was complemented by the increased deposition of metals on catalyst surface. Therefore, it was suggested that metals contributed to the loss in permeability. In this case, the vacuum residue derived from Venezuelan crude was studied at 23.6 MPa and between 683 and 703 K over the $NiMo/Al_2O_3$ catalyst.

After deposition on catalyst surface, V and Ni are gradually converted to sulfides. The overwhelming information showed that V tends to deposit on the external surface of catalyst particles, whereas the radial distribution of Ni is more uniform [10,191,79,197]. This is illustrated in Fig. 4.11 [197]. However, for a macroporous HDM catalyst (pore volume of

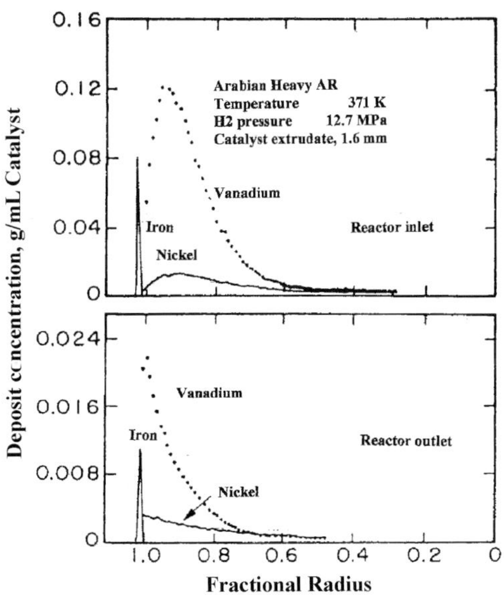

Figure 4.11: Effect of fractional radius on deposition of vanadium, nickel and iron [From ref. 197. Reprinted with permission].

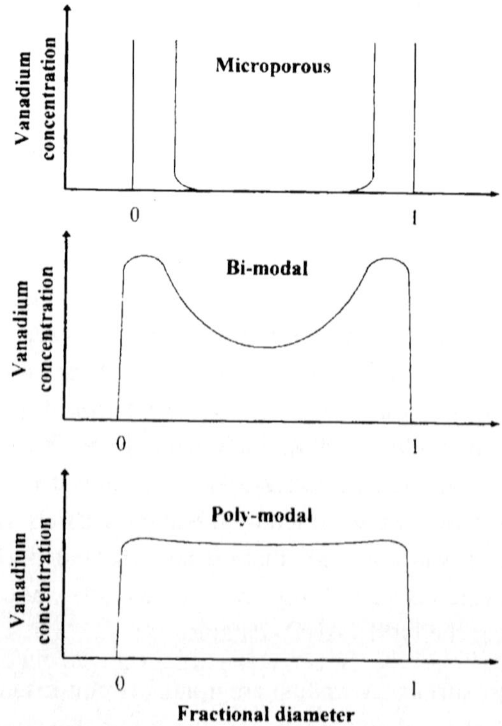

Figure 4.12: Effect of the type of porosity on radial distribution of vanadium [From ref. 199. Reprinted with permission].

0.95 cm^3/g), the difference between the distribution patterns of V and Ni was less pronounced [198]. An example of the effect of porosity on metal distribution is shown in Fig. 4.12 [143,199]. It was indicated that the presence of V in the vanadyl form is one of the reasons for the enhanced reactivity of V-containing porphyrins compared with Ni-containing porphyrins during the deposit formation on the catalyst surface. As the result of this, vanadium is deposited on the surface before it could diffuse into the catalyst interior. It has been observed that small amounts of V may deactivate catalyst because of the blocking active sites, whereas large amount of deposited V deactivates catalyst due to pore mouth plugging [145,200]. The pore mouth plugging by Ni deposits is much less evident. In fact, information suggests that the Ni deposited on the catalyst from the feed may improve catalyst performance, its HYD activity in particular [201,202]. Koyama et al. [203] proposed two regions of deactivation by metals, i.e., the initial one involving the poisoning of active sites and the other causing the decrease in effective diffusivity due to pore mouth plugging. It is believed that in both regions, the deactivating effect of the V deposits was more pronounced than that of the Ni deposits.

Arsenic in spent catalysts has been attracting attention mainly for environmental and safety reasons. Arsenic is naturally present in many crudes and usually concentrates in light fractions. Deactivation due to the presence of arsenic deposited on catalysts has been reported [204].

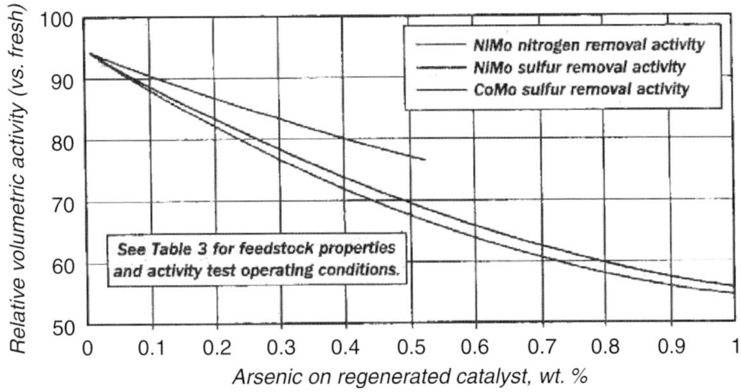

Figure 4.13: Effect of arsenic content in regenerated catalysts on relative volumetric activity [From ref. 204. Reprinted with permission].

Figure 4.13 confirms that arsenic build-up in regenerated catalyst, after successive utilization-regeneration cycles, had adverse effect on catalyst activity. Apparently, HDS was more affected than HDN. Also, deactivating effect was the catalyst structure dependent. The correlations in Fig. 4.13 were established from the results on evaluations of numerous samples of regenerated catalysts. The regenerated catalyst had good physical properties and had a low content of other contaminants.

4.4 Effect of Temperature and Hydrogen Pressure

Temperature is an important parameter, which can be used to control coke deposition. For distillate feeds, a linear increase in the coke deposition with temperature increase is anticipated. This was confirmed in the study of de Jong [173] on hydroprocessing of a VGO. Thus, as Fig. 4.14 shows, the coke build-up increased almost exponentially with increasing temperature to a maximum and then suddenly decreased. The sudden decrease in coke deposition was caused by the change in the flow patterns, i.e., the predominantly liquid system was gradually changing to gaseous system.

Figure 4.15 [205] shows trends in the coke build-up during hydroprocessing of a residue. The increased coke deposition to about 375 °C may be attributed to a slow conversion of heavy components of the feed thus prolonging their life on catalyst surface. This increases the chance for their conversion to coke. A faster conversion of resins than asphaltenes may be another reason for enhanced coke deposition. In such a case, the decreased compatibility of the feed would favor precipitation of asphaltenes. Above 375 °C, the conversion of coke precursors to lighter products successfully competed with their polymerization to coke. Apparently, above 440 °C, the coke formation was dominated by thermal effects. Other studies confirmed the trends in Fig. 4.15, although the temperature ranges were not identical. This is not surprising

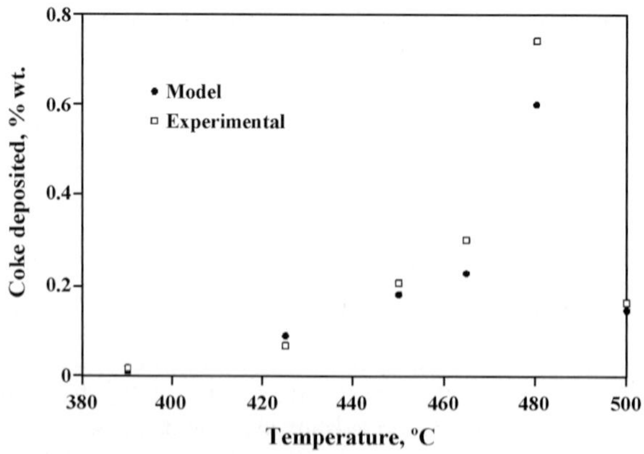

Figure 4.14: Effect of temperature on coke deposition (vacuum gas oil [VGO], $CoMo/Al_2O_3$, 3MPa) [From ref. 168. Reprinted with permission].

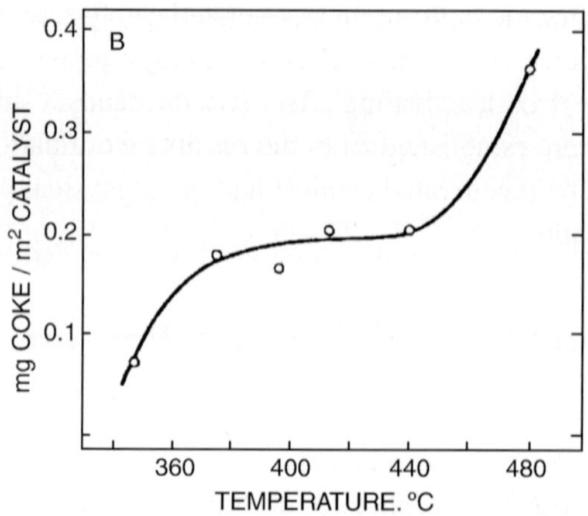

Figure 4.15: Effect of temperature on coke on catalyst [From ref. 205. Reprinted with permission].

when the differences in the origin of residues and H_2 pressure as well the type of catalyst employed used in various studies are taken into consideration.

The temperature effect on coke formation observed by Gualda and Kasztelan [145,206] differed from that in Fig. 4.15 [205]. In the former case, the coke build-up increased and reached a maximum before further increase in the rate of coke formation with temperature increase was observed. The atmospheric residues used in their studies contained ~110 ppm of V + Ni, i.e., about one third of that in the residue used in the previous studies [145,206]. On the other hand, in the case of the Kuwait atmospheric residue, the coke build-up increased linearly with increasing temperature from 633 to 693 K [143,207]. These observations again confirm that trends in the effect of temperature on the coke formation on catalyst surface depend on several experimental parameters. Therefore, it is not surprising to observe different trends in different studies.

In the study of Seki and Yoshimoto [208], the build-up of the "hard" coke, defined as the toluene insolubles on the spent catalysts ($Ni/CoMo/Al_2O_3$), was quantified. The catalysts were pre-aged during the treatment with Kuwait atmospheric residue (from 643 to 653 K, 14.0 MPa, 16 h). After pre-aging, the catalysts were washed in-situ with the light cycle oil (LCO) at 623 K for 6 h to remove "soft" coke before being used for aging tests using the demetallized atmospheric residue as the feed. The tests of 20 h duration were conducted at 643 and 703 K and 8.0 MPa. Under these conditions, the accumulation of the additional "hard" coke decreased with increasing temperature. This may be attributed to the more extensive conversion of asphaltenes to light products with increasing temperature as observed by Seki et al. [209]. Similar temperature effects on coke structures were reported by Takahashi et al. [210].

The importance of temperature for controlling coke deposition can be also established from the product analysis. This is evident from the results in Fig. 4.16 published by Marafi et al. [211]. In this study, the atmospheric residue derived from a Kuwait crude was used. Thus, decline in the content of asphaltenes and resins in products with the increasing temperature suggested that their contribution to coke formation became less important with increasing

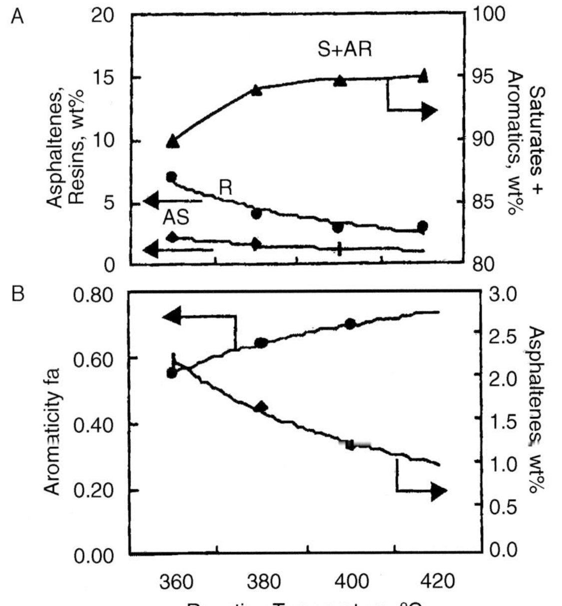

Figure 4.16: Temperature versus (A) content of asphaltenes and resins in products and (B) their aromaticity ($NiMo/Al_2O_3$, atmospheric residue [AR], 12 MPa) [From ref. 211. Reprinted with permission].

temperature. This may be attributed to the enhanced conversion of asphaltenes to light products. Consequently, the HDM rate should be increased as well.

At the same temperature, the H_2 pressure may be a critical parameter for controlling coke formation. It is however believed that the decreased coke formation caused by an increase in the H_2 pressure would favor the deposition of metals relative to that of coke. Richardson et al. [212] used Athabasca bitumen to study the H_2 pressure effect on the initial coke formation (between 1.5 and 5 h on stream) in the continuous stir tank reactor (CSTR) system and in an autoclave reactor using the commercial $NiMo/Al_2O_3$ catalyst at 703 K. After a rapid coke build-up during the first hour on stream, the coke formation did not change with the increasing ratio of the feed to catalyst. At the same time, increasing H_2 pressure from 7 MPa to more than 15 MPa decreased the amount of coke from about 17 wt.% to about 11 wt.%. In the study of Gualda and Kasztelan [145] on hydroprocessing an atmospheric residue, the amount of coke decreased from about 10 wt.% to about 4 wt.% by increasing the H_2 pressure from 2 to 15 MPa (Fig. 4.17). Moreover, the H_2 pressure had a pronounced effect on the H/C ration of coke on the catalyst. Figure 4.18 [212] shows the effect of H_2 pressure on the steady-state level of coke. It is believed that in the case of Athabasca bitumen, large asphaltenic molecules had the predominant role during the initial stages of coke formation. Thus, there was a sufficient amount of asphaltenes to form the same amount of coke even for the low feed/catalyst ratios.

Higashi et al. [213] studied the coke deposition on catalyst surface during the very early stages on stream at a low H_2 pressure using an atmospheric residue as the feed. The study was conducted in a pilot plant. They observed that the coke could not be removed and/or catalyst activity could not be recovered by increasing the H_2 pressure at the same temperature, during the later stage on stream. This indicated the permanent deactivation by coke. It is therefore essential that the coke deposition control by H_2 pressure begins at the start of the run. In this case, the loss of the HDS activity was noticed in particular. It was observed that the catalyst presulfiding was an important factor in controlling the initial coke deposition.

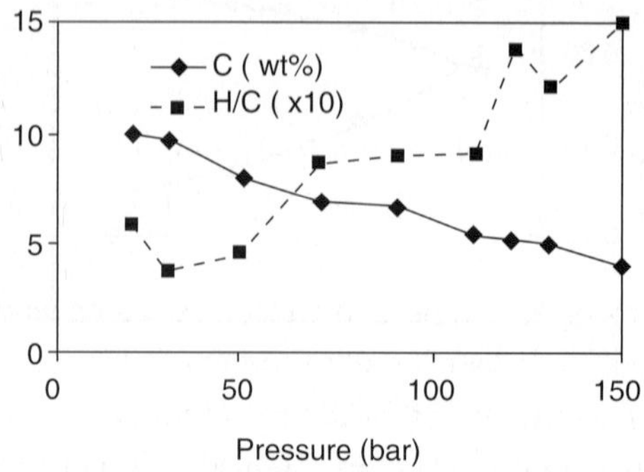

Figure 4.17: Effect of H_2 pressure on H/C ratio and amount of carbon on catalyst ($NiMo/Al_2O_3$, atmospheric residue [AR], 663 K) [From ref. 145. Reprinted with permission].

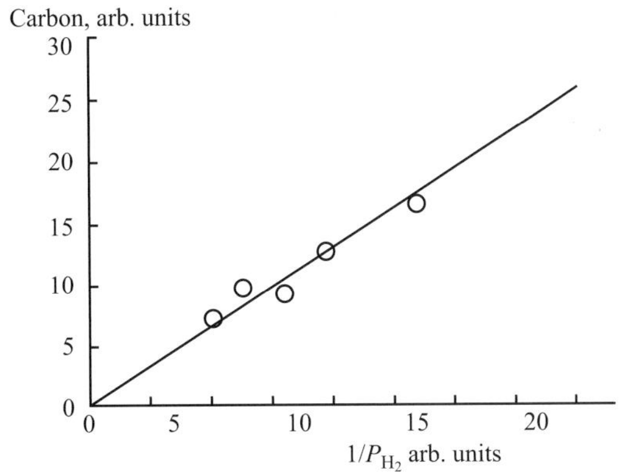

Figure 4.18: Effect of H_2 pressure (1/P_{H_2}) on carbon on catalyst [From ref. 212. Reprinted with permission].

Rather complex deactivation patterns were observed in the study of Kumata et al. [214] on the combined effect of temperature and H_2 pressure on the coke build-up. As the feed, they used the partially demetallized atmospheric residue derived from the heavy Arabian crude. The partial HDM of the feed was conducted over the typical HDM catalyst (4.2 wt.% of MoO_3 on γ-Al_2O_3). The experimental system comprised two trickle-bed reactors connected in a series. The adjacent reactor was loaded with the typical HDS catalyst of the Ni/CoMo/Al_2O_3 formulation. It was observed that at 653 K, coke was rather evenly distributed between the top of the first HDS reactor and bottom of the second HDS reactor. However, at 683 K, the coke build-up progressively increased in the same direction. When H_2 pressure was increased from 8.0 to 14.0 MPa at 653 K, the amount of deposited coke decreased from about 20 to 14 wt.%. This indicates that an optimal combination of temperature and H_2 pressure, for which coke deposition can be minimized, may be established.

4.5 Effect of Mechanical Properties of Catalyst

The desirable performance of the bed of catalyst cannot be achieved without the catalyst having an adequate mechanical strength. In the fixed-bed, the fine particles formed by attrition may be carried out with liquid streams, thus depleting the original load of catalyst. Moreover, in the fixed-bed, the fines may decrease the void space between the catalyst particles. This would affect the flow patterns of the liquid and gaseous streams leading to the development of pressure drops across the bed. Malfunctioning of the fixed-bed (e.g., development of channels), ending with the discontinuation of the operation, could be an ultimate result of these changes. Although this may not be catalyst deactivation in a true sense, the operating problems caused by fines of catalyst require attention. It was indicated earlier that if not removed from the feed, finely divided mineral matter might cause similar difficulties.

Table 4.3: Properties of spent catalyst particles from ebullated-bed reactor [From ref. 12. Reprinted with permission].

Property fouled	Spent mix	Lightly fouled	Heavily
Vanadium (wt.%)	10.6	4.4	13.8
Nickel (wt.%)	4.0	3.5	5.2
Carbon (wt.%)	16.2	15.8	16.3
Surface area (m^2/g)	68	122	55
Bulk density (kg/L)	1.09	0.97	1.21
Side crushing strength (lb/mm)	1.8	2.1	1.2
Pore volume (mL/g)	0.17	0.21	0.11
Particle length distribution (wt.%)			
<1.5 mm	25.2	14.4	40.0
1.5–0.0	42.3	23.5	37.0
3.0–0.0	32.5	61.3	23.0
>6.0	0	0.8	0

Because of the continuous motion of particles in ebullated-bed reactors, the depletion of catalyst material due to the particles attrition and/or disintegration is much more evident than in the fixed-bed reactors. Moreover, with the aim to enhance the active phase utilization, the typical diameter of the catalyst particles is 1 mm inside diameter (ID) or less. Without adequate mechanical strength, breaking of such particles could not be avoided. A vigorous mixing in ebullated-bed suggests that the fresh particles added periodically may be well mixed with the spent catalyst particles, which may need to be withdrawn. Then, a part of the particles are withdrawn with the spent catalyst without being completely utilized. Little information on these phenomena has been available until the work of Al-Dalama and Stanislaus [12] appeared in the scientific literature. The results from this study are shown in Table 4.3. On the basis of these results, it was estimated that the lightly fouled catalyst accounted for about 30 wt.% of the mixture. The catalysts were separated from the mixture by jigging technique using a mineral jig. Of particular importance is the length distribution of particles, which for fresh catalyst, was dominated by 3.0–6.0 mm particles. For the heavily fouled catalyst, more than 70% of these particles were broken to less than 3.0 mm length. There was a significant difference in surface area, pore volume and side crushing strength between the lightly and heavily fouled catalysts as well. It was established that fine particles could be carried out from the reactor together with the liquid streams. This represents a loss of activity per unit of the catalyst loaded. The lightly fouled particles withdrawn prematurely represent another source of the activity loss because of their incomplete utilization. These phenomena are physical and/or mechanical and in their nature differ from those occurring during catalyst deactivation. However, because the ultimate results are the loss of catalyst activity, they deserve attention during the catalyst design as well as during the operation.

4.6 Mechanism of Catalyst Deactivation

All evidence suggests that, during hydroprocessing of distillate feeds, the overall mechanism of catalyst deactivation is much less complex than that of the asphaltenes and metals containing feeds [10,27]. The presence of resins and asphaltenes in heavy feeds adds to the complexity of the mechanism of coke formation on catalyst surface. In this case, physical deposition (fouling) of the heavy components may dominate coke formation particularly during the early contact of catalyst with the feed. The extent of physical deposition may be influenced by the colloidal stability of heavy feeds. This suggests that for the asphaltenes containing feeds, both physical and chemical properties of the feed are important besides operating parameters such as H_2 pressure and temperature as well as the type of catalyst. Because most of the V and Ni in heavy feeds are associated with asphaltenic molecules, the mechanism of catalyst deactivation involves resins and asphaltenes, as well as metals.

4.6.1 Mechanism of Coke Formation

The mechanisms of coke formation established during the studies involving model compounds and light feeds form the basis for elucidating the mechanism occurring during the hydroprocessing of distillate feeds [10]. Consequently, this information aided in elucidating the mechanism of catalyst deactivation during hydroprocessing of heavy feeds as it was confirmed in the review published elsewhere [27].

4.6.1.1 Chemical Aspects

It has been generally known that the thermal cracking of C–C bonds begins at about 600 K. The primary products of cracking reactions are free radicals. Unless they are rapidly stabilized, free radicals can combine to large molecules and eventually to coke. The involvement of free radicals during coke formation was proposed by several authors [49,215–221]. The study published by Kubo et al. [220] provided a direct support for the involvement of free radicals. Thus, the coke formation was suppressed in the presence of a hydrogen donating liquid, which acted as the radical scavenger. Based on the bond strength, C_{AL}–C_{AL} bonds in methylene bridges are the most reactive, yielding the least stable radicals. Involvement of the C–H bonds scission in radical formation is much less evident unless the tertiary carbon is involved. For example, an aromatic structure with the isopropyl substituent attached would yield very stable tertiary radical. Free radicals can also be formed by the rearrangement of aromatic ring such as shown in Fig. 4.19 [215]. Subsequently, radicals gradually recombine to larger species and finally to coke, unless they are stabilized. In the mechanism proposed by Fetzer [222], small aromatic rings were converted to coronene, which was subsequently coupled to form either dicoronylene or even higher molecular weight polycondensed aromatic hydrocarbons. The experimental observations made by Nakamura et al. [223] and Kubo et al. [220], i.e., decreased coke formation on the addition of a hydrogen donor agent, were interpreted in terms

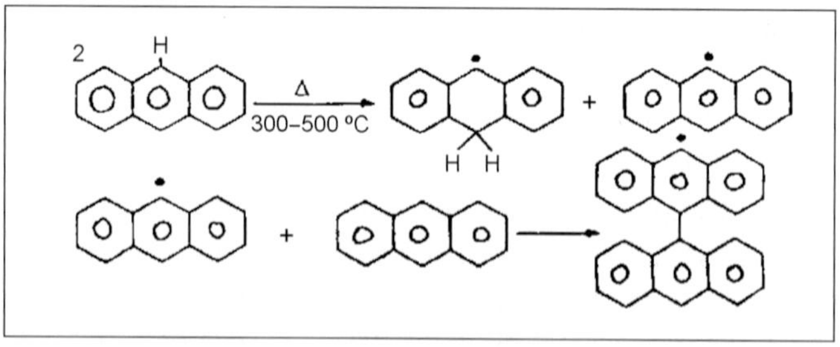

Figure 4.19: Free radical mechanism for formation of coke from anthracene [From ref. 219. Reprinted with permission].

of the free radical mechanism. Theoretically, the active surface hydrogen in the form of SH and MeH entities may stabilize radicals as well. However, at later stages on stream, this radical scavenging source may be exhausted due to the diminished hydrogen activation caused by the extensive catalyst deactivation. This is supported by the observations made in commercial units, i.e., a rapid coke build-up during final stages on stream.

The involvement of carbocations during coke formation is also possible. Carbocations are the important intermediates of some reactions, i.e., hydroisomerization (HIS), hydrocracking (HCR), polymerization, etc. If not stabilized, carbocations can combine to higher molecular weight species. The coupling of polynuclear aromatics leading to coke precursors and finally to coke was also proposed [224–230]. The rate of such reactions was enhanced in the presence of the Bronsted acid sites. This indicates the involvement of proton (via carbocation) during coke formation. The coke formation was significantly diminished after Bronsted acidity was destroyed by pretreating the catalyst support with basic species. Carbocation mechanism may be part of the overall mechanism of coke formation regardless the origin of the heavy feed. In the case of such mechanism, the type of the support may be more important factor than the type of the feed. It has been generally observed that the rate of some hydrocarbon reactions (cracking, isomerization, polymerization, etc.) was rather low unless the source of protons was available. In this regard, the catalysts supported on acidic supports (e.g., zeolites) are most suitable.

With respect to the acidity of support, carbon may represent another extreme to zeolites. Thus, it is unlikely that acidic sites are present unless carbons were subjected to special pretreatments. However, C–H bonds may be present because of hydrogen activation on carbon can proceed [231]. The results on hydroprocessing of the Kuwait AR conducted by Nakamura et al. [223] over the carbon-supported catalysts were interpreted in terms of the free radicals mechanism. In the case of the Co(Ni)/Mo(W) catalysts supported on carbon, the SH groups could be a source of the hydrogen necessary for quenching radicals unless the heavy feed involved was of a naphthenic origin. Under certain conditions, SH groups may possess a Bronsted acid character. For example, the Bronsted acid character of such groups increased with increasing temperature [36]. Then, at temperatures approaching 700 K, the SH groups could donate proton and initiate the formation of carbocations. Therefore, even for carbon-supported catalysts, the involvement of carbocations during coke formation cannot be entirely ruled out.

The development of the mechanism of coke formation benefits from the advancements in spectroscopic techniques (e.g., proton nuclear magbetic resonance [1H NMR], carbon 13 nuclear magnetic resonance [^{13}C NMR], laser desorption mass spectroscopy [LD-MS], Fourier transfer infrared [FTIR], etc.) and other analytical methods, which allowed more detailed analysis of both coke as well as corresponding feed and products. This allowed the determination of various structural parameters of the feed, products and the coke, which was deposited on the catalyst surface. With the availability of such information, the mechanism of coke formation could be defined more accurately and in more details.

The CH_2Cl_2 soluble and insoluble parts of the deposit on two spent catalysts from hydroprocessing of VGO were characterized by Sahoo et al. [232]. The structural parameters of the former, termed as a "soft" coke, were similar as those of the heavy components of the VGO feed. At the same time, the "hard" coke was more aromatic but less aromatic than the similar "hard" coke on the spent catalysts from hydroprocessing of residues. This is not surprising because the latter require more severe conditions (e.g., higher temperatures) to attain desirable level of conversions. Also, in the case of VGO, the HYD of some coke components could occur because of the less severe conditions. Then, the factors, which dominate coke formation using the distillate feeds, may differ from those for the residues.

Seki and Kumata [233,234] carried out the extensive characterization of asphaltenes and resins in the products by spectroscopic techniques. The study involved the HDM and HDS of the Kuwait atmospheric residue over the Mo/A_2O_3 and $NiMo/Al_2O_3$ catalysts, respectively. In this case, the molecular weight of both asphaltenes and resins in products progressively decreased in the course of HDM reactions. The rate of coke build-up significantly increased above 673 K. This was accompanied by the removal of alkyl chains from asphaltenes molecules. Therefore, the aromaticity of asphaltenes was increased. Such change facilitated the adsorption of asphaltenes on the catalyst surface and increased deactivation. In the presence of alkyl chains, the adsorption of asphaltenes was diminished because of the steric interference between the coke molecules and catalyst surface, provided mainly by aliphatic chains.

Fonseca et al. [235–237] recognized that the solid-state ^{13}C NMR could be a useful tool for the characterization of coke deposits on catalyst surface. The investigated $CoMo/Al_2O_3$ catalyst (0.7 wt.% CoO; 4.5 wt.% MoO_3) was used in the three-stage ebullated-bed pilot plant. The spent catalyst was withdrawn after four, 21 and 120 days on stream from the first and third reactor. The feed was the blend of Khafji vacuum residue and a diluent. Less than 69% of the coke carbons could be observed by the NMR technique employed. In the study of Hauser et al. [238,239], the solid-state ^{13}NMR with the application of the cross polarization with polarization inversion at the low or moderate magic angle spinning was used. The analysis was complemented by using the proton-gated decoupled single-pulse excitation and by the results of elemental analysis. With this approach, the limitations of the technique, used by Fonseca et al. [235–237], were minimized. The spent catalysts were extracted either by toluene or THF before their characterization. After 1 h on stream, the H/C ratios of both coke and asphaltenes in the feed were similar. This suggests that fouling of the catalyst by asphaltenes was the main cause of the coke formation. This would indicate little involvement of the catalyst surface during very early stages. However, the NMR analysis of the toluene insolubles (TIS)- and tetrahydrofuran insolubles (THFIS)-coke indicated that already after 1 h on stream, the structure of asphaltenes changed after being deposited on the catalyst. This included the loss of long chains in particular. As it is shown in Fig. 4.20, after 1 h, the TIS- and THFIS-coke structures differed from that of asphaltenes. It consisted of the less polycondensed aromatic rings with shorter but heavily branched alkyl substituents attached to them. Between 1 and 12 h, the coke deposition slowed down. In this region, a simultaneous accumulation of aromatic carbon in the coke, some HYD of aromatic rings, isomerization and dealkylation were occurring simultaneously. For the THFIS-coke, a ring condensation occurred as well.

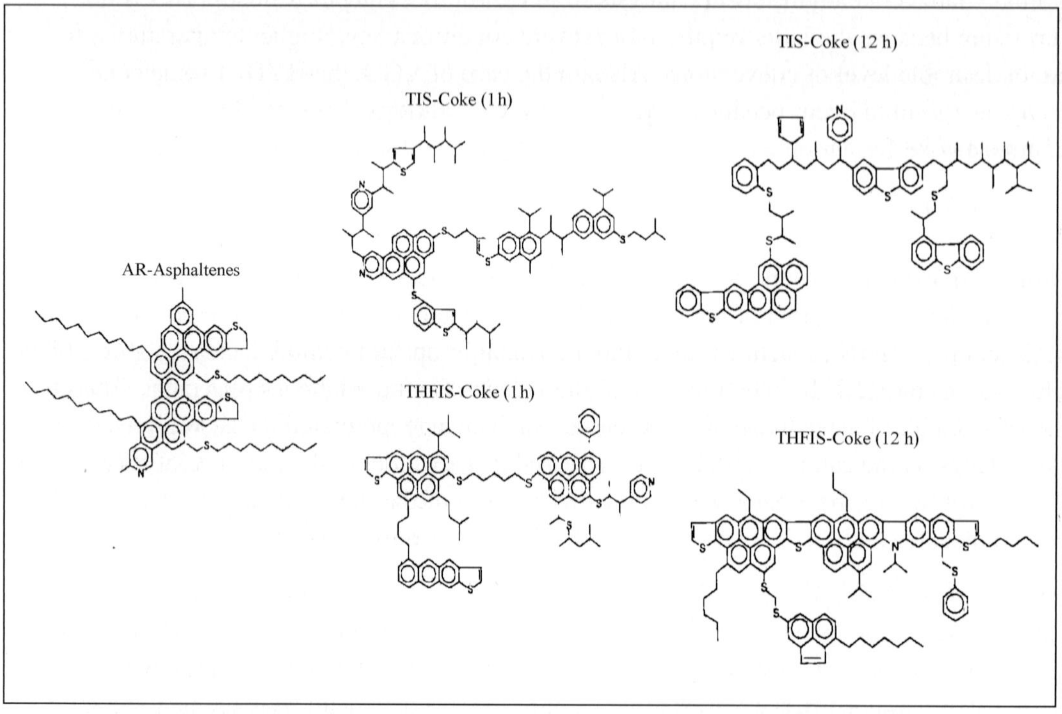

Figure 4.20: Effect of time on stream on structure of TIS and THFIS of coke on catalyst (Mo/Al_2O_3, atmospheric residue [AR], 653 K, 12 MPa) [From ref. 239. Reprinted with permission].

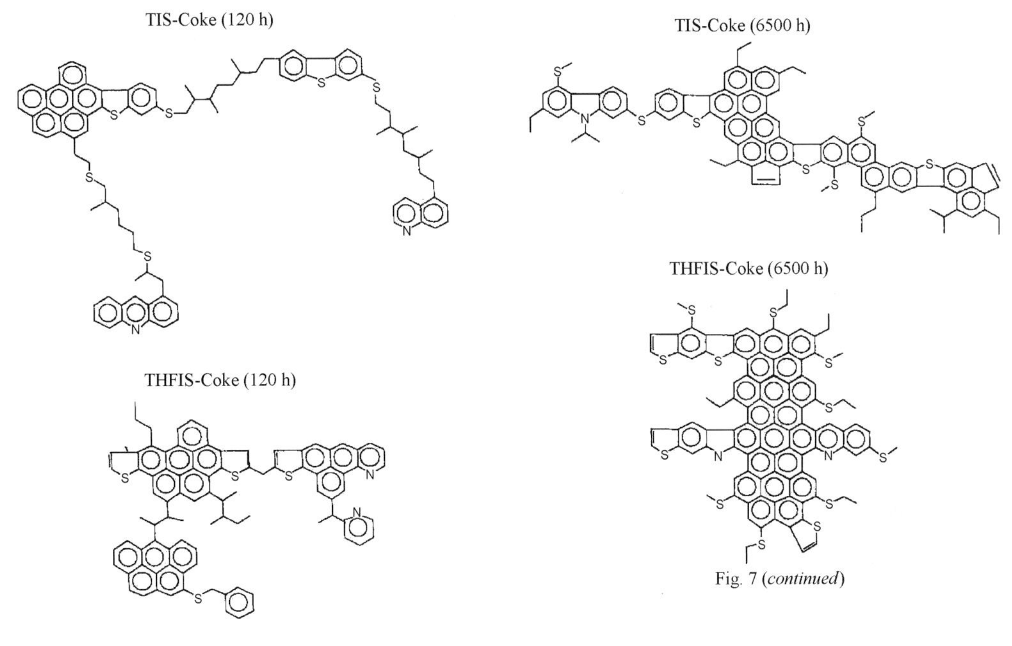

Figure 4.21: Effect of time on stream on structure of TIS and THFIS of coke on catalyst (Mo/Al_2O_3, atmospheric residue [AR], 653 K, 12 MPa) [From refs 238 and 239. Reprinted with permission].

The H/C ratio of both TIS-coke and THFI-coke was greater than one even after 240 h on stream, although the degree of alkyl substitution decreased significantly. These observations suggest that some of these changes could not occur without the direct involvement of catalyst surface. For example, a strong interaction with catalyst surface could be one reason for a low solubility of the THFIS-coke. Figure 4.21 shows the structure of coke on the same catalyst used for the HDM of the same feed in the industrial ARDS process after 6500 h. In this case, a high degree of the aromaticity of cokes, particularly that of the THFI-coke, was quite evident. The formation of such structure may be considered as the beginning of the coke graphitization on the catalyst surface. A conclusion can be drawn from Figs. 4.20 and 4.21. Thus, in the course of operation, the HYD and deHYD of coke on catalyst may occur in parallel together with some HCR, isomerization and condensation reactions.

The additional evaluation of spent catalysts from the study of Hauser et al. [238,239] was undertaken by Matsushita et al. [240] using several complementary techniques. During temperature programmed oxidation (TPO), they observed two maxima of the CO_2 formation, i.e., one at 573 K and the other at 698 K, which were formed presumably from the oxidation of a "soft" coke and a "hard" coke, respectively. In its structure, the "soft" coke may approach the structure of the TIS coke formed within the first 120 h, whereas the "hard" coke that of THFIS-coke formed after 6500 h, shown in Fig. 4.20. As expected, the latter coke had very low solubility. With time on stream, the "soft" coke was gradually converted to more refractory coke. This was supported by the decrease of the low temperature and increase of the high temperature CO_2 peak between 1 and 240 h. Also, more nitrogen and sulfur were concentrated in the refractory coke than in the "soft" coke.

The optical microscopic techniques could characterize coke deposits according to their reflectance, fluorescence and anisotropy. Micrographs usually reveal the presence of meso-phase, i.e., the spherical domains, which exhibit characteristics of liquid crystals. The meso-phase is denser, has a higher surface tension and wets catalyst surface better than the phase from which it was originated. From the structural point of view, this is consistent with the loss of long aliphatic chains from the coke precursors. These chains contributed to the steric hindrance between the catalyst surface and coke precursor and as such inhibited the wetting of catalyst surface. The mechanism of coke formation involving meso-phase as an intermediate phase was proposed by Beuther et al. [241]. With time on stream, the liquid crystals could be converted to coke whose structure was changing progressively. This involved ordering and stacking of aromatic sheets. This may be considered as the very early stage of graphitization, which tends to increase with increasing severity. Figures 4.20 and 4.21 [238,239] offer some support for this mechanism. Thus, the coke after 6500 h represents a sheet, which possesses a high aromaticity. The stacking of such sheets into platelets may have occurred particularly when the catalyst was approaching the end of its life, i.e., at this point, the active surface hydrogen was very limited.

The optical microscopy of the polished cross-sections of a series of the spent catalysts after heavy feed upgrading was investigated by Munoz et al. [242] and Gray et al. [243]. The fluorescence due to the presence of the feed components and anisotropy due to the presence of meso-phase were observed in addition to the high reflectance, which indicated the presence of domains having higher aromaticity than surrounding matrix. This was an indication of the gradual conversion of heavy components in the feed to meso-phase, which subsequently converted to the high aromaticity species. This was supported by the absence of the feed components and predominance of high aromaticity domains after more severe conditions, i.e., higher temperature and longer time on stream. These observations are in a good qualitative agreement with the other studies [230,244–246].

4.6.1.2 Physical Aspects

Physical properties of the petroleum feeds may be a contributing factor to deactivation. In this regard, heavy feeds require much more attention than light feeds. Of particular importance is colloidal stability of the system comprising oil, resin and asphaltenes phases. This stability may be affected when resins are converted at a greater rate than asphaltenes. Similar effect would have a high rate of the HYD of oil phase of the colloidal system. Compatibility is a non-issue for distillate feeds because in these systems the oil phase is predominant. The situation may be less clear for the DAO where the conditions of deasphalting and origin of the feed may have a pronounced effect on the colloidal structure and stability of the feed.

It is evident that during hydroprocessing, the asphaltene entities in the same feed may exhibit a wide range of reactivity. Thus, the most soluble part (the least polar) of asphaltenes may be the most reactive, whereas the insoluble part is the least reactive. The latter part sometimes referred to as carboids can be separated from the asphaltenes by solvent precipitation [1]. After most of the reactive portion of asphaltenes was converted to lighter fractions, the remaining carboids may physically deposit on catalyst surface and as such contribute to the coke formation. Because the content of carboids in asphaltenes from different heavy feeds is different, their coke-forming propensity will be different as well. Nevertheless, carboids, as the least reactive component of the colloidal system, may be partly responsible for the increased aromaticity of asphaltenes isolated from the products compared with that in the corresponding feed, sometimes reported in the scientific literature.

The results on solubility of asphaltenes published by Matsushita et al. [247] complement the mechanism proposed by Seki and Kumata [233,234]. The former authors introduced the solubility index defined as the ratio of the H/C ratio of asphaltenes to that of the DAO obtained from the same feed using the different solvent/feed ratios. The decreasing solubility index would indicate the loss of alkyl chains in asphaltenes (increase in aromaticity), in agreement with the observations made by Callejas et al. [181]. Thus, the paraffinic hydrogen in alkyl chains is an important contributor to the total hydrogen. This would decrease the solubility of asphaltenes in oil and enhance their deposition on the catalyst surface. Also, the precipitation of asphaltenes from the products would be enhanced. The onset of the asphaltenes precipitation can be established using the critical solubility parameters, which can be determined by the flocculation onset titration method [248]. These results suggested that fouling is much less dependent on the origin of catalyst than on the colloidal stability of the feed. It was observed that coke deposition was slowing down with the increasing solubility index, i.e., with the decreasing H/C ratio of resins (heptane solubles) [249].

The studies of Mochida et al. [250,251] showed that the compatibility problem in the graded hydroprocessing systems might be alleviated by optimizing operating parameters. Thus, an extensive deposit formation in the one-stage system could be prevented using a two-stage system. In the latter case, the first reactor was operated at 663 K under conditions favoring the HYD and asphaltenes depolymerization. The high rate of asphaltenes conversion at relatively short contact time was achieved in the second reactor, which was operated at 693 K. A large pore $NiMo/Al_2O_3$ catalyst was needed to achieve these results. These studies focused on the sludge formation in products however, similar factors are involved during the deposit formation on the catalyst surface.

4.6.2 Mechanism of Metals Deposition

In petroleum feeds, metals can be present in both inorganic and organometallic forms. The deposition mechanism involving inorganic solids should differ from that of the organometallic forms of metals. Moreover, for organometallic compounds, the different form of deposits is formed during the non-catalytic demetallization of porphyrins via the reactions with H_2 and H_2S compared with the catalytic HDM of porphyrins.

4.6.2.1 Deposition of Inorganic Solids

Inorganic solids in heavy feeds include minerals and clay-like solids, which contaminated crude either in reservoir or during the production. For bitumen separated from tar sands, part of the mineral matter originated from the caustic material used in the hot water separation process. In addition, a small amount of the finely divided particles of sand still remained in the bitumen after the separation process. During distillation, these solids accumulate in residues. During hydroprocessing of residual feeds, inorganic solids deposit on the external surface of catalyst particles. In its nature, the deposition mechanism is physical. This is confirmed by the predominant accumulation of the inorganic solids on the external surface of catalyst particles in a "skin-like" form. Indeed, a high content of Fe and Ca was noted in the "skin" on the spent catalysts from hydroprocessing of Athabasca bitumen [243]. A similar form of deposition may undergo the sulfides of V and Ni formed during the non-catalytic reaction with H_2S and H_2.

In the crudes, the solids containing alkali and alkali earth metals are predominantly in an oxidic form. They are usually associated with SiO_2, Al_2O_3 and SiO_2–Al_2O_3 in the form of clays and minerals. Under hydroprocessing conditions, the oxidic form of alkali and alkali earth metals are gradually converted to corresponding sulfides similarly as the Fe oxides. If present, finely divided water emulsions may contain chlorides of alkali metals. The problem with such solids is alleviated by dewatering of the crude oil as soon as it enters the petroleum refinery.

The industrial experience confirmed that the operating problems caused by the deposition of inorganic solids have been more evident in fixed-bed reactors than in ebullated-bed reactors. In the former case, the front of the fixed-bed will be most affected. The crust-like layer created by the deposition of such solids on the front of the bed may affect the operation by creating channels and developing pressure drops through the bed. In this regard, attention should be paid to the possible contamination of the feed by metallic particles from the corrosion of upstream equipment. As such, iron scale or fine particles usually do not penetrate deeply the catalyst porous system and do not have any strong deactivating effect. It is more a concern as contributor to the pressure drop builds up, as these particles may accumulate at the top of the bed or in the interstices between the catalyst granules. Similar problems may be caused by silicon, which originates from the anti-foaming agents. Such agents are sometimes added to the feed before coking operation, therefore they might be present in HGO. For some heavy feeds, a filtration system must be installed upstream of the catalytic reactor to avoid the operational difficulties caused by deposits of inorganic solids. Such function may be fulfilled by a guard chamber filled with a low cost solid (e.g., alumina, bauxite, clays, etc.).

4.6.2.2 Deposits of Organometallic Origin

Arsenic occurring naturally in many crudes in an organic form has been receiving little attention, although under hydroprocessing conditions, the As organic compounds are very reactive. Also, deactivating effect of arsenic after repeated utilization-regeneration cycles is quite evident as it is shown in Fig. 4.13 [204]. Thus, they are either converted to AsH_3, which ends up in gaseous products or readily decompose and remain adsorbed on the catalyst surface. In fixed-bed reactors, very steep As gradient is observed between the inlet and outlet. Once on the catalyst, As may be converted to a sulfide. Such conversion is favorable under typical hydroprocessing conditions. Arsenic may be a sever poison, particularly for HDS. This was more evident for regenerated catalysts than for the fresh catalysts [252]. In fact, in excess of 0.3 wt.%, As prevented regeneration of the spent catalyst used during hydroprocessing under moderate conditions.

The porphyrin forms of V and Ni are the main organometallic compounds in heavy feeds. They are the main cause of the metal deposits formation on catalyst surface. In crude oils, most of these metals are associated with the asphaltenes entities. The depolymerization of asphaltenes entities is considered as a very first stage of the metal deposit formation. As the result of this, porphyrins are released into the oil phase and become available for HDM reactions. Initial stage of the metal deposition coincides with the final stage of the overall HDM of metal-containing compounds, i.e., the separation of metal from the pyrrole ligand skeleton.

In most of heavy feeds, the predominance of the V-porphyrins compared with Ni-porphyrins has been noted. Therefore, the metal deposition patterns are influenced by the former to a much greater extent. Moreover, reactivity of the V-porphyrins is greater than that of the Ni-porphyrins, particularly at low conversions [253]. This is confirmed by kinetic data in Fig. 4.22 [207] from hydroprocessing of the Khafji AR. Thus, the reactivity difference was maintained for the different particle size of the catalyst. The higher reactivity of V-porphyrins is attributed to the presence of VO-group, which facilitates much stronger interaction with catalyst surface than that of the Ni-porphyrins. At high conversions, e.g., under severe hydroprocessing conditions, the rate of deposition of the Ni-porphyrins competes more successfully with that of the V-porphyrins. Another way of looking at this issue is the change in conversion between the inlet and outlet of the fix-bed of a catalyst. Thus, it is believed that the conversion will increase towards the outlet of the fixed-bed. Then, contribution of Ni compounds to the deposit formation increases in the same direction, as it was confirmed by Tamm et al. [197].

4.6.2.2.1 Vanadium containing deposits

Once on catalyst surface, transformation of the vanadyl group to a V-sulfide may be affected because of the steric hindrance. This may prevent complete sulfidation of the V=O group after

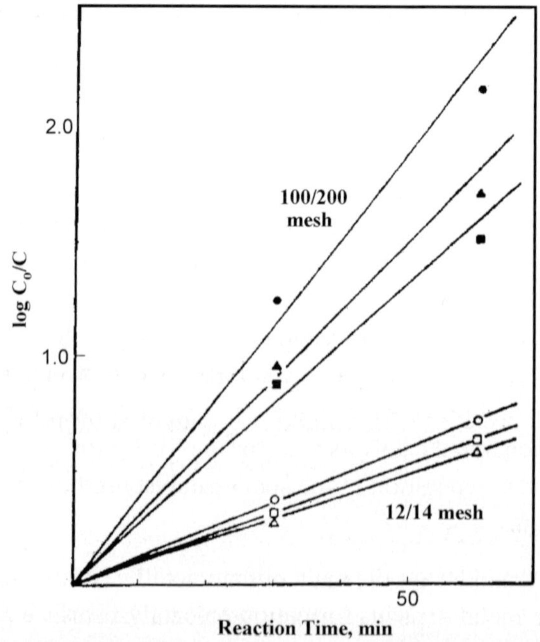

Figure 4.22: First-order plots; (○●) HDV, (△▲) hydrodenickelization (HDNi), (□■) hydrodeasphalting (HDAs) [From ref. 207. Reprinted with permission].

its separation from the porphyrin skeleton. Thus, in spent catalysts and fresh presulfided catalysts, only a partial sulfidation, with the V=O entity still being present, was observed [51,234]. The V=O entity can interact both with the uncovered support and catalytically active metals deposited on the support [254]. According to Loos et al. [255], the V=O was still present, essentially unaltered in spent NiMo/Al_2O_3 catalyst, although additional four sulfur atoms contributed to the average coordination polyhedron of V. This suggests that the sulfidation of V was incomplete compared with that of Mo and Ni/Co. The same was confirmed by Janssen et al. [256]. Thus, during the sulfidation of the Al_2O_3 impregnated with the ammonium metavanadate, the complete conversion to V_2S_3 required temperature of 1273 K, whereas at 673 K, most of the V was still present as an oxysulfide.

In an extreme case, even the unconverted porphyrins in deposits on catalyst surface may be present. This is supported by several potential forms of the interaction of porphyrins with catalyst surface, which have been identified. For example, they may include a donor–acceptor bonding, in which the Π system of the porphyrin ring is the donor and the Bronsted and/or Lewis sites are acceptors [257]. It should be noted that this information was obtained under low temperature conditions. There is little experimental evidence confirming the presence of unconverted porphyrins in coke. It is believed that because of the complex nature of deposits, a convincing identification of porphyrin structures in coke on the spent catalysts from hydroprocessing of heavy feeds would be a rather challenging task.

Asaoka et al. [257] showed that in the presence of a catalyst, there is a significant difference between the metal deposition patterns in pure H_2 and that of the $H_2 + H_2S$ mixture. In the latter case, the precursor was converted to deposits at the first contact with catalyst surface. Then, the deposits progressively penetrated into the catalyst particle interior. Also, the amount of deposit was decreasing from the inlet towards the outlet of the reactor. Spectroscopic evaluations of the deposits (formed in $H_2 + H_2S$) identified V_3S_4 as the predominant composition [257]. In V_3S_4, V was present partly as V^{4+} and partly as VO with the proportion of the latter increasing towards the catalyst particle exterior. Contrary to this, Loos et al. [255] observed the formation of V_2S_3 rather than V_3S_4. However, the latter authors used the model VO-containing porphyrin rather than the heavy feed. Kim and Massoth [258] pointed out that the structure of the V deposits formed during hydroprocessing of real feeds may differ from that formed during the treatment with model V-porphyrins. This was indicated by rather different effect of deposits on catalyst functionalities. Thus, the catalyst was much more deactivated by the real deposits than by those formed using model V-compounds. The difference between the V/S ratio of the model deposits and the real feed deposits should be noted as well.

4.6.2.2.2 Nickel and mixed deposits

The product of the reaction of Ni-porphyrins with H_2S may be at least a partially sulfided Ni. The main HDM proceeds via hydrogenolysis of the Ni–N bond releasing metallic Ni. After deposition on the catalyst surface, Ni is sulfided via established mechanism. Under typical hydroprocessing conditions, the complete sulfidation of Ni would lead to the formation of Ni_3S_2-sulfide. A partially sulfided Ni and/or an oxosulfide form of Ni may be present as well.

The radial distribution of the Ni-sulfides formed non-catalytically via reaction with either H_2 or H_2S should differ markedly from that formed catalytically via established HDM mechanism. The former shall deposit physically predominantly on the exterior of catalyst particles in a "skin-like" form, whereas the Ni-containing deposit formed as part of the HDM reactions should be distributed more evenly. It may be rather difficult to distinguish between these two types of the Ni-containing deposits on catalyst surface.

The overwhelming evidence suggests that initially, the metal deposition occurred predominantly on the bare surface of the catalyst support [259–262]. The thickness and/or size of the deposit were increasing progressively with time on stream. The multilayer deposit would consist of the mixture of V-sulfides (e.g., VS_2, V_2S_3 and V_3S_4) and V-oxosulfides as well as Ni-sulfides (e.g., Ni_3S_2). The simultaneous deposition of V and Ni supports the formation of mixed sulfides ($Ni_XV_YS_Z$). The formation of a mixed (Fe,V)S_4 sulfide was reported by Embaid et al. [263] for the Fe containing heavy feed. The ratio of the V to either Ni or Fe in the mixed sulfide deposit will change from the exterior towards the center of the catalyst particle, i.e., in the case of Ni, the V/Ni ratio will decrease as more Ni porphyrins than V porphyrins can penetrate deeper into the catalyst particle interior. At the same time, the V/Fe ratio may increase towards the particle interior because most (if not all) of the Fe deposited on the exterior of catalyst particles. Pore volume and size distribution of the catalyst may play a key role in determining this ratio.

Apparently, before separation from porhyrins, V and Ni may coordinate with sulfur of the active metal sulfide. Potential coordination with the active phase such as Co(Ni)-Mo(W)-S could lead to the change in activity of the active sites. In this regard, V is expected to have more detrimental effect than Ni. Thus, its interaction may lead to the formation of the V-Mo-S phase which is less active than the Co(Ni)-Mo(W)-S phase.

With progressive growth of the metal deposits, the pore diameter becomes less than the molecular diameter of porphyrin molecules. This prevents the access of the reactant molecules to the interior. At this stage, an abrupt loss of the catalyst activity is usually observed [264]. However, to a great extent, this point depends on the metal retention and/or metal storage capacity of the catalyst, which in turn is influenced by catalyst porosity. For example, for typical HDS catalyst, metal retention before almost total deactivation, may approach 20 wt.% or even less as it is indicated in Fig. 4.23 [265,266]. Although the sudden decline in HDS activity of HDM catalyst was observed at 50% metal retention, its activity for HDM and HDAs was still retained suggesting that the deposition of metals could continue beyond this point. It

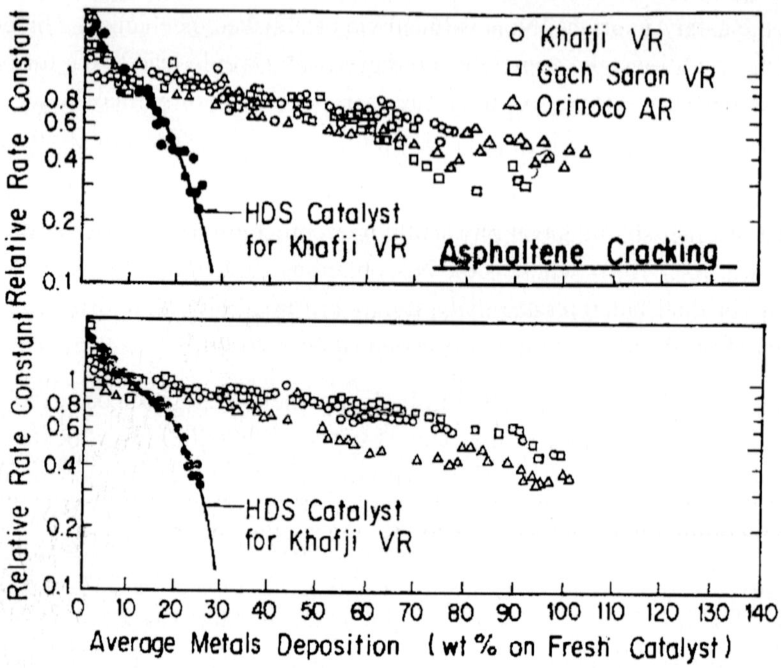

Figure 4.23: Effect of metal deposition on asphaltenes conversion and vanadium removal for hydrodesulfurization (HDS) and hydrodemetallization (HDM) catalysts [From refs 265 and 266. Reprinted with permission].

is believed that the chemical composition and structure of deposits at this level of metal retention is rather complex and will change vertically from the outer surface to the bottom of the deposit layer, which is in contact with catalyst surface. In this regard, properties of catalyst, its porosity in particular, may play an important role.

4.7 Modeling of Deactivation

Models can be used to generate the database for selecting catalysts to match the feed and a suitable reactor also, to predict long term performance of the system. Such a database can be generated quite readily. Models incorporate changes in interface, intraphase and interparticulate gradients of temperature and concentration with time on stream. The complexity of chemical structure of heavy feeds compared with light feeds suggests that the development of models to simulate hydroprocessing of the former is much more challenging. It requires the determination of performance parameters such as the change in catalyst activity for hydroprocessing reactions with time on stream, the parameters accounting for catalyst deactivation, metal storage capacity of catalyst, pore size distribution, etc. The experimental techniques for determining most of the required parameters are now available. Specific parameters, which can be determined by kinetic studies include the intrinsic and apparent rate constants, activation energies, effective diffusivities, efficiency factor, distribution parameter, Thiele modulus, etc. With such parameters available, modeling can be conducted on two levels of scale, i.e., catalyst particle level and active phase level. Modeling on a reactor level requires the information on liquid holdup, height and diameter of reactor, volume of reactor and catalyst bed, superficial liquid velocity, etc. It is noted that most of the studies were conducted on more than one level of scale.

The studies involving a wide range of feeds have shown that models are feed- and catalyst structure-dependent [10,49]. Therefore, the model developed for a particular feed may require some modifications in order to predict the catalyst performance using a different feed. Models take into account initial and steady-state deactivation by coke, which deposited on catalyst surface, as well as a more less a linear and continuous deposition of metals from heavy feeds. Attempts have been made to simulate deactivation by coke deposition occurring during very early stages of the contact of feed with catalyst. The validity of models can be verified using the data from the experiments on accelerating aging carried out in bench scale units, pilot plants and from commercial reactors. The conflicting results obtained during the accelerating aging experiments and those obtained in the commercial units have been noted [153]. Then, with respect to model development, the former results may have a limited validity. However, the accelerating aging test developed by Alvarez et al. [46] could predict the catalyst performance in a pilot plant unit quite accurately.

Kinetic parameters determined experimentally form a basis for catalyst evaluation on the active phase level. Generally, experimental data are compared with kinetic laws in either power form or Langmuir-Hinshelwood form. Reaction order is usually chosen to obtain the best fit of the experimental results for a particular kinetic model. The model developed by Long et al. [259,260] on a catalyst activity level showed very good fit of the experimental results with those predicted by the model. However, such a good fit could only be obtained by assuming that during very early stages, part of the V was deposited on the uncovered support, thus having no adverse effect on active metals' phase. Therefore, this model does not assume a uniform metal layer deposition contrary to some other models [267–269]. Moreover, metal deposits (e.g., V_3S_4) may exhibit some activity for hydroprocessing reactions. The autocatalytic effect of deposits is usually overlooked during the development of models. The autocatalysis may be, at least, partly responsible for deviation of the predicted results from those observed experimentally. This may be evident particularly during the early stages of the operation. The model tested by Melkote and Jensen [270] was among few in which the effect of autocatalysis was considered.

A detailed account of the catalyst deactivation by metals was given by Tamm et al. [197] who used five residues, metal content (V + Ni) of which varied from about 40 to almost 500 ppm. Their model confirmed that the metal deposition patterns were feedstock dependent and poisoning of active sites by metals and physical obstruction of pores by metals were contributors to catalyst deactivation.

Surface area, pore volume and pore size distribution, size and shape of catalyst particles are of the primary interest for designing catalysts for hydroprocessing of the metals and asphaltenes containing feeds. For this purpose, parameters such as the effective diffusivity, efficiency factor, distribution parameter, Thiele modulus, metal storage capacity, etc. are included in the models. In addition, the development of models on the particle scale would be incomplete without incorporating data on catalyst activity. This indicates the need of kinetic data and catalyst deactivation pattern. Therefore, it may also be appropriate to refer to this level of model development as the two-scale approach, i.e., an active phase and a single particle scales. Thus, the applicability of the models on particles scale would be somehow limited without including the effects of active phase on the catalyst performance.

The usefulness of the particle scale models for designing the catalysts for hydroprocessing of heavy feeds was demonstrated in the study of Oyekunle et al. [271,272]. These authors performed calculations for the three types of catalysts, i.e., microporous and macroporous with the predominant portion of pores having APD < 100 D and APD = 100–250 A, respectively, as well as the random pore distribution with the predominant APD between the microporous and macroporous catalysts. They used the data on hydroprocessing of the heavy Maya crude published by Ancheyta et al. [273,274]. Figure 4.24 showed a good fit of the process data with those predicted by the models. The catalyst lifetime was then estimated by using the linear regression analysis of the results in Fig. 4.24. The total activity loss was predicted to occur after 462, 316 and 150 days for the macropore, random pore and micropore systems, respectively.

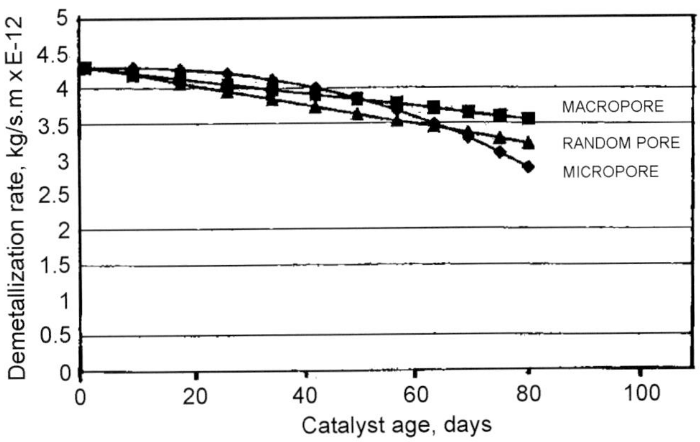

Figure 4.24: Variation of demetallization rate with time on stream for different models [From refs 273 and 274. Reprinted with permission].

The single pellet model proposed by Perreira et al. [275,276] was suitable for predicting the performance of the bimodal catalyst comprising 0.12 and 0.60 mL/g of micro- and macropore volume, respectively. A series of correlations involving the change of parameters such as distribution factor, reaction rate, pore radius, Thiele modulus, etc. with time on stream was developed. A database on the effect of the feed origin and of the catalyst properties on HDM was established. The results supported the advantages of bimodal catalysts compared with unimodal catalysts. The model was suitable for selecting and designing the tailor-made catalyst to match the type of catalyst with the feed properties. This includes the size and shape of catalyst particles as well as the catalyst porosity.

Among other models, the concept introduced by Toulhoat et al. [277,278] involved the estimate of the "ultimate storage capacity" of the single catalyst particle, defined as the mass of metals accumulated in the HDM catalyst pellet, relative to the unit mass of the fresh catalyst, until the catalyst activity became zero. There are a number of other models the validity of which was tested using various feeds. For example, the models developed by Dautzenberg et al. [279] and Hannerup and Jacobsen [54,280] are based on the pore-plugging by metals. The pore-plugging model developed by Oyekunle and Hughes [281] predicted performance of the HDM catalyst lasting 1 year providing that the catalyst possessed a suitable pore size distribution.

Modeling on reactor level requires all parameters used for modeling on a catalyst activity and single particle scale. In addition, new parameters such as liquid holdup, bed height, bed volume, catalyst volume, pellet size, superficial liquid velocity, etc. are necessary. For fixed-bed reactors, most of the models assume trickle flow of the liquid and gaseous streams co-currently from the top to the bottom through the void space between the catalyst particles. The wetting of catalyst surface ensures that reactions occur predominantly in the liquid phase. Moreover, the plug flow transport through the reactor, usually assumed for such models, ensures an ideal mixing, in which both radial and axial dispersion or back mixing are neglected. Without such simplification, the development of models would be rather complex and difficult. The complexity of modeling on reactor level is confirmed by the results published by Tamm et al. [197], which indicate significant differences in the catalyst deactivating patterns between the inlet and outlet of the reactor. On the basis of microscopic evaluations, these authors observed that the V and Ni deposition profiles exhibited either U or M shape, i.e, the maximum of metal deposition was located either at the catalyst particle surface or inside the catalyst particle. This was dependent on the location of catalyst particles in the fixed-bed. Thus, the M shape profile was observed in the reactor inlet, and the U shape in the reactor outlet. The maximum of metal deposition shifted inward with the decreasing temperature, decreasing H_2 pressure, decrease in pellet diameter and increase in the average pore diameter.

The model developed by Khang and Mosby [282] was based on pore-filling by metals and its suitability was verified using the macroporous catalyst for applications in the trickle-bed reactor. Deactivation process could be expressed by two adjustable Thiele modules, i.e., one assuming the bulk diffusivity and kinetics, and the other effecting diffusivity and kinetics in deactivated pores. The model was suitable for predicting deactivation curves for the HDS and HDM reactions in a good agreement with the experimental data before less than 50% of pores were filled with metals. The model data showed a reasonably good agreement with the results obtained in pilot plant such as the HDM activity and metal loading between the inlet and outlet of the fixed-bed reactor. A modified form of the model developed by Khang and Mosby [282] was used by Togawa et al. [122] to analyze five sets of the deactivation data obtained from the commercial operation employing ARDS process using a Kuwait residue as the feed. After modification, the model was applicable; up to 80% of the pores were filled with metals compared with 50% observed by Khang and Mosby [282]. An improved simulation using the modified model was also reported by Kam et al. [283] on the catalysts.

The integrated mechanistic reactor model developed by Kam et al. [284] considered the initial rapid catalyst deactivation by coke deposition. In this case, a distinction was made between the "soft" coke formed during very early stages and the "hard" coke formed in the steady state of catalyst deactivation. At the same time, during the start of run period, the deactivation by metals was much less evident, whereas in the middle of run, and particularly before the end was approached, the metal deposition was the dominant mode of deactivation. Figure 4.25 compared the simulation data with those obtained in pilot plant during more than 400 days on stream. These results are for the second reactor of the four-reactor process. For HDS, a good agreement between the predicted and measured results was obtained for less than 4,000 h on stream, whereas the large discrepancies were observed for the HDAs results. Similarly, a good prediction was made for HDV and hydrodenickelization (HDNi). However, the deviation of the HDM data became more evident beyond 4,000 h on stream. In an effort to further advance their model, Kam et al. [284,285] assumed that a part of the V and Ni were removed from

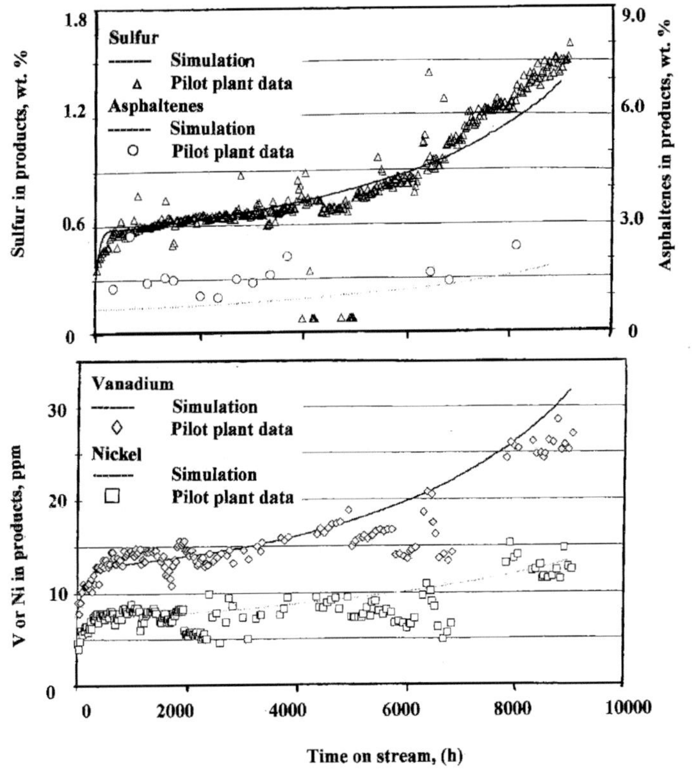

Figure 4.25: Comparison of pant data with simulation data for hydroprocessing atmospheric residue [From ref. 284. Reprinted with permission].

heavy feed via non-catalytic reactions. After applying this assumption, the prediction of the catalyst performance beyond 4000 h significantly improved.

Figure 4.26 shows the prediction of catalyst performance using this model proposed by Takatsuka et al. [286]. The properties of the catalysts are shown in Table 4.4. It was evident, that initially, catalyst A was more active for HDS than catalyst C, as it was indicated by the lower content of sulfur in the products. However, the activity of the former declined with time on stream. The same catalyst was the least active for V removal. The best performance of the catalyst B resulted from the optimal combination of pore volume, particle diameter and surface area.

The percolation model based on the Bethe network [287–289] was used for the simulation of reactor performance. It involves more than 50 mathematical equations used to describe the events occurring on an active phase level, as well as the transport-phenomena taking place on a particle level. The trickle flow reactor, in which liquid and gases flow co-currently from the top to the bottom through the void space between catalysts particles were used for modeling on the

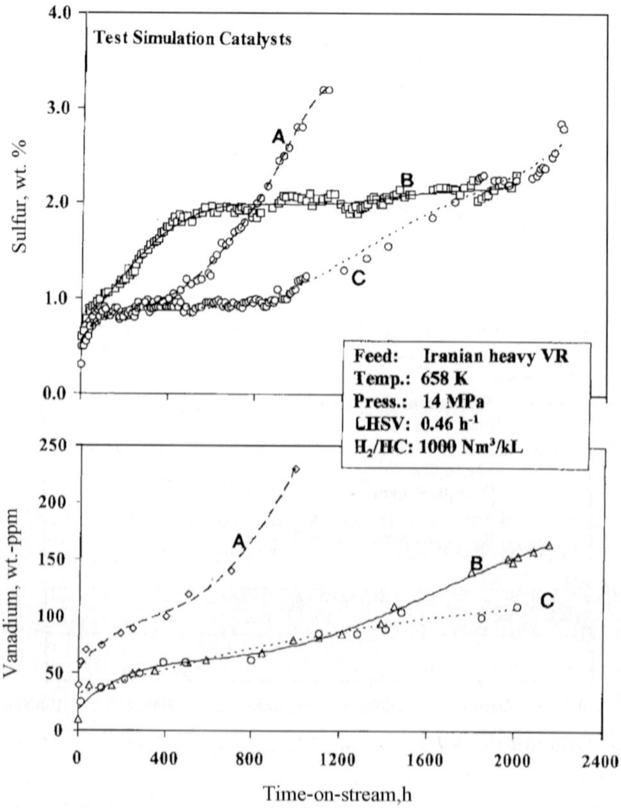

Figure 4.26: Simulation of catalyst aging (pore volume of catalyst A, B and C was 0.44, 0.60 and 0.79 mL/g, respectively) [From ref. 286. Reprinted with permission].

reactor level. Under such conditions, the wetting of catalyst surface ensured that the reactions occurred in the liquid phase. The variables such as liquid holdup, superficial liquid velocity, bed height and diameter, reactor volume and catalyst volume were incorporated in the mathematical equations in addition to all parameters used on the active phase and single particle level. The plug flow transport through the reactor assumed for the study ensured an

Table 4.4: Properties of catalysts [From ref. 286. Reprinted with permission].

	Catalyst		
	A	B	C
	Ni-Co-Mo	Co-Mo	Co-Mo
Pore volume (mL/g)	0.44	0.60	0.79
Surface area (m^2/g)	280	155	341
Particle diameter (mm)	1.16	1.34	1.59

ideal mixing in which radial and axial dispersions as well as back mixing could be neglected. There was a reasonably good agreement between the predicted and experimental data for the top layer of the catalyst. Theoretical part of this study includes several correlations in which the values of Thiele modulus were arbitrarily chosen to identify the effect of the presence and absence of diffusion on the V distribution profiles on the particle level. With such correlations available, the meaning of Thiele modulus becomes much clearer.

The thermal monitoring for iso-performance desulfurization of oil residues (THERMIDOR) model was developed by the French Institute of Petroleum to simulate operation of the HYVAHL process comprising two guard reactors operating in the perturbating mode upstream from the series of four fixed-bed reactors [290]. The model considered both the grain scale and the bed scale. The shape of catalyst particles (sphere, cylinders, polylobe, etc.) were considered at the grain level using the fractional radius of the catalyst particle as space variable, whereas the fractional axial position was used to describe changes between the inlet and outlet of the reactor. More than 40 mathematical equations, which were considered represent one of the most comprehensive approaches used for the model development. These equations were beyond the scope of this review. The pseudo Langmuir-Hinshelwood law was used to express the rate of hydroprocessing reactions as first-order in disappearing reactant (e.g., sulfur, metals and asphaltenes) using the assumption that about 10% of asphaltenes present in the feed were responsible for coke build up. Figure 4.27 [290] shows the experimental and simulated deactivation curves for Boscan heavy feed using the macroporous HDM catalyst. Remarkable agreement between the experimental data and those predicted by the model should be noted. The THERMIDOR software incorporates Fick law to express the local molar fluxes in the liquid phase. The Stokes-Einstein law and Andrade law were combined to describe the effect of temperature on the feed viscosity and molecular diffusivities. The random spheres, random needles and random coins models were used for the representation of the catalyst porous media. Other equations derived as part of the THERMIDOR project enabled simulations of the additional parameters, e.g., the change in the inlet and outlet temperature with time on stream, longitudinal profiles of temperature, Tamm factor, surface area, porosity, etc. with time on stream. Another model suitable for the simulation of a multistage system performance (e.g., ARDS) was developed by Kodama et al. [291] was used by Al-Adwani et al. [292] to simulate performance of the four reactors in a series, such as in the ARDS process. Apparently, this model was supposed to be less dependent on the catalyst and feed properties than the other models. The model could predict the effect of LHSV on the set-point temperature for achieving certain level of HDS. For a given degree of HDS, the set-point temperature increased with the increasing LHSV, i.e., the decreasing contact time. In general agreement, the model predicted that the degree of HDS parallels that of HDM.

Few attempts have been made to develop the models to simulate performance of the ebullated-bed reactors. In this regard, it has been recognized that the artificial neural network may be a suitable modeling tool [293,294]. The neural networks comprise computing systems

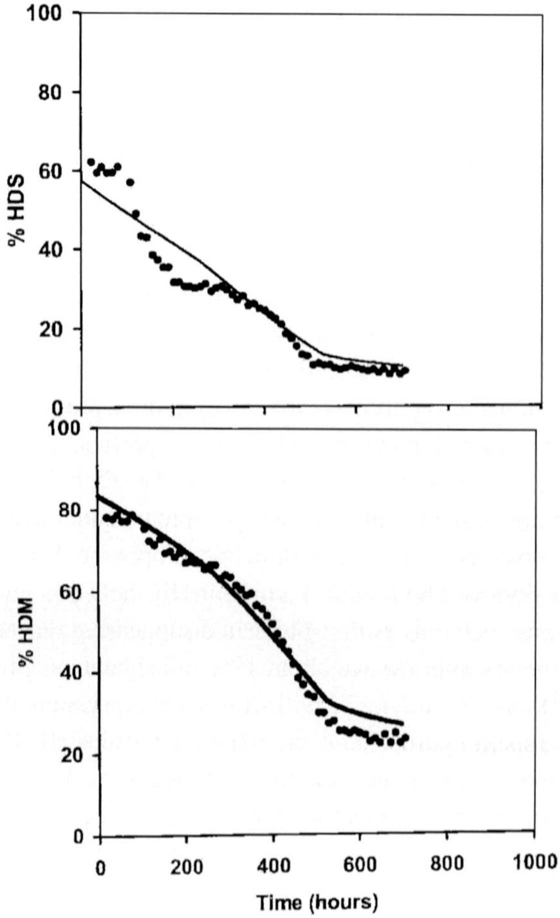

Figure 4.27: Experimental (dots) and simulated (continuous) data for hydrodemetallization (HDM) catalyst (Boscan crude) [From ref. 290. Reprinted with permission].

composed of many simulation elements interacting with bandwidth channels and/or connections to process the information by responding to the external inputs. Initially, it can be built using the historical plant data as well as the results of laboratory and pilot plant research and those found in the literature. Gradually, the models can be upgraded by incorporating more reliable and advanced information. The neural network architecture developed by Kam et al. [293] consisted of the three neural layers and eight specified input nodes and nine hidden nodes to predict the expanded bed height. A comparison of the normalized bed height data, obtained from a commercial reactor with those predicted by the artificial neural network, showed a good agreement. The absolute average deviation value indicated the applicability of the model to the expanded bed systems as employed in the commercial H-Oil reactors. Furthermore, the predicted slate of products and their yields were in the reasonable agreement with the plant data.

CHAPTER 5

Environmental and Safety Aspects of Spent Hydroprocessing Catalysts

The survey of wastes generated by US petroleum refineries, conducted in 1992 under auspices of the American Petroleum Institute (API), grouped refinery wastes into six categories, i.e., aqueous wastes, oily sludge, waste chemicals, contaminated soil, other wastes, and spent catalysts [295]. As part of this survey, almost all participants indicated a steady decline in generation of refinery wastes. As hazardous solid wastes, spent hydroprocessing catalysts come under the controlling terms of the US Environmental Protection Agency (EPA), Basel Convention, and Organization for Economic Cooperation and Development (OECD) rules. As such, the spent catalysts cannot be exported to third world countries. Furthermore, the generators have a legal obligation to ensure that their spent catalysts are properly disposed of or safely recycled. Since spent hydroprocessing catalysts have been classified as hazardous wastes, their safe handling and disposal in an environmentally acceptable way is an issue of a great and continuous concern for petroleum refiners.

The process of handling spent hydroprocessing catalysts begins with their removal from reactors at the end of the operation. The procedures described in operating manuals are applied by refiners during the catalyst withdrawal from reactors and subsequent treatment on refinery site. Special precautions have to be taken in the case that spent catalysts are transferred to a treatment, storage, and disposal facility. All these actions and/or activities are governed by a set of environmental and safety regulations. The objective is to minimize an adverse effect of spent catalysts on the quality of ambient air and ground water. In the following sections, all phases of spent catalysts handling and movement will be discussed in line with the existing regulations.

Refineries have to spend a sizeable portion of their cash flow in order to comply with environmental regulations. As a consequence, some refineries had no choice but to shut down the operation. Because environmental regulations are still evolving, refineries may be experiencing such pressures continuously. A competitive advantage may be gained by refineries or countries with a lower environmental awareness. Therefore, some global approach to deal with environmental and safety issues in petroleum refining industry is needed to prevent an unfair competition.

5.1 Regulatory Affairs

In order to protect human health and environment, any handling of the waste solids disposed from an industrial operation must be monitored within the framework of existing environmental and resource regulations. The summary of all relevant acts controlling the quality of ground water was given by Nielsen [296], whereas that of the ambient air by Wang et al. [297]. In USA, the largest body of environmental regulations has been promulgated by the US EPA. In this regard, the major regulatory programs include:

(1) The Resource Conservation and Recovery Act (RCRA) that includes the Hazardous and Solid Waste Amendments (HSWA).

(2) The Comprehensive Environmental Response, Compensation and Liability Act (CERCLA) or Superfund, including amendments.

(3) The Toxic Substance Control Act (TSCA).

(4) The Safe Drinking Water Act (SDWA) and amendments.

(5) Clean Air Act (CAA) setting up National Ambient Air Quality (NAAQ) standards.

(6) The Surface Mining Control and Reclamation Act (SMCRA).

All stages of handling solid and hazardous wastes, starting with generation and transportation, as well as all activities performed by the operators of treatment, storage, and disposal facilities (TSDF) are subject to the RCRA regulations. The subtitle C of the RCRA defines what is considered a hazardous waste and what is not. It also defines the types of facilities required for spent catalyst handling and movement which comply with regulations. For example, one requirement includes the installation and operation of the ground water monitoring system for continuous monitoring of the performance of TSDF. The permits may be granted only to those TSDF operators, who are in the compliance with the RCRA regulations. The ground water monitoring system may be necessary during the closure and postclosure period of the TSDF.

The CERCLA, better known as Superfund, was established to deal with abandoned waste sites which may pose threat to public. One of the objectives of the Superfund has been to develop strategy and set priority for cleaning up the worst existing hazardous waste sites. In this case, responsible parties have to cover the cost of clean up wherever possible. The Hazardous Waste Trust Fund (HWTF) may be used if the responsible parties cannot be identified. It is believed that the sites contaminated with spent catalysts can still be found in various parts of the world. However, their size is not expected to be large provided that only spent catalysts are involved.

The TSCA was introduced with the aim to prohibit or to regulate the production, processing, distribution, and disposal of chemical products, which pose a risk to human health or the

Table 5.1: National primary drinking water standards for selected contaminants [From ref. 296. Reprinted with permission].

Contaminant	Maximum contaminant level (ppm)
Antimony	0.006
Arsenic	0.010
Barium	2
Cadmium	0.005
Chromium	0.1
Copper	1.3
Lead	0.015
Selenium	0.05
Fluoride	4.0
Nitrate (as nitrogen)	10.0
Nitrite (as nitrogen)	1.0
Benzene	0.005
PAHs	0.0002
Toluene	1.0
Xylenes (total)	10

environment. The TSCA provides the EPA with authority to demand the premarket testing and/or premanufacturing notices with details of procedures, the amount of products, properties of products, methods of distribution, etc. Some fresh hydroprocessing catalysts (e.g., containing fluoride) may require regulation under this act.

The objective of Clean Water Act (CWA) is the protection of the quality of surface water that affects ground water and vice versa. This requires water quality management plans with the aim to control disposal of hazardous waste on land and to protect quality of the surface and ground waters. The SDWA deals specifically with the quality of drinking water. It ensures the safe supply of drinking water from public water supply systems. Several provisions refer specifically to the ground water quality. The promulgation of drinking water quality standards is required by Section 1412 of the SDWA. To meet requirements, the National Primary Drinking Water Standards (NPDWS) and National Secondary Drinking Water Standards (NSDWS) were developed by EPA. The former set maximum contaminant level and are legally enforceable. The NSDWS are non-enforceable guidelines regulating contaminants, which cause either cosmetic or esthetic effects in drinking water. Almost 90 species are listed among NPDWS contaminants [296]. Those contaminants, which may be somehow relevant to spent hydroprocessing catalysts are shown in Table 5.1. In addition, Ag, Al, Fe, Mn, Zn, and sulfate are listed among the NSDWS contaminants.

It is evident that regulatory programs, such as RCRA, CERCLA, TSCA, and CWA as well as their amendments, provide a wide range of means for monitoring regulatory compliance. In most cases, the focus is on the quality of ground water. This may require the sampling program in proximity of the sites at which ground water contamination may have a potential to occur. Detailed analysis of the samples is necessary to ensure that the site is properly operated. The authority provided by the regulatory programs may be exercised either to force corrective actions or even discontinuation of the operation.

The CAA was created to address air pollution problems in addition to environmental laws described by RCRA. Besides NAAQS, the major programs under CAA include New Source Performance Standards (NSPS) and National Emission Standards for Hazardous Air Pollutants (NESHAP). They regulate criteria pollutants (e.g., particulate matter, SO_2, CO, NO, O_3, Pb, etc.) and designated pollutants (e.g., total reduced sulfur and sulfuric acid mist). Because O_3 is criteria pollutant, VOC as O_3 precursors are also regulated by NAAQS. Thus, O_3 and an aerosol are formed when NO and VOC are exposed to sunlight. This mixture is described as photochemical smog. Respiratory systems can be affected when humans are exposed to O_3 and the aerosol.

Details of the primary and secondary NAAQSs for regulated criteria pollutants, including integrated time required to calculate the standard, were given by Vallero [298]. The primary NAAQS are the levels of air quality, which according to EPA are required, with an adequate margin of safety to protect public health. The secondary NAAQS are levels necessary to protect public welfare from any known or anticipated adverse effects.

5.1.1 Classification of Spent Hydroprocessing Catalysts

The announcement made by EPA on May 2002 reinstated its previous position that under RCRA regulations spent hydroprocessing catalysts removed from dual-purpose hydroprocessing reactors must be classified as hazardous waste [299–301]. According to EPA, dual function of hydroprocessing comprises hydrotreating and hydrorefining. The former includes the removal of impurities, such as sulfur, nitrogen, metals, and other impurities from petroleum. Spent catalysts removed from such operations are designated as hazardous wastes K-171. According to EPA terminology, hydrorefining is conducted under more severe conditions than hydrotreating. The former involves upgrading heavier feeds, i.e., residual fuel oil and heavy gas oil. The catalysts removed from such operations are designated as hazardous wastes K-172. Interestingly enough, EPA did not make a listing determination for spent catalysts from hydrocracking operations and at the time of the EPA announcement, such catalysts were not designated as hazardous wastes. However, according to EPA, the main objective of hydrocracking is the conversion of large molecules to volatile products with minimal parallel hydrotreating and hydrorefining. Yet, in order to clarify the issue, EPA declared that spent catalysts meeting the listing description for K-171 and K-172 are those which are removed from expanded and/or ebullated bed reactors, e.g., H-Oil and LC-Fining reactors, which according to refining terminology, are classified as typical hydrocracking reactors, although in these systems, the removal of impurities occurs in parallel with hydrocracking. Also, hydrorefining is generally considered as the last step before specifications of commercial fuel are attained. Therefore, it may be conducted under less severe conditions than hydrotreating. This suggests that there may be some inconsistencies between the terminology used by regulatory authorities and petroleum refiners. To reconcile the language discrepancies, it may be concluded that all catalysts removed from all commercial hydroprocessing operations have to be classified as hazardous wastes.

Some definitions of solid waste by EPA under RCRA may be subjected to interpretation. For example, refining industry represented by API has been considering solid waste as the material that is being discarded by virtue of being disposed of, abandoned or thrown away, rather than being recycled [302]. Therefore, the material that is being recycled (e.g., regenerable spent catalysts) should not be classified as solid waste contrary to the spent non-reusable catalysts that are being sent for disposal. Such exclusion of spent hydroprocessing catalysts from hazardous listing is necessary to provide a cost-effective management option for recycling. This may also encourage recycling option over disposal.

For transportation purpose, the waste is categorized in three different lists, i.e., green, amber, and red [303]. The green list covers non-hazardous waste that is only subject to normal commercial transactions. The "amber" list covers wastes that exhibit hazardous characteristics or contain hazardous components. In the case of transportation, such waste requires notification to all authorities and parties involved. Also, a tracking document is required for the movement of waste listed as "amber". The "red" list hazardous wastes require the same notification as the "amber" list wastes. However, for such wastes, a written consent is required from all concerned authorities before the transport can take place.

5.1.2 Transportation of Spent Catalysts

In OECD countries, the Chief Executive Officer of the company is legally responsible for all environmental issues including transportation and disposal of waste [304]. It is believed that directors and other lower levels officers in the company may also play certain role in these actions through the laws of joint and several responsibilities.

The shipment of spent hydroprocessing catalysts is regulated by the Transportation of Dangerous Goods Act (TDGA). In addition, the Basel Convention prohibits the international shipment of hazardous waste between developed and developing countries [311]. In the US, such shipments are regulated by EPA regulations that require a waste receiving country to certify its willingness to accept hazardous waste. In the European Community (EC), the transportation of hazardous waste including spent catalysts is controlled by two different regulations, i.e., Authorization-Dangerous-Road (ADR) code and International-Maritime-Dangerous-Goods (IMDG) code [305]. The latter code covers shipping of spent catalysts overseas, whereas the ADR applies to the road transportation within the EC.

The transboundary shipment of the RCRA hazardous waste, such as spent catalysts, must follow the export notification procedures. This involves a notification describing the material, mode of transportation, shipping company, regenerator/reclaimer, ports of exit and entry, and other details. In the case of a US exporter, this notification is sent to the responsible authorities in the receiving country through EPA and Department of International Affairs. The refiner can only proceed with the spent catalyst shipment after approval from the government of receiving country was granted. This also should ensure that the regenerator or reclaimer in the receiving country are licensed and have necessary certifications as required by environmental authorities.

Various countries may have their own regulations in effect. It is however believed that these regulations are for most part in compliance with international laws and acts. For example, in Canada, the transboundary shipment of spent hydroprocessing catalysts is controlled by the regulations on the Export and Import of Hazardous Goods [306]. As expected, these regulations are in line with the international codes established for similar purposes. The regulations define the conditions that have to be fulfilled before spent catalysts can be imported in, exported out or transited through country or a province. All parties involved are required to notify appropriate authorities in advance, i.e., one year before the proposed shipment. For shipments from a country, the generator/exporter has to complete the notice, whereas the importer (e.g., recycler and disposer) is required to provide authorities with the notice. In the case of shipment within the country, the notice should also be completed by the carrier. Always, the Material Safety Data Sheet (MSDS) has to accompany the shipment.

Because of hazardous characteristics, all safety precautions have to be taken during transportation of the spent hydroprocessing catalyst. The method of packaging must prevent contact with air and water as well as the leakage of gaseous and liquid constituents of hazardous nature [306]. It is preferable that packaging is performed by companies with special certification from the environmental authorities. The flammability and leachability characteristics dictate that spent hydroprocessing catalysts cannot be shipped in supersacks or in a bulk. The use of metal containers is the most suitable packaging method provided that they have undergone tests for resistance to impact and tightness. Such tests, which are regulated by United Nations texts, require appropriate labeling of containers [307]. The preference of catalyst bins compared with drums for catalyst transportation has been noted [308]. The bin can replace 10 regular 200 L drums. The former can contain up to 2000 kg of catalyst.

It was suggested that the self-heating nature and release of toxic material during transportation can be minimized by the formation of an organic seal over the spent catalyst while in the container [309]. The seal consists of a gelatinized starch. The shipment of fresh and regenerated catalysts requires less attention, but they should still be properly labeled and any contact with water should be avoided.

Modes of the transportation included van trailers, dump trucks, railcars, and sea containers. Generally, the mode of transportation is agreed upon by refiner and the partner who may be either a catalyst regenerator or reclaimer. It sometimes depends on the refiners' loading capabilities. In some situations, the handling and loading equipment may be provided by the partner. The partner should carefully monitor shipments from the initial point to the final destination. Because of hazardous nature, the companies involved in the transportation of spent hydroprocessing catalysts must be certified by the regulatory authorities (e.g., in US by EPA). In some cases, the transportation of spent catalysts may be a part of the agreement signed between petroleum refinery and a company with total catalyst management (TCM) certification. Such an agreement may cover all stages of spent catalysts handling.

5.1.3 Recycling and Disposal of Spent Catalysts

Various issues relevant to disposal of spent catalysts addressed in preceding sections are in line with the summary of governing regulations given by Lavers [310], i.e.:

(1) No waste may be exported to the third world countries without adequate facilities.

(2) No waste may be taken to a facility that is not properly equipped to recycle that waste.

(3) No waste that cannot be recycled, and must therefore be destroyed, must travel any further than it is absolutely necessary (the proximity principal).

(4) The country or state/province where the waste requiring destruction is generated is responsible for its destruction.

(5) Hazardous waste for recycling may be shipped to authorized facilities anywhere in the world, but only under a highly regulated system.

In the case of spent hydroprocessing catalysts, destruction may involve the process in which most of useful metals in spent catalysts are recovered for reuse either for preparation of catalysts or in other industrial applications. Recycling may involve the shipment of spent catalysts to a company capable of regenerating/rejuvenating spent catalysts for reuse. The disposal of spent hydroprocessing catalysts is only considered after all other utilization options became unattractive.

It was reinstated several times that special procedures have to be applied during all stages of handling and utilization of spent hydroprocessing catalysts because of their designation as hazardous toxic wastes. The petroleum refiner, as the producer of spent hydroprocessing catalysts, has usually limited capabilities and experience in safe handling of spent catalysts.

Such a TCM expertise is owned by the companies who obtained certification from regulatory authorities. In some situations, refiner has no choice but use the services of these companies. They include the certified owners of TSDFs operating under RCRA regulations. It might be desirable that this process is conducted in a close cooperation with refiner who must be ready to respond timely whenever necessary, i.e., a change in regulatory acts. The costs of these services have been continuously increasing. Therefore, the final decision on the fate of spent catalysts should only be made after all available options were carefully considered.

Significant efforts have been made to bring the storage and disposal of spent catalysts under control [311]. In spite of these efforts, there might still be cases of irresponsible dumping of spent catalysts. This was supported by the survey conducted by the API in 1982. The survey concluded that only 70% of all spent non-regenerable catalysts were sent either to metal reclaimers or were disposed of in commercial landfills [312–314]. It is probable that the remainder of spent catalysts was disposed of on unauthorized sites. It is not certain that all such sites were subsequently cleaned to ensure environmental compliance. It is hoped that sufficient time has passed to realize that today that unauthorized dumping of spent catalysts defies any logic.

Spent catalyst wastes could be disposed of into a landfill only if it could be proven with certainty that both landfill facility and spent catalysts meet non-hazardous criteria. Thus, landfill does not remove or destroy any hazardous materials that may pose long-term health and environmental hazards unless they are properly managed. According to RCRA, not only the approved dump-site owner is liable, but also the owner of the buried hazardous waste. This environmental responsibility and/or liability continues for the life of the dump-site as well as during the postclosure period until it is proven that the site no longer poses any danger to the environment.

Historically, in 1982, commercial landfills were operating under the RCRA interim status permits. However, the RCRA amendments issued in 1984 required all interim status facilities to meet ground water and insurance requirements. For a continued operation, minimum technology of a double liner and leachate collection system was required to be installed by 1988. In this regard, some refineries responded by replacing all surface impoundments with the above ground tankage [315]. After closure, the spent catalyst and contaminated soil were removed and the impoundment filled with an uncontaminated material. In view of the potential future liabilities, such costly approach was deemed to be necessary, particularly in regard with the re-authorization of the RCRA, which requires that most of the surface impoundments are either retrofitted or closed. Originally, the concept of joint liability required that if something went wrong with an unsecured landfill within 20 years of disposal (e.g., ground water contamination), the company would be asked to cover cost of the entire clean-up [316]. However, subsequently, this concept was modified and replaced with the unlimited liability.

According to CERCLA or "Superfund" which was promulgated by EPA, refining industry may be forced to a cleanup wherever spent catalysts were disposed of in the past.

The RCRA subtitle C requires the installation and operation of the ground water monitoring system for evaluation of the performance of TSDFs unless the owners of such facilities can demonstrate that there is no and/or low potential of migration of hazardous species to ground water [296,297]. This includes establishment of a ground water sampling and analysis program as well as a ground water quality assessment plan. Regulations require at least one up-gradient and three down-gradient wells for obtaining samples of ground water. The RCRA provides details of the maintenance of the wells and water sampling procedures [296]. Monitoring is required during the operation of the facility, during its closure, and during its postclosure period if necessary. Postclosure monitoring, usually lasting 30 years after closure, is required in the case that hazardous solid was not removed from the facility after the closure. This may apply to landfills and surface impoundments that were closed but waste solid left in.

In view of the complications discussed above, it may be wise to conduct a thorough assessment of the site before a landfill facility is constructed. A similar assessment was made by a company before constructing a new metal reclamation plant [317]. In this case, the company commissioned a third party environmental audit, which also included the previous site operation. The audit comprised of the evaluation of the potential impact on ground water and soil contamination as well as the review of the permit status and files to assess potential liability of the site. As part of the audit, a shallow, confined, permeable deposit was discovered about 10 m under previous site. This zone was confined by clays and silts. The water samples taken from it revealed that all metals were at the background level or at the level of the nearby river. As the result of the audit, the new plant was constructed on top of very tight formation of clays and silts, having very low permeability.

5.1.4 Handling of Spent Catalyst on Refinery Site

When the decline in catalyst activity cannot be offset by adjustment of operating parameters (e.g., temperature), the operation has to be discontinued. In some cases, such actions have to be taken because of a high-pressure drop caused either by accumulation of solids on the front of fixed bed or due to the attrition of catalyst particles and formation of clumps caused by temperature excursions. The problems may be alleviated by skimming the front of the bed and replacing it with fresh catalyst. In more complicated cases, whole bed may need to be unloaded, screened, and reloaded. If no other action can restore the operation, catalyst has to be removed from reactor, which is then reloaded with either fresh or regenerated catalyst. The catalyst removed from reactor at this point is referred to as spent catalyst. It should be noted that such situations might be encountered during the operation of fixed bed reactors rather than moving bed reactors. In fact, avoiding problems with high-pressure drops was the main reason for the development of the latter reactors.

To ensure safety and environmental acceptance, it is desirable that spent catalyst is de-oiled and dried prior to unloading. This can be achieved by replacing the flow of the feed with a lighter fraction while H_2 is still flowing. The absence of discoloration of the washing liquid caused by dissolution of the catalyst carry-overs may indicate the completion of the washing stage. The accelerated solvent extraction method used for catalyst characterization has a direct relevance to catalyst de-oiling [318,319]. According to this method, the spent catalyst de-oiling, using hydrocarbon solvents, is conducted at elevated pressures (e.g., ~10 MPa). Under such conditions, de-oiling was complete within 5 min compared with conventional Soxhlet extraction requiring 6–12 h. This suggests that de-oiling efficiency can be optimized by operating conditions. After de-oiling, the drying of the spent catalyst can be performed in the flow of H_2 after the flow of washing liquid was discontinued. The last phase of drying involves the replacement of H_2 with an inert gas and cooling to room temperature. After de-oiling, drying, and cooling under inert gas, the safety during catalyst unloading is ensured.

Some fixed bed reactors, equipped with catalyst dump nozzle, are shown in Fig. 5.1 [320]. For spent catalyst withdrawal, the nozzle is opened under the purge of N_2. Then, spent catalyst flows into catalyst bins that were also purged with N_2. Dry ice (CO_2) is added to the filled containers to expel remaining air. The removal of the last amount of spent catalyst may require the personnel to enter the reactor. It is essential that in such situation the personnel carries all necessary safety equipment to prevent serious injuries.

Apparently, there is no catalyst unloading procedure that could be commonly adapted by all refiners. Generally, refineries apply their own procedures unless the catalyst withdrawal is part of the agreement between the refiners and a partner (e.g., companies involved in regeneration, transportation, storage, etc.). There may be a need for an approved procedure, which could be commonly adapted by all refineries. In this regard, several patents describing the catalyst unloading techniques have been noted [320,321]. In every case, the primary focus is on the reducing a self-heating character of spent catalysts. Otherwise, a spontaneous combustion of spent catalyst may result in the release of toxic species, such as SO_X, NO_X, HCN, NH_3, etc. [322].

Attempts have been made to develop a pretreatment method that would minimize hazardous nature of spent catalysts during catalyst unloading and all stages following after. In this regard, the process developed by Kashima Engineering Co. in Japan enables the catalyst unloading under air [323]. The process passivates self-heating nature of spent catalyst during reactor shutdown by applying a proprietary mixture of chemicals. The mixture contains compounds that deposit a film on the surface of spent catalysts. This film slows down oxygen penetration considerably. Figure 5.2 [323] shows a generalized shutdown procedure. Initially, feed-rate is reduced by about two third, while the reactor starts cooling down. Then, the carrier oil is introduced to displace the mixture of the feed and products. Once carrier oil is in the total

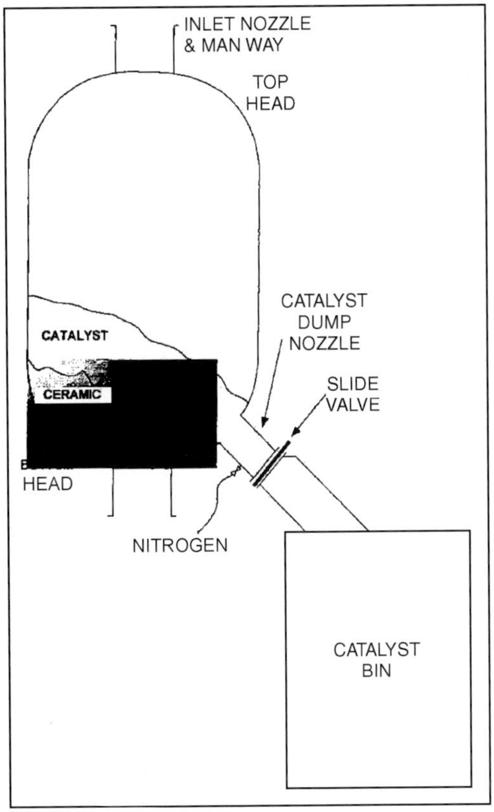

Figure 5.1: Schematic of spent catalyst withdrawing system from fixed bed reactor [From ref. 320. Reprinted with permission].

cycle, a chemical inhibitor is injected and circulation continues until it is necessary. At about 140 °C, the carrier oil is replaced by N_2 to achieve a drying and further cooling of spent catalyst. A similar method involves treating spent catalyst (while still in reactor) with a mixture comprising oxygen-containing hydrocarbons having carbon number of at least 12 [324].

After unloading, a decision has to be made regarding the fate of the spent catalyst, although most of the petroleum refineries have already a necessary system in place, usually based on many years of practical experience. This may include the agreement with a partner, i.e., certified company to perform regeneration, metal reclamation, and storage of hazardous waste. For refiner, the situation can be simplified in the case that the involvement of partner during the catalyst unloading, packaging, and shipping is part of the agreement. The decision making process has been described elsewhere [303]. Otherwise, it is responsibility of refiner to take all necessary precautions during the entire presence of spent catalyst on the refinery site to ensure safety and environmental acceptance.

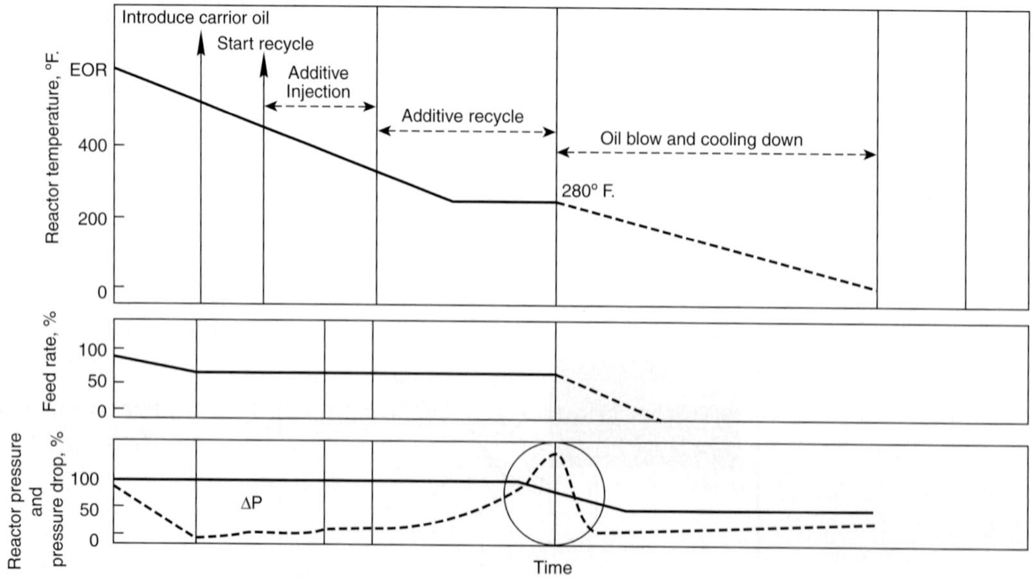

Figure 5.2: Summary of procedure for catalyst passivation and reactor shutdown [From ref. 323. Reprinted with permission].

Lassner et al. [325] summarized the actions that should be taken prior to shipment of waste materials from refinery site. They include a series of tests to determine whether or not the material is hazardous (e.g., toxic, corrosive, leachable, and flammable). It has been generally observed that more analytical work was necessary to prove that spent catalyst is non-hazardous than if it was [326]. The analysis should also indicate the presence of impurities, which could have an adverse effect during the catalyst handling. The MSDS should be prepared for each spent catalyst. The MSDS should list necessary precautions and emergency procedures for the catalyst packaging and shipping.

There may be circumstances in a refinery when the fate of spent catalysts was not yet determined and a temporary storage was necessary. Also, some refiners may store spent catalysts on site awaiting when better treatment techniques become available. In this case, carefully maintained storage using polypropylene super-sacks may be adequate provided that activities, such as welding, cutting, etc., are not conducted in the proximity. In some cases, refineries may have an access to a specially engineered landfilling comprising of separate lined cells capped and isolated from each other and the environment [306]. Another option which may deserve some attention involves placing metallic drums filled with spent catalyst in the non-operating mines. But those are still only temporary solutions. Dumping spent catalyst, although only temporary, near, and/or on the refinery site is perhaps the worst alternative. Nevertheless, information suggests that there is a 90-day deadline for moving hazardous waste from the site [326]. Otherwise, refiner has to go through lengthy administrative procedures to retain spent catalyst on the site beyond the deadline.

5.1.5 Cradle to Grave Approach to Spent Catalyst Management

There are several experienced and certified companies that offer a reclamation solution to generators of spent catalyst wastes at the time of the fresh catalyst procurement. They provide a "cradles to grave" approach for catalyst management, including loading, unloading, transport, regeneration, recycling, and disposal of spent catalysts. For example, CRI International Inc., in association with a group of affiliated companies, such as Criterion Catalyst Company, Catalyst Recovery Group, Catalyst Technology, Inc., and CRI-MET, provide all catalyst related services to refiners under one roof [327,328]. A similar consortium, including Albemarle, EURECAT, ACI Industries Inc. (Belgium), and London Chemicals & Resources Ltd (UK), also provides similar TCM services. The main benefits of the total integrated catalyst management system to refiners are:

(1) Reduced unit down time during catalyst change out in refinery.

(2) Deciding on regeneration/disposal options after utilization cycle may involve a team of personal experienced in handling spent catalysts and in regulatory affairs.

(3) Relief from transportation issues related to shipment of spent catalyst to a recycling facility.

(4) Relief from safety and environmental issues related to spent catalyst handling and disposal.

The strategy for TCM provided by Albermarle Catalysts was recently presented by Eijsbouts et al. [308,329]. It includes cooperating companies, such as Universal Oil Products (UOP) (hydroprocessing), EURECAT (presulfiding and regeneration), and Petroval (transportation). If necessary, companies with the additional services and expertise can be added to the consortium. This also includes personal with experience in regulatory affairs. Figure 5.3 [329] gives a detailed account of services provided by the TCM consortium. It is evident that the TCM group can play an advisory role and provide services during all stages of catalyst utilization, since fresh catalyst enters petroleum refinery. They participate in the final decisions, i.e., when further catalyst recycling is not feasible and metal reclamation or disposal remains the only option.

5.2 Hazardous Characteristics of Spent Hydroprocessing Catalysts

According to EPA, a hazardous waste is defined as one posing a substantial or potential hazard to human health and environment if mishandled. There are two basic criteria that are used to identify hazardous solids, i.e.:

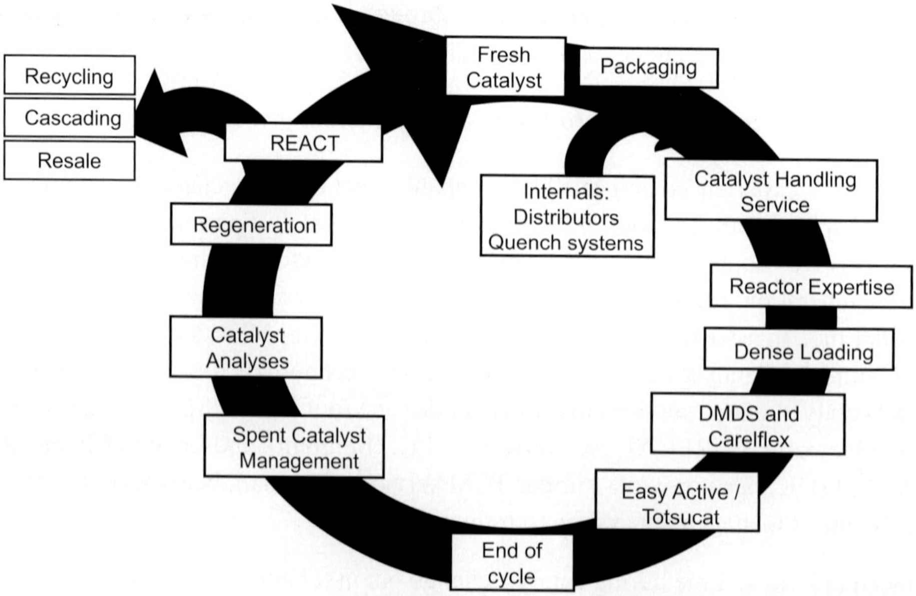

Figure 5.3: Stages of total catalyst management system [From ref. 329. Reprinted with permission].

(1) Characteristics, which can be defined in terms of physical and chemical properties, cause the waste to be hazardous.

(2) Properties defining hazardous characteristics must be measurable and quantified by testing protocols and must be detectable by hazardous waste generators (e.g., petroleum refineries).

Potentially hazardous constituents that are relevant to spent catalysts, listed User's Guide, are shown in Table 5.2 [306]. The list includes all metals that are considered by EPA as hazardous (e.g., Pb, Cd, Hg, Cr, Se, Ba, Ag, and Cu). The constituents in Table 5.2 can be divided into two groups, such as those present in the fresh catalysts as well as those present in spent and regenerated catalysts. Other possibilities are to classify the constituents either as inorganic and organic or combustible and non-combustible. It is believed that Mo and W, which are the principal active metals in hydroprocessing catalysts, should also be added to the list in Table 5.2 in anticipation that most of the heavy metals might be regulated in a near future. The efforts to develop more active and stable catalysts may require the addition of other constituents to the list.

Table 5.2 [306] covers all metals that are included in the NPDWS listed in Table 5.1 [296]. The organic constituents listed in these tables, i.e., benzene, toluene, PAHs, etc., may be of a

Table 5.2: Constituents of potentially hazardous solids [From ref. 306. Reprinted with permission].

Compounds of Be, V, hexavalent Cr, Co, Ni, Cu, Zn, As, Se, Te, Ag, Cd, Sn, Sb, Ba, Hg, Pb, and Ta
Inorganic acids
Inorganic sulfides
Inorganic fluorine compounds, excluding CaF_2
Inorganic cyanides
Phenols
Ethers
Aromatic compounds; polycyclic and heterocyclic
Organic nitrogen compounds, especially aromatic and aliphatic amines
Organic sulfur compounds
Substances of an explosive character
Organohalogen compounds

concern for spent catalysts that were not properly de-oiled and dried. Such cases of spent catalysts cannot be completely ruled out. The complete list of regulated organic compounds, given by Vallerot, was part of the groups in Table 5.2 [298]. Inorganic acids (e.g., sulfuric) require attention during a prolonged exposure of spent catalyst to air and water, such as in landfills.

For hydroprocessing catalysts, the major inorganic constituents that are added to the catalysts during operation include V and Ni. Depending on the origin of the feed, other contaminant metals, such as As, Fe, Ti, Ca, Na, clays, etc., may also be added. Details of the composition of organic and inorganic deposits in spent hydroprocessing catalysts were presented in the Chapter 4 on catalyst deactivation. It should be noted that the metals that are part of the spent catalysts (e.g., V, Ni, Co, Mo, W, Fe, etc.) are not included among the NPDWS constituents in Table 5.1 [296]. In spite of this, all necessary precautions have to be taken to prevent release of these metals to ground water and, thus, to avoid future liabilities.

The performance of hydroprocessing catalysts can be improved by modifying γ-Al_2O_3 support with various additives. In this regard, the catalysts modification, using fluoride and phosphate compounds, resulted in a significant enhancement in the activity and stability of catalyst. All handling stages of spent hydroprocessing catalysts must take into consideration the presence of the fluoride and phosphate species. So far, little attention has been paid to the release of fluorine and phosphorus containing compounds in spite of their hazardous nature. It is desirable that such compounds are added to the list of toxic emissions if their release from spent catalysts can be confirmed.

Hazardous characteristics of spent hydroprocessing catalysts are listed in Table 5.3 [16,306]. The RCRA characteristic hazardous waste is defined in 40 CFR 261 subpart C as one that fails

Table 5.3: Hazardous characteristics [From ref. 306. Reprinted with permission].

Explosive
Flammable
Liable to spontaneous combustion
Corrosive
Toxic
Liberation of toxic gases in contact with air and water
Capable, by any means, after disposal, of yielding another material

the test ignitability (40 CFR 261.21), corrosivity (40 CFR 261.22), reactivity (cyanide and sulfide, 40 CFR 261.23), and toxicity characteristics leaching procedure (Toxicity Characteristic Leaching Procedure [TCLP], 40 CFR 261.24) [280]. The EPA added spent hydrotreating catalyst (K171) and spent hydrorefining catalyst (K172) to its hazardous waste list in August 1998 [16] because of their self-heating behavior and toxic chemicals content. The spent hydrocracking catalysts from the dual hydroprocessing operations were added to the list in 1999 [16–18].

It is evident from the hazardous characteristics listed in Table 5.3 [306], that hazardous nature of spent catalysts is always associated with the potential release of toxic species on the exposure either only to air or a combined exposure to air and water. The characteristics, such as explosive, flammable, and liable to spontaneous combustion dictate that all precautions are taken to minimize the contact with air particularly in the case of spent catalyst that was not adequately de-oiled and dried. Toxic species may be released either to atmosphere or to ground water causing harm to humans on exposure. Special regulations and regulatory affairs, which have been in effect to deal with all stages of handling spent catalysts, e.g., unloading, storage, transportation, disposal, etc., were discussed above.

The analytical methods used for determining hazardous characteristics were discussed by McKenna and Pickering [330]. For example, corrosivity can be simply determined by measuring the pH of soils. For cyanide reactivity test, samples are held in a hermetic system and acidified to a pH of 2. The generated gases are scrubbed and analyzed. For sulfide reactivity, samples are acidified in an enclosed system. The generated gases are scrubbed and analyzed. The objective of the cyanide and sulfide reactivity tests is to determine the potential release of HCN and H_2S.

The ignitability of spent catalysts can be determined using the Pensky-Martens method, which has been used widely in petroleum refining for determining the flammability of fuels. According to this method, a sample is heated while being exposed to the atmosphere directly above the sample, to an open flame. The lowest temperature, at which the vapor above the

Table 5.4: Properties of spent catalysts treated in N_2 at 350 °C [From ref. 331. Reprinted with permission].

Analysis (wt.%)	Catalyst A	Catalyst B
Loss on heating	15	1
Carbon	18.50	7.64
Hydrogen	1.31	0.46
Nitrogen	0.38	0.14
Sulfur	5.62	6.09
H/C	0.75	0.72
N/C	0.018	0.016

sample ignites, is taken as the flash point. A flash point occurring at low temperatures would indicate that spent catalyst was not adequately de-oiled and dried. Therefore, potential for ignition of such spent catalyst would be high.

5.2.1 Exposure to Air

A number of the reactions releasing toxic species and those which generate heat occur during the exposure of spent catalyst to air alone and/or in the presence of water. Both organic (coke) and inorganic portions of spent catalysts are involved, although in rather different ways. It should be noted, that for most part, these reactions proceeds at very low rates. Therefore, their effect may only be evident after a prolonged exposure.

5.2.1.1 Reactions of Air with Coke

On exposure to air, the flammability and ignitability of spent catalysts are of primary concern. As it was indicated above, a spent catalyst may ignite if a sufficient vapor pressure of flammable gases builds up over its surface. The flammable gases may include H_2 and volatile hydrocarbons entrapped in the coke. As it was suggested earlier, the flammability can be minimized by spent catalyst pretreatment, e.g., de-oiling and drying by flashing in H_2 followed by inert gases. The actions that may be taken on the refinery site with the aim to minimize hazardous nature of spent catalysts were discussed above.

Figure 5.4 [331] shows TPO of differently pretreated spent catalyst A and catalyst B. The properties of these catalysts, after being treated in N_2 at 350 °C, are shown in Table 5.4. A large difference in the loss on heating between these catalysts should be noted. Curve 1 depicts the behavior of catalyst A (as received) during the temperature programmed pyrolysis (TPP) in N_2, whereas curve 3 the same catalyst during the TPO in 2% O_2. It is evident from curve 3 that the sufficient vapor pressure of the combustible volatile matter was developed already at

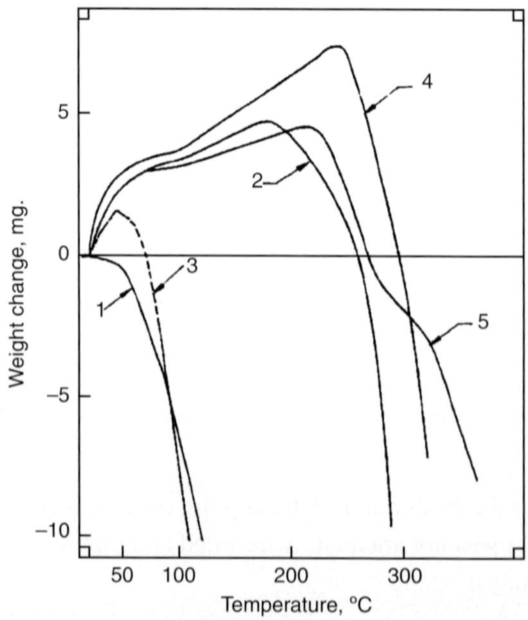

Figure 5.4: Effect of pretreatment on ignition temperature of spent catalysts [From ref. 331. Reprinted with permission].

about 50 °C. Consequently, the catalyst particle ignited as indicated by a sudden decrease in the weight. The curve 2 shows TPO of the catalyst A after being pretreated in N_2 at 200 °C until weight stabilization. For this catalyst, a significant increase in the O_2 chemisorption should be noted. Moreover, the ignition temperature increase from about 50 °C for unpretreated catalyst to almost 250 °C for the pretreated catalyst was noted. Further increase in the ignition temperature was observed when catalyst A was extracted by tetrahydrofuran (THF) followed by the pretreatment in N_2 (curve 4 in Fig. 5.4). The catalyst B was much less deposited with coke (Table 5.4) [331]. The TPO of this catalyst (curve 5) indicates on the involvement of the inorganic sulfur during the catalyst ignition. For this catalyst, the abrupt weight decrease temporarily slowed down before subsequent steep weight loss. It was postulated that the weight loss in the first region was dominated by combustion of inorganic sulfur, whereas in the second region by that of the coke. Because of a thick coke layer on catalyst A, the presence of the two burning regions was not observed. It should be noted that the ignition in Fig. 5.4 occurred in spite of only 2 vol.% O_2 used in the oxidizing gas. However, for pretreated catalysts, it is the concentration of the reactive surface oxygen groups, rather than the O_2 concentration in gas phase, which is the dominant factor causing the ignition. Thus, a sufficient vapor pressure of volatile combustibles would have to be developed for the O_2 in gas phase to get involved in combustion reactions. Curve 3 shows that only for unpretreated catalysts, the ignition can be caused, involving the gas phase O_2.

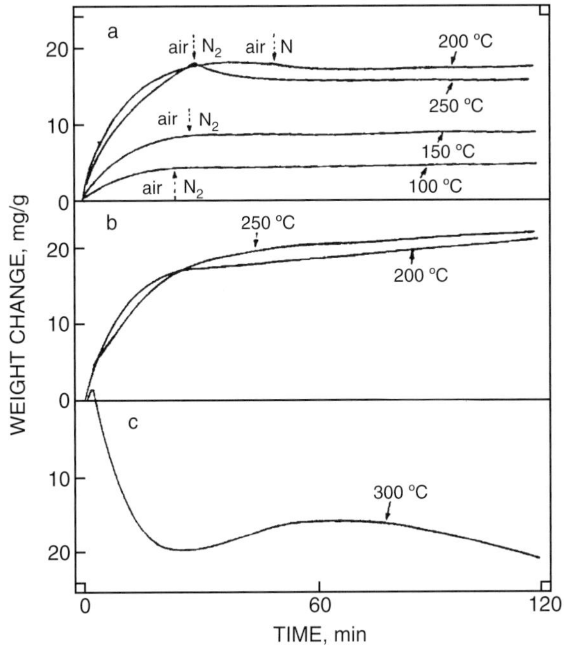

Figure 5.5: Isothermal oxidation of spent catalyst particles [From ref. 332. Reprinted with permission].

The results in Fig. 5.5 [332] were obtained in air rather than in 2 vol.% O_2 as it is the case of Fig. 5.4 [331]. In contrast to Fig. 5.4, these experiments were conducted isothermally at indicated temperatures. Prior to the experiments, the spent NiMo/Al_2O_3 catalyst was extracted by THF followed by pumping at 200 °C overnight. After approaching a steady state, the flow of air was replaced by the flow of N_2 (Fig. 5.5a). In this case, a slight decline in the weight at 200 and 250 °C indicated the beginning of the decomposition of oxygen-containing surface complexes. No weight loss was observed during the continuous oxidation (Fig. 5.5b). Under these conditions, a decomposition of the complexes was more than offset by the formation of additional complexes. A rapid weight loss at 300 °C indicated the ignition of catalyst particles aided either by the most volatile component of coke or by the oxidation of metal sulfides. This is supported by a slight weight gain during the later stages of oxidation shown in Fig. 5.5c.

With regard to the exposure of spent catalysts to air, some important conclusions may be drawn from the results in Figs 5.4 and 5.5 [331,332]. First of all, it is a continuous chemisorption of O_2, which can be accelerated with increasing temperature, that deserves attention. At sufficient concentration of the oxygen-containing complexes, the catalyst particles may ignite, particularly in proximity of hot objects. Also, some activities, e.g., welding, cutting, etc., in a proximity of the storage of spent catalysts may increase the potential of ignition. It is again emphasized that these facts have to be taken into consideration even during a temporary storage of spent catalysts.

5.2.1.2 Reactions of Air with Catalyst

The composition of fresh, fresh-sulfided, and spent hydroprocessing catalysts were discussed earlier. On exposure to air under dry conditions, the mineral matter of spent catalyst undergoes oxidation even at ambient temperatures. Apparently, during the storage, this process is very slow. However, in view of the unlimited liability, even a slow oxidation deserves attention in the case that it may lead to the release of hazardous constituents. Thus, the oxidation of inorganic sulfur in solid waste (e.g., coal rejects) has been noted after more than 50 years exposure to air. Consequently, leaching of metals and decrease in pH of the ground water in proximity were noted. It is believed that a brief discussion of these events may be necessary.

During the exposure to air, the sulfided form of metals is gradually converted to an oxidic form. This process may be very slow and may involve a gradual replacement of sulfur atoms by oxygen atoms to form oxysulfide intermediates before a complete oxidation was achieved. For MoS_2, this may be depicted using the following set of reactions:

$$MoS_2 + O_2 = MeSO + SO$$

$$MoSO + O_2 = MoO_2 + SO$$

$$2SO + O_2 = 2SO_2$$

$$MoO_2 + 0.5O_2 = MoO_3$$

The sum of all these reactions provides a reaction for the overall conversion of MoS_2 to MoO_3, i.e.:

$$MoS_2 + 3.5O_2 = MoO_3 + 2SO_2$$

However, the probability of a complete oxidation of MoS_2 during storage (e.g., in landfill) is rather low. The conversion of MoS_2 to $Mo(SO_4)_2$, e.g.:

$$MoS_2 + 2O_2 = Mo(SO_4)_2$$

cannot be ruled out, although a slow diffusion of O_2 through and/or consumption by coke layer would not be favorable for this reaction to proceed, suggesting that under typical exposure of spent catalysts to air this reaction may be kinetically limited. Other metal sulfides, which are part of the spent hydroprocessing catalysts, may undergo a similar gradual transformation to corresponding metal oxides via metal oxosulfide intermediates. For example, the oxidation of Ni_3S_2 may proceed as follows:

$$Ni_3S_2 + O_2 = Ni_3SO + SO$$

$$Ni_3SO + 1.5O_2 = 3NiO + SO$$

$$2SO + O_2 = 2SO_2$$

$$Ni_3S_2 + 3.5O_2 = 3NiO + 2SO_2$$

The overall conversion of Ni_3S_2 to $NiSO_4$ would involve the following reaction:

$$Ni_3S_2 + 4.5O_2 = 2NiSO_4 + NiO$$

Because of a limited stability, the SO species is expected to compete successfully with metal sulfides or oxysulfides for O_2 and, as such, will be converted to more stable SO_2 and, if sufficient O_2 is available, the oxidation may proceed to SO_3.

In the presence of water, additional reactions during the exposure of spent catalysts to air may be anticipated. For example, the reactions, such as:

$$SO_2 + H_2O = H_2SO_3$$

$$H_2SO_3 + 0.5O_2 = H_2SO_4$$

may be part of the overall oxidation mechanism. The formation of H_2SO_4 may be one of the reasons for classifying spent hydroprocessing catalysts as corrosive wastes. Moreover, the presence of H_2SO_4 may have a dramatic effect on spent catalyst because of the potential reaction releasing H_2S from the unconverted metal sulfides, e.g.:

$$MoS_2 + H_2SO_4 = Mo(SO_4)_2 + H_2S$$

In similar reactions, H_2S would be also released from other transition metal sulfides, which are part of the spent hydroprocessing catalysts. The potential release of H_2S is one of the hazardous characteristics of spent hydroprocessing catalysts. Information suggests that preoxidation of waste materials, using H_2O_2 before disposal, resulted in a significant decrease in the content of sulfidic sulfur [333]. Consequently, the release of H_2S to environment could be minimized. Nevertheless, the formation of H_2SO_4 would increase leachability and corrosivity (due to lowered pH). The potential H_2SO_4 aided release of toxic species, such as HCN, deserves attention as well.

The study of Afonso et al. [343] provides some support for the above rational. Thus, these authors placed metal boxes containing spent $NiMo/Al_2O_3$ into the ground (30–50 cm depth) for exposure to the environment during 40 to 70 days. After the exposure, the boxes were removed to determine the change in the composition of both mineral and organic matter of spent catalyst. After 14 months exposure, Al(III) and phosphate species became insoluble, while sulfur was almost quantitatively converted to sulfate. The migration of Ni and Mo into liquid phase increased as well. The Fe content in liquid phase increased due to corrosion of the metal container. At the same time, the solubility of coke in dichloromethane was reduced drastically, whereas the coke solubility in methanol increased. The increased solubility in methanol was attributed to the formation of O-containing groups, such as acids, ethers, esters, and phenols. After six months, the holes in metal boxes, presumably formed by corrosion with H_2SO_4 released from spent catalyst, were noted. Consequently, the contamination of ground water with Ni, Fe, Mo, and sulfate was observed.

The active phase of the hydroprocessing catalysts comprises of Co(Ni)-Mo(W)-S entities [53]. To various degrees, this phase may be still present in spent catalysts. As it was indicated earlier, in this phase, promoters, such as Ni and Co, decorate MoS_2 and/or WS_2 crystallites. Then, the O_2 will access the promoting atoms more readily compared with the sulfide component of the active phase. This may be depicted by the general reaction such as:

$$\text{Ni(Co)-Mo(W)-S}_X + O_2 = \text{NiO(CoO)} + \text{Mo(W)-S}_X$$

The oxidation of the residual Mo(W)-S_X entity would proceed according to the reactions discussed above.

The presence of fluoride and phosphate may limit utilization options of spent hydroprocessing catalysts [335]. A potential release of fluorine and phosphorus containing species from spent catalysts received little attention. Because of their modifying effect on the support, it is believed that both fluorine and phosphorus are associated with the Al_2O_3. In the absence of any experimental data, only a speculative reaction may be proposed. Thus, for the fluorine, the release of HF could be anticipated if sufficient concentration of H_2SO_4 builds up on the exposure of spent catalyst to air and water. Similarly, species containing Al_2O_3 and a phosphorus could be converted to $Al_2(SO_4)_3$ as a more stable salt compared with Al phosphate.

Various concentrations of arsenic in petroleum have been noted [336]. During hydroprocessing, a portion of the arsenic may deposit on catalyst surface, most likely in a sulfidic form, whereas another part may be released in refinery gases as AsH_3. A similar set of the oxidation reactions as postulated above for other metal sulfides can also be proposed for As sulfides. It should be noted that a significant increase in the solubility of arsenic caused by the conversion of arsenic sulfides to corresponding oxides, which may occur during the storage of spent catalysts, should be of a concern because of the toxic nature of the As-containing species. Consequently, leaching of arsenic to ground water could not be avoided unless catalyst is handled in accordance with the regulations.

Table 5.5: Analysis of Toxicity Characteristic Leaching Procedure (TCLP) leachates (ppm) of spent catalysts and Environmental Protection Agency (EPA) regulatory levels [From ref. 337. Reprinted with permission].

Metal	Catalyst 1	Catalyst 2	Catalyst 3	EPA regulatory levels
As	53	<0.1	<0.1	5
Ba	<0.1	<0.1	nd	100
Cd	nd	<0.1	0.1	1
Co	1	2	293	–
Cr	<0.1	<0.1	nd	5
Fe	3	83	15	–
Pb	0.3	0.3	0.4	5
Hg	nd	nd	nd	0.2
Mo	249	9	310	–
Ni	657	455	4	–
Ag	nd	nd	nd	5
Se	<0.1	<0.1	nd	1
V	<0.1	205	<0.1	–
Zn	0.2	<0.1	0.3	–

5.2.2 Leachability

The methods, such as ASTD-3987 and the TCLP, are normally used to assess leachability of the wastes dumped in landfills. The results in Table 5.5 are used as an example to illustrate the hazardous nature of spent hydroprocessing catalysts [337]. The regulatory levels of trace elements (last column in Table 5.5) and volatile organics are reported in Federal Register [16], which has been issued and periodically updated by the US EPA. It should be noted that the permitted levels of some of these metals in drinking water (Table 5.1) are lower by a factor of 100 or more of the regulatory levels. Thus, these levels assume a dilution factor of 100 or more, once the metal is leached out from the hazardous solid. For the analysis, catalysts were used without decoking. The leachability of the catalysts was determined the TCLP procedure. The method involves shaking catalyst particles in the solution of a buffer for 18 h. Subsequently, solid particles and suspended fines were filtered off to obtain leachates for analysis. The test simulates potential release of metals to the ground water during the contact with water. Potential release of PAHs deserves attention as well. In this case, the same leachate that was used for the determination of trace elements is being used. The set of analytical methods, used for the PAHs analysis, is described in details in the EPA's Federal Register [16–18] as well.

The catalyst 1 in Table 5.6 [337] was the extrudate form of the $NiMo/Al_2O_3$ catalyst used for hydroprocessing a distillate from primary upgrading of a heavy feed. The chestnut bur form of

Table 5.6: Analysis of spent hydroprocessing catalysts (as received) [From ref. 337. Reprinted with permission].

Analysis (wt.%)	Catalyst			
	1	2	3	4
Carbon	7.4	15.0	7.7	56.3
Hydrogen	1.2	3.3	0.8	1.1
Nitrogen	0.40	0.14	0.14	0.30
Sulfur	7.13	12.9	6.2	4.60
Iron	0.08	7.0	0.05	0.49
Vanadium	0	8.6	34 ppm	2.7
Nickel	2.7	3.9	64 ppm	1.2
H/C	1.9	1.5	1.2	0.2

the $NiMo/Al_2O_3$ catalyst 2 was used for upgrading an atmospheric residue using a fixed bed reactor. The extrudate form of the $CoMo/Al_2O_3$ catalyst 3 was used in a commercial hydrodesulfurization (HDS) unit. The $NiMo/Al_2O_3$ catalyst 4 was obtained from a fixed bed reactor after upgrading of a heavy feed. This catalyst was received as a lump weighing about 4 kg. It is believed that at the end of operation, the catalyst was deactivated to such an extent that the fixed bed essentially fused. Therefore, a chisel and hammer had to be used to produce particles (minus 20 mesh) for evaluation. These particles were non-extractable by hot THF in a Soxhlet apparatus.

According to Table 5.5 [337], catalyst 1 is classified as hazardous solid because for this catalyst, the concentration of As in the leachate was about 10 times greater than allowed by regulatory levels. Most likely, arsenic originated from the distillate feed during the continuous operation lasting more than two years. For other spent catalysts, the content of the trace metals were lower than regulatory limits. However, high concentrations of the active metals (e.g., Mo, Ni, and Co) in leachates also suggest that these catalysts should be handled as hazardous solids in spite of the fact that they are not included among the NPDWS in Table 5.1 [296]. Again, there is no reason to believe that this will not change in a future.

5.3 Pretreatment of Spent Catalysts for Disposal

Spent catalyst wastes could be disposed of into a landfill only if it could be proven with certainty that the landfill met non-hazardous criteria. In the USA, the disposal and treatment of spent refinery catalysts is governed by the RCRA, which holds not only the approved dump-site owner liable, but the owner of the buried waste as well. This environmental responsibility continues for the life of the dump-site. Spent catalysts sent to landfills should be properly treated for stability to reduce leaching of the harmful materials that could pollute the environment. Methods such as ASTM-3987 and the TCLP are normally used to assess leachability of the wastes dumped in landfills.

The treatments, which enhance environmental acceptance of spent catalysts prior to their disposal in landfills, will be illustrated on several examples from the scientific and technical literature. In every case, the aim was to minimize leachability of spent catalysts. These concerns are reflected by a growing number of the related articles that have been appearing in the scientific literature during the recent years. This is not surprising in view of the cost of potential liability associated with landfilling which may exceed $200 per ton of a spent catalyst. In this regard, several methods based on immobilization, vitrification, and encapsulation of toxic substances have been evaluated with various degrees of success.

It has been generally observed that a significant decrease in the leachability of hazardous solids may be achieved using thermal treatment during which the spent catalyst is fused. For example, the process patented by Phoenix Environmental [338] involves heating spent catalyst in the flow of O_2 until the solid becomes a molten bath. After solidification, the molten bath has a spinel structure that can bond harmful metals and convert them into a non-leachable form. As it was shown by Kox and Van der Vlist [339], this method has been successfully applied to spent hydroprocessing catalysts. It is expected that as part of this method, the effluents formed during heating is captured because it may contain some volatile contaminants.

Treatment methods, such as encapsulation, vitrification, and stabilization/solidification are commonly used to immobilize the hazardous waste materials and to make them non-leachable [340–344]. For example, as it was indicated by Trimm [345], a decrease in leachability may be achieved by encapsulation of spent catalysts using organic substances. The method involves thermal treatment of spent-decoked catalyst with substances, such as bitumen, paraffin wax, and various polymers. After cooling, the catalyst material is well sealed in the thermoplastic film. However, long-term effects of this method on leachability are unknown. Also, after thermal treatment, some of these films may be efficient adsorbents of O_2. It is therefore believed that the ignitability of spent catalysts pretreated by this method may require attention. Besides organic encapsulants, the encapsulation using various inorganic solids, such as clay, concrete, vitreous silicates, glass composite, etc., may also be achieved. In this case, the encapsulant converts a leachable form of metal into non-leachable, usually by fusion.

As it was indicated, the immobilization of metals is one of the options that can decrease the leachability of spent catalysts. In this case, catalysts are treated with other solids (e.g., silica, clays, Portland cement, etc.) at temperatures ensuring the slugging (softening) of the mineral matter. On cooling, the slag retains glassy properties. This prevents the release of metals in contact with water. The immobilization by incorporating hazardous metals in glasses has been

Table 5.7: Toxicity Characteristic Leaching Procedure (TCLP) of pretreated and unpretreated solids [From ref. 350. Reprinted with permission].

	Concentration in leachate (ppm)		
	Unpretreated	Pretreated	Regulatory levels
Pb	3 to 3700	<1	5
Cd	1 to 1596	<1 to 07	1
Cr	5 to 660	<0.3	5
Se	1 to 300	<0.0025	1
As	≤4190	<1.75	5
Ni	5 to 250	<0.5	5
Ba	>400	<10	100

successfully applied to radioactive wastes [346–348]. For a similar purpose, Shi and Fernandez-Jimenez [340] used alkali-activated cement comprising Ca-silicate hydrate and a zeolite. The leachability of the material solidified with the alkali-activated cement was much lower than that of solids, which were hardened with Portland cement.

The Maectite process, patented by Sevenson Environmental Services Inc. [349], is capable of converting reactive metals contained in solid wastes into non-leachable minerals in the apatite and barite group. These minerals are resistant to acidity and degradation by geological and chemical conditions, such as those found in landfills and natural settings. The leachability of the unpretreated hazardous waste and that pretreated using the Maectite process is shown in Table 5.7 [350]. It is evident that on pretreatment, the concentrations of all trace metals were well below the regulatory levels. It is believed that the Maectite process can be also successfully applied to spent hydroprocessing catalysts. A similar method known as Sealosafe has been developed by Stablex Corporation for the treatment of hazardous wastes to make them non-leachable [351]. The process involves adding a mixture of calcium containing cement powder and an alumino-silicate powder to the waste material dispersed in water and converting them to an impermeable solid. Cement-based solidification/stabilization/immobilization methods have also been widely studied for treatment of hazardous wastes containing arsenic and heavy metals [352–355]. Sun et al. [341] used Portland cement as a matrix to encapsulate vanadium and nickel present in spent fluid catalytic cracking (FCC) catalyst and found that Portland cement was an effective means of stabilization. In another study, these authors used marine clay as a matrix to stabilize the heavy metals present in spent FCC catalysts and produce value-added high strength bricks through a sintering process [355,356] similarly as they did for the spent $CoMo/Al_2O_3$ catalyst [357]. The results showed that the resulting bricks had high compressive strength ranging between 20.0 and 92.0 N/mm^2 for marine clay-spent FCC catalyst samples and low-water absorption values ranging between 4.8 and 18.5 wt.% as well as low metal leaching rates with metal concentrations ranging between 0.1 and 6.6 mg/L for vanadium and between 9.5 and 52.8 ug/L for nickel. Both chemical fixation and encapsulation mechanisms were proposed by these authors for stabilization of V and Ni. In the chemical fixation, the heavy metals form chemical bonds with the surrounding matrix to become part of its crystal structure, while in the encapsulation, a physical barrier is formed around the heavy metals preventing them from leaching into the environment.

The effect of spinel based construction ceramics (e.g. γ-Al_2O_3, hematite, kaolinite) in the stabilization of nickel-containing waste sludge was evaluated by Shih et al. [358]. The study demonstrated the feasibility of transforming nickel-laden sludge into spinel phases with the use of readily available and inexpensive ceramic raw materials and successful reduction of metal mobility under acidic environments. Solidification and stabilization of toxic cadmium ions in sand-cement clay mixture were achieved recently by Shawabkeh [359]. Clay-based stabilization of the heavy metals (V, Mo, and Ni), present in spent hydroprocessing catalysts, was also reported by Stanislaus et al. [360]. In this study, the spent catalyst in the form of a fine powder was mixed with clay, gatch (containing SiO_2–Al_2O_3 and CaO), sand, and water and heated at high temperatures in the range 1150–1130 °C. Leaching of the heavy metals from the resulting material was very low (<1.0 mg/L).

A combined process for the recovery of metals of interest from spent hydroprocessing catalysts and immobilization of the unleached metals remaining in the residue by stabilization has been reported by Sun et al. [361]. The spent catalyst that contained 16 wt.% C, 7.3 wt.% S, 10.9 wt.% Mo, 4.0 wt.% Co, and 4.6 wt.% V was first calcined in air at 500 °C to remove carbon and sulfur and to convert the metal sulfides to metal oxides. The metals were then leached in two stages using concentrated NH_4OH in the first stage and 10% H_2SO_4 in the second stage to recover 87% of Mo and 77% of Co. The residue containing the unleached metals was mixed with marine clay and fired at temperatures around 1000 °C to produce commercial value bricks. The leachability of the heavy metals from the produced bricks was very low. It is believed that the SiO_2 in the clay together with the γ-Al_2O_3 in the residue formed an impermeable ceramic material to prevent leaching.

In another related study, Sun et al. [362] used ladle furnace slag (LFS) rich in CaO and SiO_2 to stabilize small amounts of the heavy metals (V, Mo, and Co) remaining in the γ-Al_2O_3 residue after recovering a major portion of the metals by leaching. The combined waste material containing mainly Al_2O_3, SiO_2, CaO, and small quantities of heavy metals (V, Mo, and Co) was melted in a high temperature electric furnace at 1130 °C for 3 h to form a glass-ceramic product. The TCLP leaching values of the metals from the glass-ceramic material were lower than the allowable level, which indicates immobilization of the metals in the glass-ceramic matrix.

A combined leaching, recovering, and residue immobilization approach is applied by Hydromet Corporation Ltd. [363] in Australia for treating metals-containing wastes. When the recovery/recycle option is possible, the targeted metal is recovered by suitable chemical leaching, separation, and purification methods and the residue is treated to immobilize the remaining contaminants to make them acceptable for landfill disposal. Where either and/or economic reasons prevent a recovery/recycle route, the waste is treated to immobilize the contaminant process. Basically, the Hydromet's immobilization process involves the addition of chemical reagents to convert all of the metal constituents into extremely insoluble compounds.

CHAPTER 6

Regeneration

For the purpose of this review, regeneration is considered to be the process for only removal of the coke deposited on catalyst surface with the aim to restore as much as possible of the original catalyst activity. This also requires the recovery of surface area and porosity. It is generally believed that, to be suitable for regeneration, the content of contaminant metals (e.g., V and Ni) in spent catalyst should not exceed 5 wt.%. Otherwise, at least 80% recovery of the original activity to make regeneration attractive may not be achieved. Such level of catalyst activity recovery can be readily achieved for the spent catalysts after hydroprocessing atmospherics distillates and lighter feeds. It is difficult to reach this level of the activity recovery when both coke and metals, which were deposited from the feed, are present on the catalyst surface. The removal of metals with the aim to restore activity is discussed as part of the Chapter 7 dealing with catalyst rejuvenation.

The most widely used method for activity recovery involves the coke removal via oxidative regeneration using diluted air and/or air and steam mixture. Potentially, other oxidative agents can also be used. The oxidative regeneration technique has been used on the commercial scale for several decades. Attempts have also been made to use steam and CO_2 as oxidative agents. The coke removal may be achieved via reductive regeneration using H_2. However, this method has not yet reached a commercial stage. Attempts to remove the deposits from catalyst surface by extraction using various solvents have also been noted. However, the potential of this option seems to be rather limited. It was indicated earlier that after calcining to obtain necessary mechanical strength, an enhancement of active metals dispersion may be achieved via postcalcining treatment with chelating agents [329]. The same approach may also be applied to regenerated catalysts.

6.1 Regenerability of Spent Catalysts

In petroleum refinery, a decision has to be made on the fate of spent catalyst after unloading from reactor. This may require testing on a laboratory scale. A regeneration test should reveal the potential level of recovery of catalyst activity. Generally, at least 80% recovery of hydrodesulfurization (HDS) and hydrodenitrogenation (HDN) activity of the fresh catalyst is the acceptable minimum [364], although some refineries prefer 90% activity recovery [365]. Otherwise, the spent catalyst is classified as non-regenerable. Compared with the pore volume and size distribution, the surface area recovery may not be a precise indication of the activity recovery. The crushing strength and particle size distribution of the laboratory-regenerated catalysts should conform to minimum specification requirements as well. The conditions of the laboratory tests must be optimized to ensure reliability of the results. Otherwise, a situation may be encountered in which laboratory tests indicate regenerability of the catalyst contrary to the industrial regeneration results.

Because of the complexity involved, the refiner may request either catalyst supplier or catalyst regenerating company to perform laboratory testing of spent catalysts prior to final decision is made. The TCM approach discussed earlier is another potential option available to refiner. Otherwise, testing on a pilot plant scale may provide more reliable information. According to Smart [364], such testing may comprise four parts, e.g.:

(1) Measurement of activity, kinetic rate constants, and activation energy.

(2) Measurement of catalyst response to the change in H_2 pressure, space velocity, and temperature.

(3) Estimate of relative catalyst stability and lifetime using an accelerating aging test.

(4) Evaluation of the regenerability of catalysts and of the other factors governing regeneration.

Of course, all these testing phases are usually complemented by determination of the chemical and physical properties of catalysts.

It was indicated in Chapter 5 that the structure and composition of the deposits on spent catalysts surface depend on the origin of feed, type of catalyst, and operating conditions. A spent catalyst used for the hydroprocessing of atmospheric distillate feeds can be readily regenerated because the content of contaminant metals in the deposit is negligible. Thus, several regeneration-utilization cycles can be achieved with such catalysts. In this case, the recrystallization of active phase and/or the loss of active metals due to diffusion to the support may not be avoided. During the prolonged exposure to operating conditions, sometimes lasting several years, this could be the main cause of deactivation during hydroprocessing of the atmospheric distillates. Most likely, this would be the cause of a permanent deactivation.

Apparently, the spent catalysts from hydroprocessing of vacuum gas oil (VGO) and heavy gas oil (HGO) can also be regenerated, although they were exposed to more severe conditions than those applied during hydroprocessing of atmospheric distillates. More severe conditions would result in an enhanced catalyst deactivation due to sintering. Therefore, the number of regeneration-utilization cycles is expected to be lower than that for the spent catalysts from the upgrading of atmospheric distillates. Temperature and H_2 pressure are among the most important parameters influencing the regenerability of spent catalysts [366]. In fact, the influence of operating conditions may be much more pronounced than that of the origin of vacuum distillates and type of catalyst. In one case of hydroprocessing of a VGO, catalyst was exposed to a temperature runaway under low H_2 pressure. For this catalyst, more severe regeneration conditions were necessary for achieving a desirable level of the coke removal.

The content of contaminant metals in residues derived from conventional crude was generally low. It is, therefore, believed that spent catalysts, particularly those used in the downstream layers and/or reactors of a multistage system, can also be suitable for regeneration. However, for such spent catalysts, the number of regeneration-utilization cycles will be less than that for the spent catalysts from hydroprocessing of lighter feeds derived from a conventional crude. Similarly, for vacuum residues (VR), the regeneration of catalysts from the fourth or even third stage reactors of the multistage processes (e.g., atmospheric residue desulfurization [ARDS] and HYVAHL processes) can be performed without difficulties. This was confirmed by the results published by Al-Dalama and Stanislaus [12] showing a significant recovery of the surface area and porosity of catalysts taken from the third and fourth reactor of the ARDS process, compared with a little recovery for the catalysts taken from the first and second reactors. If not reused directly, i.e., in the same reactor, the spent-regenerated catalysts can be cascaded to different catalytic operation. This option becomes attractive in larger refineries with several catalytic reactor units in operation. Various options for cascading regenerated catalysts in petroleum refinery as well as other potential uses of regenerated catalysts are evaluated in several following chapters.

A high content of contaminant metals complicates upgrading of residues derived from heavy crudes. In this case, spent catalysts may be non-regenerable, particularly if upgrading was conducted in the ebullated bed reactors. The situation may be elevated by pretreating the heavy feeds, e.g., using solvent deasphalting method. The quality of deasphalted oil (DAO) produced is influenced by the type of solvent employed. There is always a trade-off between the yield and quality of DAO. For example, the contamination with metals increases with increasing yield of DAO. This will have an adverse effect on the regenerability of spent catalysts used for hydroprocessing of the DAO, i.e., the number of regeneration-utilization cycles will decrease. In some cases, the quality of DAO derived from a heavy feed may approach that of a residue derived from a conventional crude.

Information on the regeneration of catalysts used for hydroprocessing of heavy feeds and at the same time deactivated only by coke is limited. The results shown in Fig. 6.1 [367,368] were obtained during the regeneration of a VGO. They indicate that almost complete carbon removal could be readily achieved. In the same studies, several catalysts used for hydroprocessing of lighter feeds were also included. The main differences included the higher temperatures required for the removal of sulfur from the spent catalyst used for hydroprocessing of the VGO than that for the lighter fraction. Moreover, the regeneration lasted longer because of the

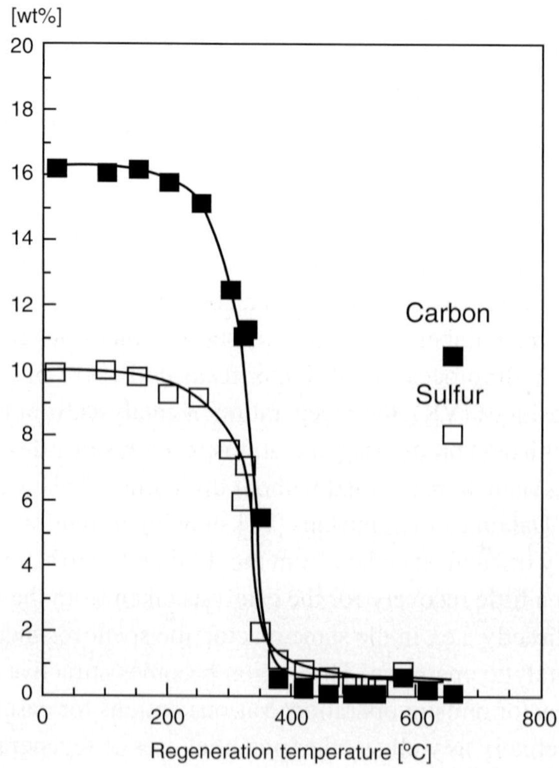

Figure 6.1: Burn-off profiles for removal of carbon and sulfur during regeneration of spent catalyst [From refs 367 and 368. Reprinted with permission].

greater amount of the deposited coke. A lower reactivity of coke, because of more severe conditions applied during hydroprocessing of VGO, may be a contributing factor as well.

It was indicated that there may be some exceptions where regeneration of the catalyst used for hydroprocessing of the metals and asphaltenes containing feeds can be justified, i.e., the catalysts from the third and fourth reactors either of the ARDS process or HYVAHL process [12]. It is believed that in some cases, even the second reactor catalyst may be at least partly utilized. For example, the catalyst unloaded from the bottom of the bed can be (after regeneration) cascaded to the top bed of the same bed for the next utilization cycle.

For both light and heavy feeds, arsenic may accumulate on the catalyst surface during the operation. It has been established that under hydroprocessing conditions organo-arsenic compounds in the feed readily decompose and deposit on the catalyst surface [7,307]. Because of the severe poisoning of the HDS active sites, the oxidative regeneration of catalysts may not be a suitable method when the arsenic content approached about 0.4 wt.%. Such spent catalyst must be classified as non-regenerable.

6.2 Oxidative Regeneration

The most established regeneration method involves the oxidative removal (by burn-off) of carbonaceous deposits, generally referred to as coke, which deposited on catalyst surface during the operation. To ensure safety, the diluted air has been used as the oxidation medium during the early stages of regeneration. The last stages of regeneration process, i.e., after the most reactive components of coke were removed, may be completed using air. Figure 6.1 [367,368] shows that under properly controlled regenerating conditions, almost complete removal of carbon deposited on catalyst can be achieved. The oxidative regeneration using air is also possible provided that the regenerator temperature is carefully monitored. In this case, either the design of regenerator or an installation of cooling system may be necessary. However, the additional cost of these changes must be compensated by improved efficiency of regeneration.

6.2.1 Mechanism of Oxidative Regeneration

A significant complexity of the regeneration mechanism is indicated by the simultaneous involvement of organic coke, contaminant metals associated with coke and those on catalyst surface as well as the inorganic portion of the spent catalyst. Because of catalyst porosity, two main regions, i.e., chemically controlled and diffusion controlled, are always part of the overall regeneration mechanism. In addition, variable conditions (e.g., O_2 concentration, temperature, particle size of spent catalyst, method of spent catalyst pretreatment, etc.) can have a pronounced effect during regeneration.

6.2.1.1 Oxidation of Coke

The distribution of gaseous products provides essential information on the mechanism of oxidative regeneration. Various experimental techniques used for this purpose were compared elsewhere [13]. In the presence of O_2 under typical regeneration conditions, both inorganic part and organic part of spent catalyst are oxidized. The involvement of the latter becomes more evident with the gradual removal of coke. Typical analyses of the coke on spent catalysts are shown in Table 6.1 [13]. They indicate that the oxidation of coke will involve the following general reaction in which the formation of all major gaseous products is shown:

$$\underbrace{H_mC_nS_xN_y}_{\text{Coke}} + (0.5m + n + x + 0.5y)O_2 = \underbrace{H_2O + CO + SO_2 + NO + CO_2}_{\text{Products}}$$

Depending on the experimental conditions, the formation of N_2O, NO_2, and SO_3 can be anticipated as well. The difference between these reactions and the oxidation of coke during storage discussed in the previous section should be noted. Thus, under regeneration conditions, the spent catalyst is contacted with the oxidation medium at much higher temperatures

Table 6.1: Analysis of coke on spent hydroprocessing catalysts (wt.%) [From ref. 13. Reprinted with permission].

	A	B	C
MoO_3	15.0	16.0	16.0
CoO	–	4.0	4.0
NiO	4.0	–	–
Carbon	18.8	11.4	2.8
Hydrogen	1.5	2.0	–
Sulfur	7.1	4.4	6.1
Nitrogen	0.5	0.5	0
Surface area (m^2/g)	26	65	126

compared with that during storage. In other words, the rate of these reactions under regeneration conditions is greater by several orders of magnitude.

It has to be noted that CO_2 is a major product even under limited availability of O_2. This follows from the established mechanism of hydrocarbon auto-oxidation, which involves the chemisorption of O_2 leading to the formation of surface complexes, solid peroxy radicals, and hydrogen peroxides [331,332]. This is depicted on the mechanism shown in Fig. 6.2. The less stable surface complexes decompose to yield CO and CO_2, once the temperature is increased as part of the stepwise regeneration process. This is clearly demonstrated in Figs 5.4 and 5.5 [331,332] showing the temperature programmed oxidation (TPO) profiles and the isothermal oxidation profiles of several spent hydroprocessing catalysts, respectively. Thus, at lower temperatures, the weight of catalyst particles increased due to the chemisorption of O_2 accompanied by the formation of surface complexes. The complexes rapidly decompose as soon as the critical temperature is reached. It is believed that at least during the early stages of regeneration, the gas phase reaction, i.e.:

$$CO + 0.5O_2 = CO_2$$

is the minor contributor to the CO_2 formation because of the limited availability of O_2. Therefore, most of the CO_2 originated from the decomposition of surface complexes is depicted in Fig. 6.2 [332].

Based on Fig. 6.3 [322], the yield of CO_2 was more than 13 times greater than that of CO. However, during the last stages of regeneration conducted with air at increased temperature, the gas phase oxidation of CO to CO_2 may be an important source of the latter. The hydrogen removal from coke during regeneration is generally overlooked although it is an important contributor to the overall mechanism. The involvement of hydrogen can be quantified by

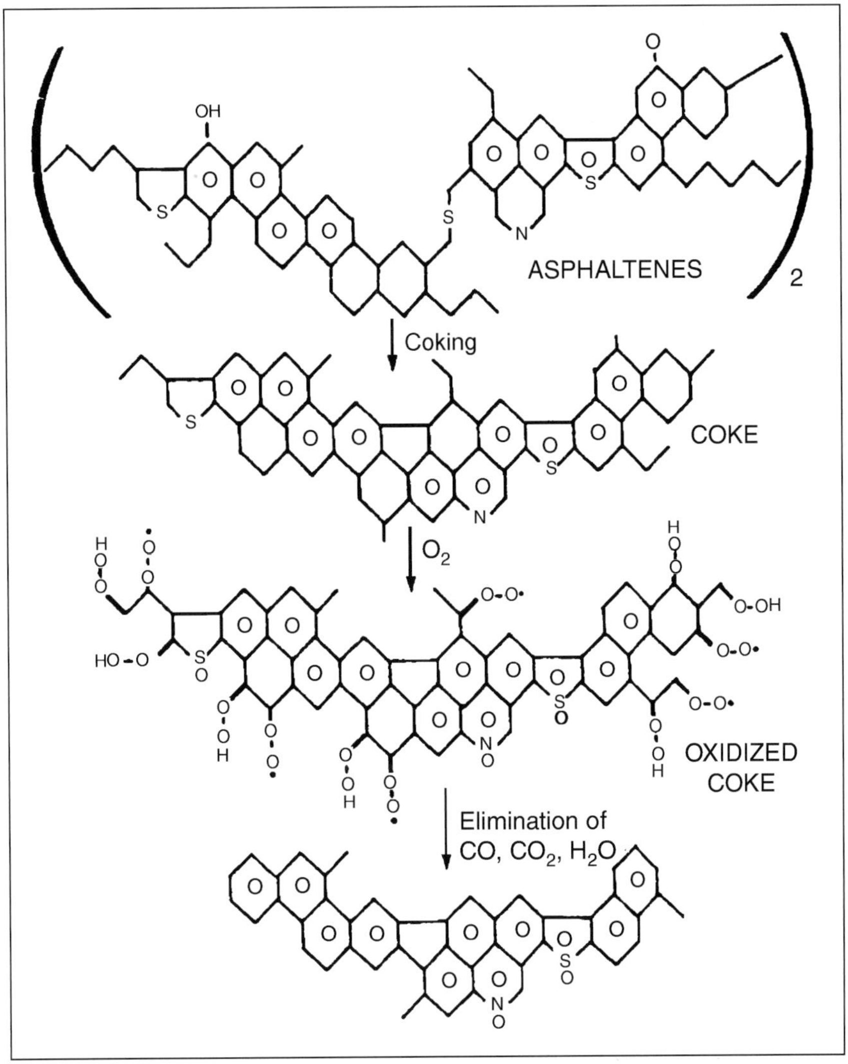

Figure 6.2: Mechanism of oxygenated complexes formation during auto-oxidation of coke [From ref. 332. Reprinted with permission].

measuring yields of H_2O in similar manner as that of CO_2 and CO shown in Fig. 6.3. Figure 6.4 shows that during the stepwise regeneration, most of the hydrogen was already removed at 350 °C. At the same time, less than 20% of carbon was removed. It is believed that most of this carbon originated from the oxidation of alkyl substituents and hydrocarbons trapped in the coke, i.e.:

$$C_nH_m + (n + 0.5m)O_2 = nCO_2 + 0.5mH_2O$$

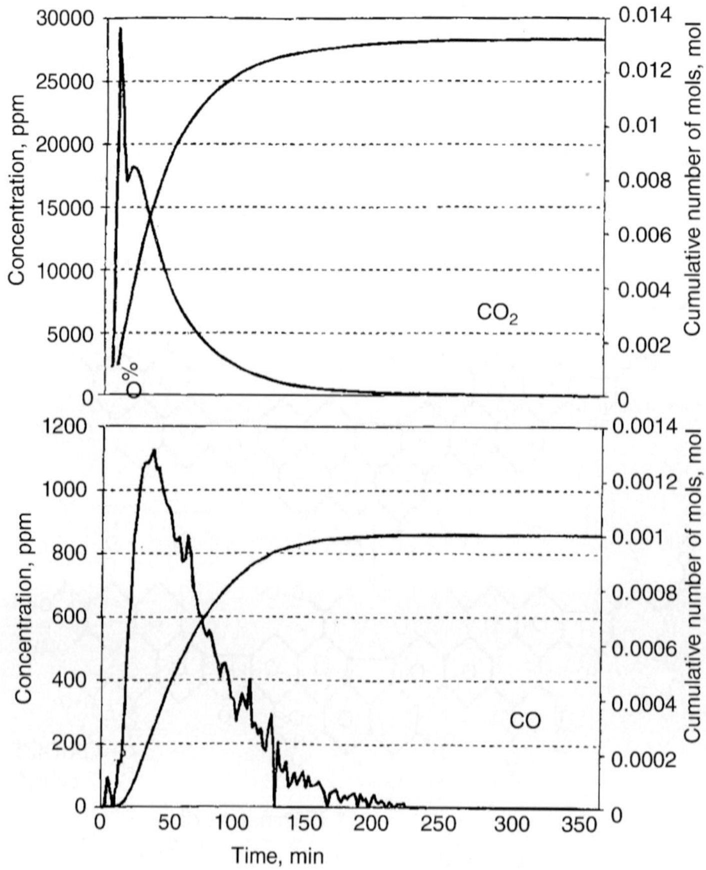

Figure 6.3: Concentrations and yields of CO and CO_2 during temperature programmed oxidation (TPO) of spent catalyst in 2% O_2 at 350 °C [From ref. 322. Reprinted with permission].

On the basis of the results shown in Figs 6.3 and 6.4, two distinct regions and/or mechanisms of oxidative regeneration may be applicable, i.e., the one dominated by removal of hydrogen and the other dominated by removal of carbon. The latter involves much less reactive coke left behind because most of the reactive moieties rich in hydrogen were already removed. Mechanistically, the results in Fig. 6.4 can be interpreted in terms of the removal of the most of aliphatic carbon and all hydrogen associated with it. Some aliphatic structures, i.e., alkyl substituents and methylene bridges, which are part of coke molecules, can be identified in coke model shown in Figs 4.20 and 4.21 [238,239]. This would leave behind the highly aromatic and non-reactive residue. Figures 4.20 and 4.21 show a continuous change in coke structure during various stages of the operation. This suggests that the burning profiles (e.g., Figs 6.3 and 6.4) of the spent catalysts taken at different stages on stream of hydroprocessing operation will be different.

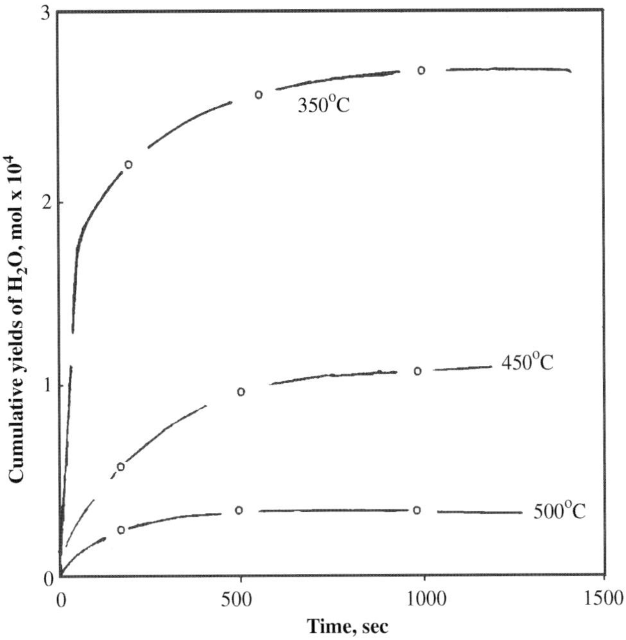

Figure 6.4: Effect of temperature on evolution of H_2O during isothermal burn-off of spent $CoMo/Al_2O_3$ catalyst.

The oxidation of S and N heteroatoms, which are part of the heteroring structures of coke, precedes the formation of SO_2 and NO. This is depicted in Fig. 6.2 [332] by the formation of the sulfoxide and nitroxide entities. These results in the weakening of the C–S and C–N bonds in the oxidized structures compared with the corresponding bond in heterorings. Such situation favors decomposition of the transition complexes to produce SO and NO. Very low stability of the former species favors rapid oxidation to SO_2 even under a limited availability of O_2. The higher oxidation state oxides, such as SO_3 and NO_2 can also be present, particularly after a diluted air is replaced with air. In fact, a gas phase oxidation of SO_2 to SO_3 and NO to NO_2 can proceed while burn-off gases are exiting regeneration zone.

All carbon and nitrogen oxides produced during the oxidative regeneration are of an organic origin. At the same time, the structure of spent catalysts suggests that both organic coke and catalyst contribute to the SO_2 formation. Two different sources of SO_2 can be clearly distinguished in Fig. 6.5 [369]. In this case, the oxidic form of $CoMo/Al_2O_3$ (catalyst A) was used for hydroprocessing the mixture of hexadecane + phenol. The same reaction was repeated with the sulfidic form of the $CoMo/Al_2O_3$ (catalyst B). Then, the sulfided $CoMo/Al_2O_3$ catalyst was used for hydroprocessing VGO (catalyst C). The TPO profiles of the spent catalyst C indicate two regions of the SO_2 formation. The lower region coincides with that of catalyst B and is attributed to the oxidation of metal sulfides. On the other hand, the higher

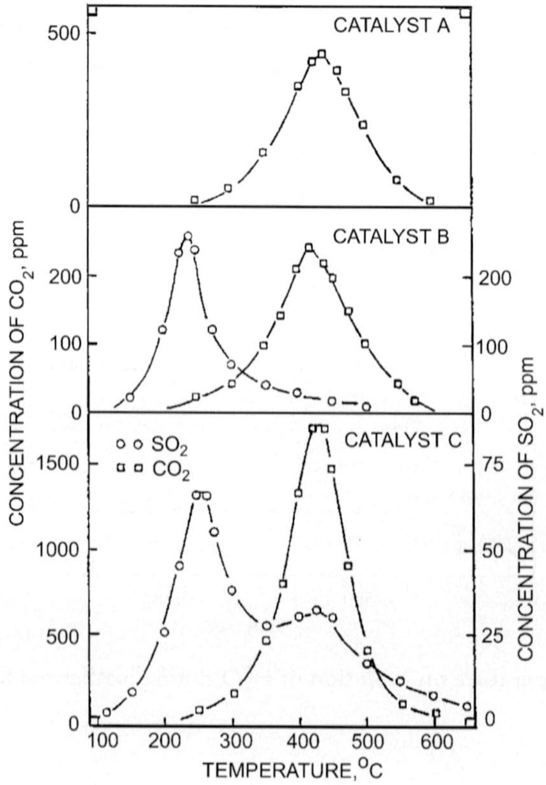

Figure 6.5: Formation of SO_2 and CO_2 during temperature programmed oxidation (TPO) of spent catalyst [From ref. 369. Reprinted with permission].

temperature region coincides with the formation of CO_2 and can be almost certainly attributed to the oxidation of organic sulfur in coke. The catalysts in Fig. 6.5 [369] were crushed to 100–200 mesh before TPO experiments to minimize the interference of diffusion. The trends observed in Fig. 6.5 are in agreement with the studies published by Zeuthen et al. [370,371].

Matsushita et al. [240] evaluated the spent catalysts from the study of Hauser et al. [238,239] using the TPO technique. The spent catalysts were used for hydroprocessing the atmospheric residue derived from a Kuwait crude. They were taken at different stages on stream during the same operation. The TPO profiles (Fig. 6.6) revealed the two maxima of CO_2 formation, i.e., one at about 573 K and the other at about 700 K. These maxima were attributed to the oxidation of a "soft" coke and a "hard" coke, respectively. It should be noted that in the study of Seki and Yoshimoto [208], the "hard" coke was defined as toluene insoluble portion of coke. The "hard" coke studied by Hauser et al. [239] was much more refractory and was adsorbed strongly on the catalyst surface. In its structure, the "soft" coke in the Hauser et al. [238] study may approach the structure of the toluene insolubles portion of coke formed

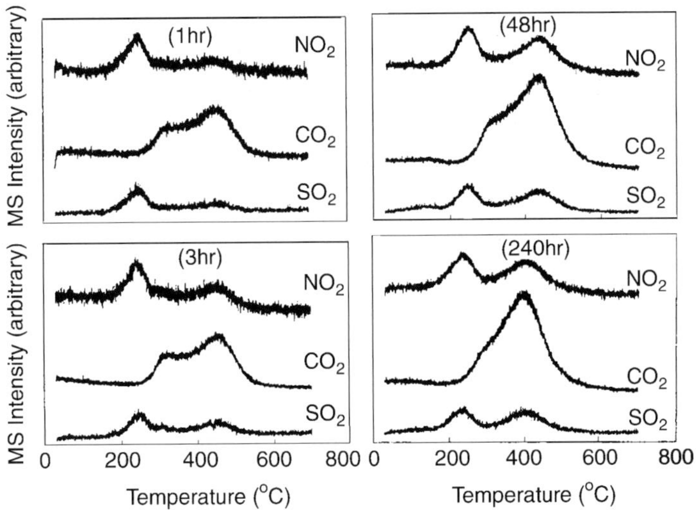

Figure 6.6: Effect of time on stream on temperature programmed oxidation (TPO) profiles of spent catalysts (Mo/Al_2O_3, atmospheric residue [AR], 653 K, 12 MPa) [From ref. 240. Reprinted with permission].

within the first 120 h, whereas the "hard" coke that of THF insolubles coke formed after 6500 h. The chemical transformations occurring during the conversion of a "soft" coke to "hard" coke are shown in Figs 4.20 and 4.21 [238,239]. As expected, the latter coke had very low solubility. It is believed that the "soft" coke was formed predominantly on the uncovered support. With time on stream, the "soft" coke was gradually converted to more refractory coke. This was supported by the decrease of the low temperature and increase of the high temperature CO_2 peak between 1 and 240 h shown in Fig. 6.6 [240]. After 120 h on stream, most of the coke was refractory and was insoluble in THF. More nitrogen and sulfur were concentrated in the refractory coke than in the "soft" coke. These observations agreed with the structural mechanism of coke formation proposed by Hauser et al. [238,239]. The distinction between the "soft" coke and "hard" coke in Fig. 6.6 may be affected by the involvement of active metals during TPO. Thus, the transition metals catalyzed oxidation of coke, i.e., oxygen transfer from active metals to carbon has been proposed by several authors [181,208,369]. Then, the coke in the vicinity of active metals would exhibit higher reactivity (e.g., be "softer") than its actual reactivity at the end of hydroprocessing operation. With respect to the overall mechanism of oxidative regeneration, the hydrogen removing reactions during the oxidation of "soft" coke should be much more pronounced than those during the oxidation of "hard" coke.

The study of Callejas et al. [181] gives another dimension to the coke structure issue. During the TPO, these authors observed the emergence of another coke peak with maximum at about 700 °C. This peak appeared for the catalyst after 1100 h on stream, whereas there was no such peak after 100 h. The intensity of this peak further increased with increasing time on stream beyond 1100 h. The NMR data confirmed the absence of any aliphatic structures in this portion of coke. It is proposed that most of this coke was removed via metal catalyzed gasification of carbon involving oxygen transfer via redox cycle.

For every catalyst in Fig. 6.6 [240], the SO_2 profiles show the low temperature and high temperature peaks that originated from the oxidation of metallic sulfides and organic sulfur, respectively in agreement with Fig. 6.5 [369]. Similarly, two peaks were observed for the formation of NO_2. It should be noted that under the conditions of analysis, all nitrogen oxides were oxidized to NO_2 before they reached the detector of analytical system. Most likely, the low temperature NO_2 peak did not originate from the oxidation of coke. Thus, this peak occurred at lower temperature than the low temperature CO_2 peak. It is postulated that this peak originated from the gas–solid oxidation of ammonia and N-containing base observed that the oxidation of ammonia proceeds as follows [372,373]:

$$2NH_3 + 2O_2 = N_2O + 3H_2O$$

Under the conditions of analysis used in the study of Hauser et al. [239], the N_2O produced as primary product was converted to NO_2, i.e.:

$$N_2O + 1.5O_2 = 2NO_2$$

It should be noted that the formation of N_2O during catalyst regeneration was observed by Furimsky et al. [374]. Although they were obtained under isothermal conditions (450 °C in 4% O_2 + He), the results in Fig. 6.7 indicate the occurrence of two N_2O peaks. The first peak, originated from a more reactive nitrogen containing species, may be attributed to the oxidation

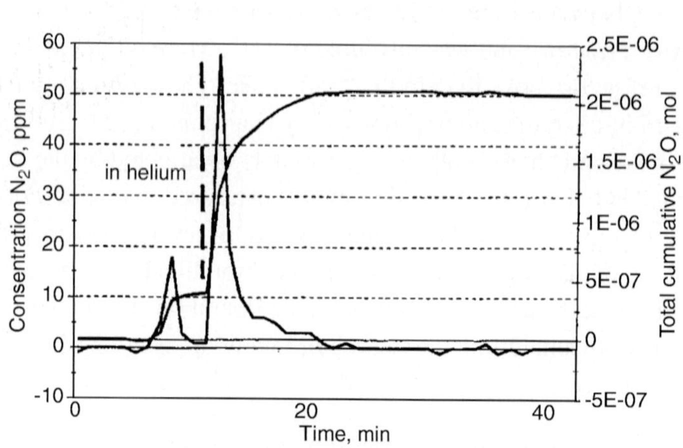

Figure 6.7: Formation of NO and N_2O during temperature programmed oxidation (TPO) of spent catalyst in 4% O_2 at 450 °C [From ref. 374. Reprinted with permission].

of ammonia strongly adsorbed on catalyst surface, whereas the second peak from the oxidation of nitrogen in coke. It was reported that much more of N_2O was formed when nitrogen was bound in the five-membered heterorings than that in the six-membered heteroring [375,376]. Because of its impact on environment, the occurrence of N_2O in the burn-off gas from oxidative regeneration of spent catalysts needs to be addressed.

A part of the nitrogen is removed simultaneously with the inorganic sulfur, as it is indicated by an overlap of the SO_2 and low temperature NO regions [12,240]. Most likely, this part of the NO originated from the ammonia and other N-containing compounds adsorbed on and/or in the proximity of catalytically active phase, rather than on the bare support. It was suggested that the last part of carbon and nitrogen may be gradually converted to metal carbides and nitrides [377,378]. These phases are catalytically active for hydroprocessing reactions. Therefore, the last amount of carbon and nitrogen left in catalyst after regeneration may not be detrimental, although their removal requires prolong exposure to the oxidation medium. This was supported by the results published by Noguchi et al. [379] who reported that the activity of the spent catalyst was almost completely restored as soon as the amount of carbon on the catalyst was less than 3 wt.%.

Some primary reactions, which are part of the mechanism of conversion of nitrogen in coke on catalyst, are similar as those occurring during the oxidation of other carbonaceous solids [380]. Thus, at low O_2 concentrations (e.g., ~2 vol.%), NH_3 and HCN are among important products. This is supported by the product distribution obtained during the isothermal oxidation of spent catalyst from an industrial hydroprocessing operation, shown in Table 6.2 [322]. The evolution of NH_3 and HCN coincided with that of CO_2 and CO, therefore, it was associated with the oxidation of coke. Increasing the O_2 concentration from 2 to 4 vol.% resulted in the formation of N_2O in addition to NH_3 and HCN. At 4 vol.% O_2 concentration, temperature increase from 350 to 450 °C resulted in the significant decrease in the yield of

Table 6.2: Distribution of N-containing compounds during burn-off [From ref. 322. Reprinted with permission].

	2% O_2		4% O_2	
	350 °C	450 °C	350 °C	450 °C
N total[a]	20.7	13.0	23.9	8.8
HCN	10.8	0.6	11.3	0.9
NH_3	8.3	0.2	5.1	0.4
NO	1.5	5.2	1.3	5.1
N_2O	0	7.0	6.2	1.6
C total[a]	63.9	35.0	98.3	0.8

[a] *N total and C total in wt.% of N and C in catalyst, respectively.*

NH_3, HCN, and N_2O. At the same time, NO became the major N-containing product. Based on these observations, the initial stage of the mechanism of nitrogen conversion during the oxidative regeneration of spent catalysts may involve thermal decomposition of coke, i.e.:

$$\text{Coke} \rightarrow \begin{cases} NH_3 \\ HCN \end{cases}$$

There is a significant database to support the O_2 aided elimination of NH_3 and HCN from carbonaceous solids [374–376]. There is the evidence supporting the preferential formation of N_2O from NH_3 and NO from HCN. These reactions occur in the gas phase, although the formation of these products via surface reactions cannot be ruled out. Then, the mechanism of nitrogen oxides formation in gas phase during catalyst regeneration can be refined as:

$$2NH_3 + 2O_2 = N_2O + 3H_2O$$

$$2HCN + 2.5O_2 = 2NO + 2CO + H_2O$$

$$N_2O + 0.5O_2 = 2NO$$

$$NO + 0.5O_2 = NO_2$$

This mechanism suggests that during oxidative regeneration, there are two sources of N_2O and NO, i.e., one involving the oxidation of the NH_3 strongly held on catalyst surface and the other originating from the decomposition of N-containing moieties in coke yielding precursors to the formation of NH_3 and HCN, which are subsequently oxidized in gas phase to nitrogen oxides.

6.2.1.2 Involvement of Metals

The coke in spent catalysts after hydroprocessing of residues contains metals, such as V and Ni. Depending on the feed, other metals, i.e., Fe, Ti, Ca, As, etc., can also be present. It was suggested that V in coke may be either in a partially oxidized or in a sulfided form, whereas most of the Ni was in a sulfided form [27]. Under conditions of oxidative regeneration, these metals will be converted to the corresponding oxides, i.e.:

$$VO_xS_y + 1.5yO_2 = VO(x + 0.5y) + ySO_2$$

$$NiS_x + 1.5xO_2 = NiO0.5x + xSO_2$$

Obviously, all other metals that deposited on catalyst surface during the operation should be oxidized in similar manners giving the corresponding oxides of Fe, Ti, Ca, etc.

Overwhelming information confirms that during hydroprocessing operation, most of the catalytically active metals are in a sulfided form [53–55]. Thus, the active phase comprises Co(Ni)-Mo-S structure, although other sulfided species, such as MoS_2, WS_2, Co_9S_8, Ni_3S_2, and partially sulfided oxides of these metals, can also be present. The oxidation of these metal sulfides to corresponding oxides is illustrated using the following general reactions:

$$MoS_2 + 3.5O_2 = MoO_3 + 2SO_2$$

$$WS_2 + 3.5O_2 = WO_3 + 2SO_2$$

$$Co_9S_8 + 12.5O_2 = 9CoO + 8SO_2$$

$$Ni_3S_2 + 3.5O_2 = 3NiO + 2SO_2$$

It is believed that the oxidation of these sulfides may proceed to the completion because the latter stages of regeneration are usually conducted in air and at elevated temperatures (e.g., ~450 °C). However, conversion of the metal sulfides may proceed through intermediate oxosulfidic species, to metal sulfate structures, before this stage of oxidation is attained. In this regard, some support is provided by Fig. 6.8 [381] showing the TPO profiles of the fresh-sulfided catalysts. Thus, for a commercial catalyst A ($CoMo/Al_2O_3$), a small SO_2 peak occurring at about 700 K approached the temperature range in which the sulfates of transition metals usually decompose, giving the corresponding metal oxide and SO_2. Iwamoto and Kagami [382] reported that the formation of $Al_2(SO_4)_3$ was also possible particularly in the presence of V. They suggested that the V oxides catalyzed the oxidation of SO_2 to SO_3 leading

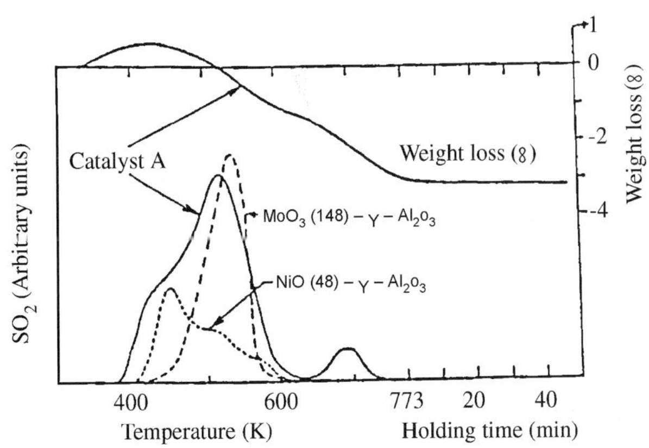

Figure 6.8: Formation of SO_2 and weight loss during temperature programmed oxidation (TPO) of presulfided NiMo/Al_2O_3 catalyst [From ref. 381. Reprinted with permission].

to the formation of H_2SO_4 and finally to $Al_2(SO_4)_3$ involving the following set of tentative reactions:

$$V_2O_5 + 2SO_2 = V_2O_3 + SO_3$$

$$SO_3 + H_2O = H_2SO_4$$

$$Al_2O_3 + H_2SO_4 = Al_2(SO_4)_3$$

The formation of $Al_2(SO_4)_3$ was evident particularly on the exterior of catalyst particles. This is in the agreement with the preferential accumulation of V on the exterior of the spent catalyst particles generally observed. Similar observation was made by Yoshimura et al. [383].

It has been reported that regeneration of the spent $CoMo/Al_2O_3$ catalyst proceeded more readily than that of the spent $NiMo/Al_2O_3$ catalysts [384]. This was attributed to the formation of Ni carbonyl. If so, Ni carbonyl could arise from the secondary reaction involving CO and Ni, i.e.:

$$Ni + xCO = Ni(CO)x$$

An additional study is necessary to confirm the occurrence of this reaction during the oxidative regeneration of spent hydroprocessing catalysts.

Using a set of thermodynamic data, Yoshimura et al. [385] concluded that the oxidation tendency of Mo sulfide was much greater than that of Co sulfides. The latter were oxidized to both $CoSO_4$ and Co oxides, in contrast to MoS_2, which was only oxidized to Mo oxides. The decomposition of $CoSO_4$ below 973 K had little probability to occur. Therefore, this portion of Co became unavailable for the formation of the Co-Mo-S phase. After repeated utilization-regeneration cycles, the loss of Co to $CoSO_4$ formation may be one of the reasons for the decline in catalyst activity. In this regard, experimental evidence was provided by Mahadjev et al. [386]. The information on behavior of Ni under similar conditions as used by Yoshimura and Furimsky [387] indicated the formation of $NiSO_4$, which could not be removed completely during the subsequent catalyst resulfidation. It is therefore desirable that during regeneration the conditions should be established under which the formation of the sulfates of promoting metals is minimized.

An important contribution to the understanding of the mechanism of oxidative regeneration was made by Yoshimura and Furimsky [387], who studied the oxidation of the sulfided Co/Al_2O_3, Mo/Al_2O_3, and $CoMo/Al_2O_3$ catalysts, using varying concentrations of O_2 in the oxidation mixture. For the Mo/Al_2O_3 catalyst, a significant decrease in the amount of Mo^{4+} was observed after oxidation below 533 K. However, most of the Mo still remained in a sulfidic form. At 653 K, most of the Mo^{4+} was converted to Mo^{6+}. At the same time, the amount of sulfidic and oxosulfidic sulfur decreased substantially. For the Co/Al_2O_3 catalyst, almost all Co was converted to Co^{2+}, while sulfur remained as sulfate below 423 K. In $CoMo/Al_2O_3$ catalyst, the Co sulfide was more oxygen tolerant than in the Co/Al_2O_3 catalyst as confirmed by the presence of Co sulfide in the former catalyst even after oxidation at 488 K. The gradual conversion of sulfidic form of sulfur into oxosulfidic and sulfate forms in the $CoMo/Al_2O_3$ catalyst was more pronounced than in the Mo/Al_2O_3 catalyst. When air was used, the Co sulfate was usually concentrated on the exterior of catalyst particles forming a layer that slowed down the diffusion of O_2 into the particles interior. These effects were diminished when air was replaced with 0.5% O_2. The extended X-ray absorption fine spectroscopy (EXAFS) data confirmed a gradual replacement of the Mo–S and Mo–Mo bonds with the Mo–O bonds.

An intimate contact of coke with catalyst surface suggests a potential involvement of the latter during coke oxidation. This would however require the presence of an oxidic form of the metals in spent catalyst. This is supported by the results shown in Fig. 6.9 [387]. Each point in Fig. 6.9 was obtained during the isothermal burn-off performed at the corresponding temperatures. Thus, at 200 °C, about 20% of carbon (as CO_2) was removed compared with almost 80% of sulfur removal during the oxidative burn-off of the spent sulfided catalyst. The CO_2 and SO_2 profiles in Figs 6.5 and 6.6 [369,240] provide another confirmation that a large portion of metals sulfides was oxidized before the onset of the oxidation of carbon. However, this observation may not be identical in the case of the spent catalyst heavily deactivated with

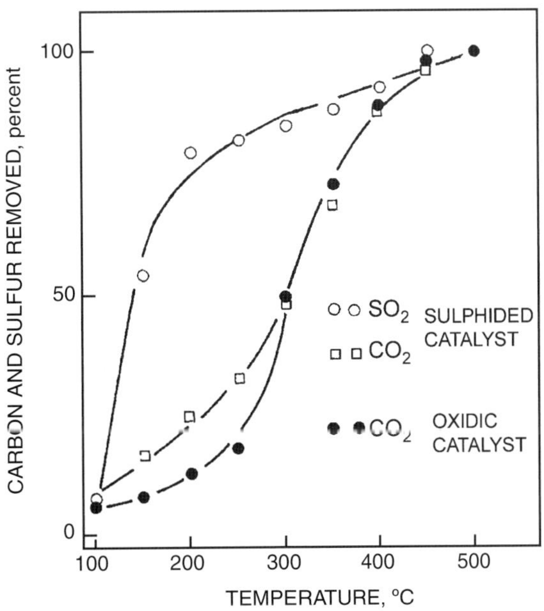

Figure 6.9: Effect of temperature on cumulative amount of carbon and sulfur removed during stepwise burn-off [From ref. 387. Reprinted with permission].

coke. Then, almost certainly, the oxidation of carbon with metal oxides involving a redox cycle becomes part of the overall regeneration mechanism. Of course, the extent of the redox cycle involvement will be different in different stages of regeneration process. This can be shown on the following general reactions:

$$Me_xO_y + C = Me_xO_{(y-1)} + CO$$

$$Me_xO_y + CO = CO_2 + Me_xO_{(y-1)}$$

$$Me_xO_{(y-1)} + 0.5O_2 = Me_xO_y$$

There is an extensive evidence confirming participation of the oxides of transition metals (e.g., Mo, Ni, Fe, etc.) during the oxidation of carbon via redox cycle [13,388]. It has been generally observed that the rate of carbon oxidation could be enhanced significantly in the presence of transition metal oxides. Therefore, the portion of coke removed via redox cycle would appear to be less refractory, e.g., a "hard" coke removed in this way would appear to be like a "soft" coke.

A direct interaction of the catalyst surface with coke was supported by the H_2O evolution during temperature programmed pyrolysis (TPP) of the $NiMo/Al_2O_3$ catalyst [389]. In this case, the model coke was produced by deposition of the catalyst with either pyrrole or pyridine followed by pyrolysis. The H_2O produced during the TPP of the deposited catalyst could only arise from the interaction with catalyst surface. Thus, under the conditions applied during the TPP, including a special catalyst pretreatment prior the TPO runs, the catalyst was the only source of oxygen in the system.

Yoshimura et al. [390] compared the NiW/Al_2O_3 catalyst with the $CoMo/Al_2O_3$ catalyst used for hydroprocessing of the same feed. The coke on the former spent catalyst was softer, i.e., it had a higher H/C ratio than that on the spent $CoMo/Al_2O_3$ catalyst. Yet, it was more difficult to oxidize the coke on the NiW/Al_2O_3 catalyst. This was attributed to a lesser redox efficiency of the Ni and W metals compared with that of the Co and Mo metals.

A direct evidence for the involvement of metal oxides during the formation of CO_2 and SO_2 is provided in Fig. 6.10a [391], showing the TPP profiles of CO_2 and SO_2 obtained in the flow of N_2. In this case, the spent catalyst was extracted by toluene. Obviously, the yields of CO_2 and SO_2 are much lower than those obtained during the TPP in oxidizing medium shown in Fig. 6.10b. Nevertheless, during TPP in N_2, CO_2, and SO_2 could only be produced during the interaction of not fully sulfided surface with either organic matter or inorganic sulfided portion of the spent catalyst.

It is unlikely that during industrial regeneration, a 100% removal of the carbon, nitrogen, and sulfur is achieved. Always small amounts of these elements are left behind without having any

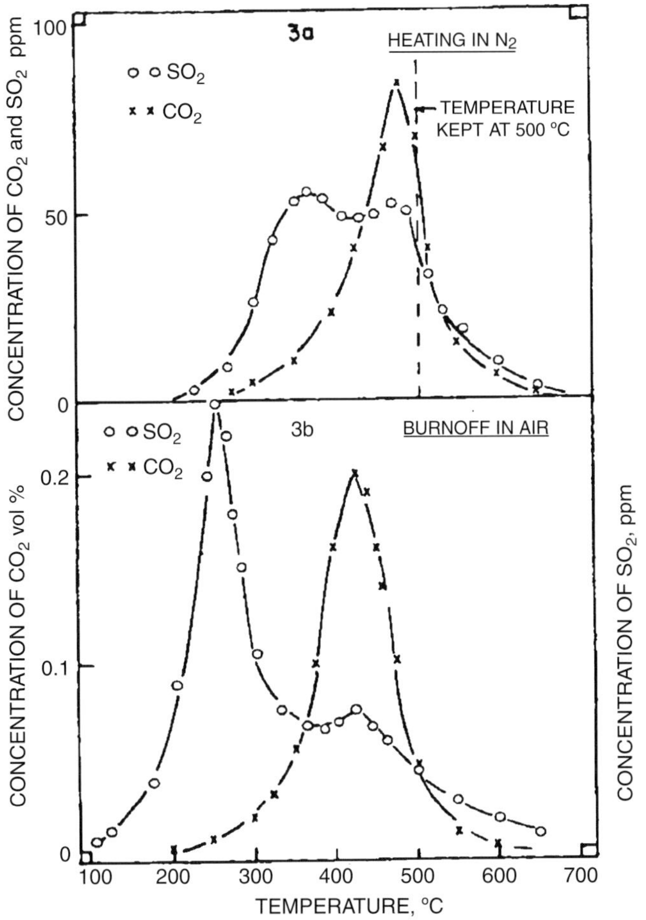

Figure 6.10: Formation of CO_2 and SO_2 under N_2 and in air for spent extracted catalyst [From ref. 391. Reprinted with permission].

adverse effect on catalyst activity recovery. There is little information on the structure of carbon and nitrogen in regenerated catalysts, although it was proposed that the metal carbides and oxycarbides as well as nitrides and oxynitrides are possible phases. The conditions under which such structures can be formed were described elsewhere [377]. The residual sulfur in regenerated catalysts was the focus of study conducted by Yoshimura et al. [381] who confirmed that the last amount of sulfur in regenerated catalyst was a sulfate form. Bogdanor and Rase [392] detected $NiSO_4$ in the regenerated $NiMo/Al_2O_3$ catalyst. This sulfate remained in the catalyst after subsequent sulfiding. Apparently, the temperature employed during sulfiding was not high enough to achieve the sulfate decomposition. This was in agreement with the observations made in several studies where a sulfate presence was confirmed in fresh-sulfided catalysts [393–395]. Although the process may be rather slow, the sulfate formation during the exposure of sulfided catalysts to air cannot be ruled out. The sulfate formation during the exposure to air was also postulated in Chapter 5 for spent catalysts.

A significant complexity of the mechanism of the oxidative regeneration of spent catalysts should be noted. This results from rather complex nature of coke and metals, which deposit on catalyst surface during the operation. Thus, only recently, the chemical structure of such coke was described in more details [238,239]. The interaction of coke with catalyst surface during regeneration is an important part of the overall mechanism. Because of the structural complexity and continuous change, the involvement of the inorganic part of spent catalysts during regeneration needs to be further investigated. Therefore, some assumptions and/or even speculation while discussing the mechanistic aspects of the oxidative regeneration of spent hydroprocessing catalysts could not be avoided.

6.2.2 Kinetics of Oxidative Regeneration

The estimate of kinetic parameters may be desirable to obtain a more quantitative picture of regeneration process. There are several methods that can be used for quantifying the removal of coke during regeneration. The method based on the change in conversion with time on stream requires several burn-off experiments performed under the same conditions but at different duration. Apparently, the most suitable method involves the quantitative on-line analysis of all burn-off products in the course of the experiments. An example of such results is shown in Fig. 6.11 [374] correlating cumulative yields of CO_2 with time at different

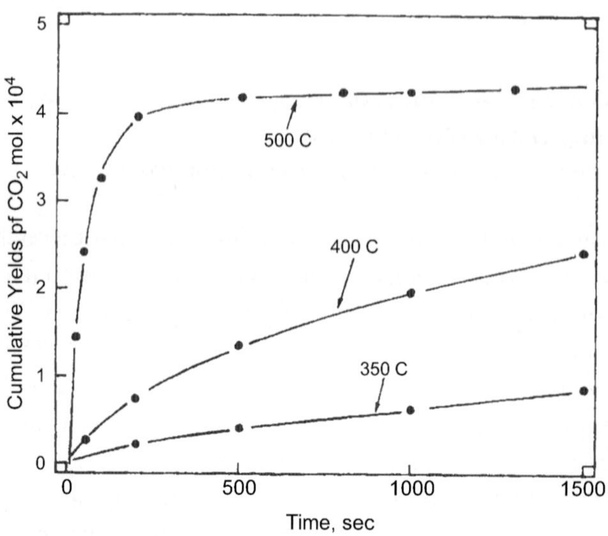

Figure 6.11: Cumulative yields of CO_2 from isothermal burn off of spent catalyst in air [From ref. 374. Reprinted with permission].

temperatures. In view of the results shown in Fig. 6.3 [322], the kinetics of carbon removal from catalyst are dominated by the formation of CO_2 as the contribution of CO appears to be rather minor. This, of course, does not take into consideration the rate determining initial step, which may involve a surface reaction. The kinetic data obtained by measuring weight loss during the burn-off are less reliable. Thus, under such conditions, the contribution of coke and mineral matter to the overall weight loss cannot be decoupled.

The opposite trends in the effect of temperature on hydrogen removal from coke compared with carbon removal have been noted (Fig. 6.4). It is evident that at early stages of burn-off at 350 °C, the overall kinetics are dominated by hydrogen removal similarly as it was indicated above for the overall mechanism of regeneration. However, at 450 °C and above, the involvement of hydrogen removing reactions becomes unimportant.

Catalyst regeneration is usually influenced by diffusion phenomena. This may be seen in Figs 6.3 and 6.11 (at 500 °C), which indicates the presence of two burn-off regions, i.e., chemically controlled, occurring during the early stages of burn-off and predominantly diffusion controlled, occurring during the final stages of burn-off. The former involved the coke deposited on the exterior of catalyst particles, whereas the diffusion controlled burn-off involved the coke deposited in pores. During the transition stage, both chemical and diffusion controlled kinetics influence the burn-off. The slow burn-off during the later stages resulted from the diminished accessibility of O_2 in the pores. Thus, the diffusion of O_2 into the catalyst interior was obstructed by the burn-off products (CO, CO_2, SO_2, and NO_X) exiting from the pores. Consequently, carbon-coke interface is developed in the interior of catalyst particles. At low temperatures and/or at low O_2 concentrations in oxidizing gas, the burn-off is chemically controlled. Under such conditions, carbon-coke interface may not be developed because the slow reaction with coke ensures enough time for O_2 to diffuse into the catalyst particles interior, rather than being consumed as it is the case at higher temperatures. Therefore, at higher temperatures, the formation of two regions in the coke layer, differing markedly in the H/C ratio, may be evident. In this case, the external region would have a lower H/C ratio.

6.2.2.1 Chemically Controlled Kinetics

Assuming first-order kinetics with respect to carbon removal, the following general equation describes the carbon burning process:

$$dC/dt = -kC \quad (6.1)$$

At constant partial pressure of O_2, k includes also partial pressure of O_2. After integration for initial conditions, i.e., $C - C_o$ at $t = 0$, the form of this equation is changed to:

$$\ln(C/C_o) = -kt \quad (6.2)$$

This form of equation enables the use of the relative carbon concentration (e.g., C/C_o) instead of absolute carbon concentration [396]. After further rearrangement, this equation can be used to follow the carbon removal in the high-rate burn-off region (C_1) and low-rate burn-off region (C_2), i.e.:

$$C_1 = C_{10}\,e^{-k1t} \quad (6.3)$$

$$C_2 = C_{20}\,e^{-k2t} \quad (6.4)$$

where total carbon content $C = C_1 + C_2$ and relative carbon concentration with time:

$$C/C_o = C_{1o}/C_o\,e^{-k1t} + C_{2o}/C_o\,e^{-k2t} \quad (6.5)$$

$$C/C_o = (1 - f_2)\,e^{-k2t} + f_2\,e^{-k2t} \quad (6.6)$$

where f_2 is the fraction of initial carbon associated with C_2 at $t = 0$. Then, after all C_1 carbon was removed, the equation becomes:

$$C/C_o = f_2\,e^{-k2t} \quad (6.7)$$

To be used for kinetic measurements, this set of equations requires determination of the carbon content in spent catalyst at different time intervals during burn-off. To get a realistic description of regeneration process, such estimate would have to be done for different stages conducted at gradually increasing temperatures, until most of the carbon was removed from spent catalyst. Such approach was undertaken by Alwarez et al. [46] who studied several spent catalysts varying widely in the level of deactivation. Their experimental data could be correlated using two parallel first order independent power laws (Eqns (6.5) and (6.6)) up to 90% conversion of coke. However, for the spent catalyst containing 4.1 wt.% V and about 36 wt.% of coke, the second order kinetic law gave the best fit of the experimental results. Established information (e.g., Figs 6.9 and 6.11) shows that on an industrial scale, regeneration process has to be conducted in at least two stages. For safety reasons, the first stage has to be conducted either with a diluted or in air but under all precautions taken to avoid temperature excursions due to the presence of highly reactive components of coke. For this stage, the removal of hydrogen from coke is much more important than that of carbon. Based on the above sequence of the equations, this process may be described by the following equation:

$$H = H_o\,e^{-k3t} \quad (6.8)$$

where k_3 is the rate constant for the burn-off of hydrogen, whereas H and H_o are the hydrogen content in coke at t and $t = 0$, respectively. Then, the following kinetic equation may apply for the overall coke removal during the chemically controlled first stage:

$$\text{Coke removed} = H/H_o\;e^{-k3t} + C_{1o}/C_o\;e^{-k1t} \quad (6.9)$$

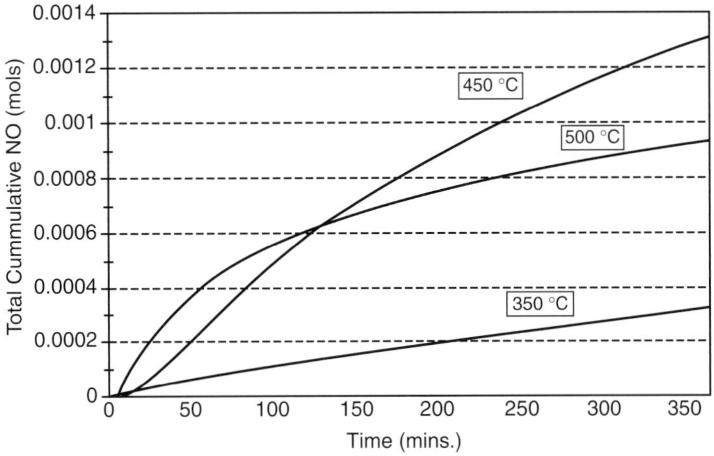

Figure 6.12: Cumulative yields of NO from stepwise isothermal burn-off of spent catalyst [From ref. 322. Reprinted with permission].

The contribution of nitrogen to coke removal is negligible in the case that the first burn-off stage is conducted at temperatures below 350 °C. This was confirmed by the results shown in Fig. 6.12 [322]. Therefore, omitting nitrogen from kinetic equations may have little adverse effect on the value of kinetic parameters for the overall coke removal. At the same time, a high content of oxygen in coke on spent catalysts used for hydroprocessing of some synthetic liquids (e.g., coal derived liquids, bio-liquids, etc.) suggests that the inclusion of oxygen removal in kinetic equations may improve the accuracy of the kinetics of the overall coke removal during regeneration. However, the content of oxygen in conventional petroleum feeds is rather low. Consequently, the content of oxygen in the coke on spent catalysts should be low as well and, as such, should play rather minor role during regeneration. However, the involvement of oxygen may not be negligible for spent catalysts exposed to air prior to regeneration (e.g., during storage, transportation, etc.).

The experimental set up to study kinetics of catalyst regeneration used by Yoshimura et al. [369,387] allowed the estimate of the product yields during very early stages of regeneration. Thus, in the course of regeneration, the measurements could be taken within few seconds intervals. The typical product yield-time trajectories are shown in Figs 6.11 and 6.13 [374] for CO_2 and in Fig. 6.4 for H_2O. The former indicate the removal of the carbon, whereas Fig. 6.4 the removal of hydrogen during coke burn-off. Using Figs 6.4 and 6.11, an instantaneous reaction rate could be determined by choosing any interval on the curve and the amount of the corresponding product produced during the same interval. At the same time, Fig. 6.13 is suitable for determining the initial rate of regeneration. The kinetic parameters estimated in

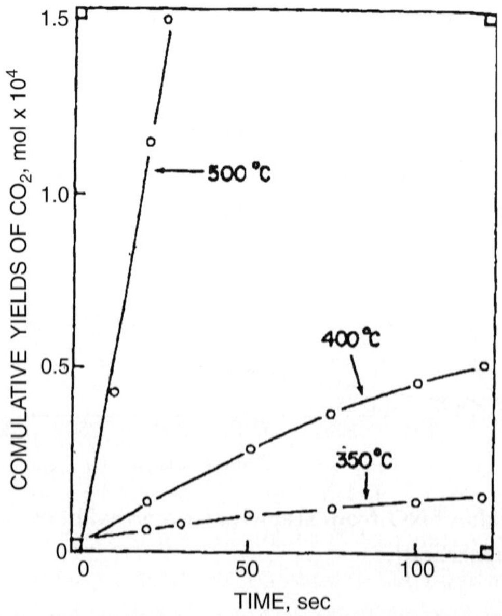

Figure 6.13: Initial cumulative yields of CO_2 from stepwise isothermal burn-off of spent catalyst in air [From ref. 374. Reprinted with permission].

this manner are much more accurate than those determined by the other techniques, e.g., measuring weight loss during regeneration.

Table 6.3 [387] compares kinetic parameters determined from the weight loss during thermal gravimetric analysis (TGA) and those obtained from the cumulative yields of products determined in the chemically controlled burn-off region. The TGA results were affected by the O_2 chemisorption that had an opposite effect on the weight loss. This resulted in the

Table 6.3: Rate constants from TGA and from yield of products [From ref. 387. Reprinted with permission].

Temperature (°C)	Rate constant (1/min atm)			
	TGA		Yields	
	Oxidic	Sulfided	Oxidic	Sulfided
300	–	–	0.094	0.97
350	0.095	1.64	–	–
400	0.37	2.00	1.08	3.01
450	0.65	2.46	–	–
500	1.52	2.60	10.30	18.90

underestimate of the rate constants. This was particularly evident for the catalyst that was not presulfided prior to hydroprocessing test. As it was indicated earlier, reliable values of the kinetic parameters can only be obtained from the determination of products with time on stream. These values are much higher than those from the TGA tests. This may be attributed to the turbulent flow present in the fixed bed reactor compared with the laminar flow present in the TGA system. The former ensured more efficient contact of O_2 with spent catalyst. Also, the rate constants for sulfided catalyst were consistently higher than those for the oxidic catalyst. Most likely, this was caused by the difference in reactivity of coke. Thus, because of a higher hydrogenation activity, the coke on sulfided catalyst should be less refractory, i.e., softer, compared with the coke on the oxidic catalyst. This was supported by the higher H/C ratio of coke on the former spent catalyst.

The rate constants in Table 6.3 [387] were expressed in the form of Arrhenius plots (Fig. 6.14) [391] for the estimate of activation energies. It should be noted that these rate constants were determined during very early stages of burn-off, i.e., during less than 50 s. Therefore, they represent chemically controlled burn unaffected by diffusion. Activation energies for the oxidic catalyst were consistently higher than those for the sulfided catalyst. For the latter catalyst, the activation energy obtained from TGA results was affected by the weight loss

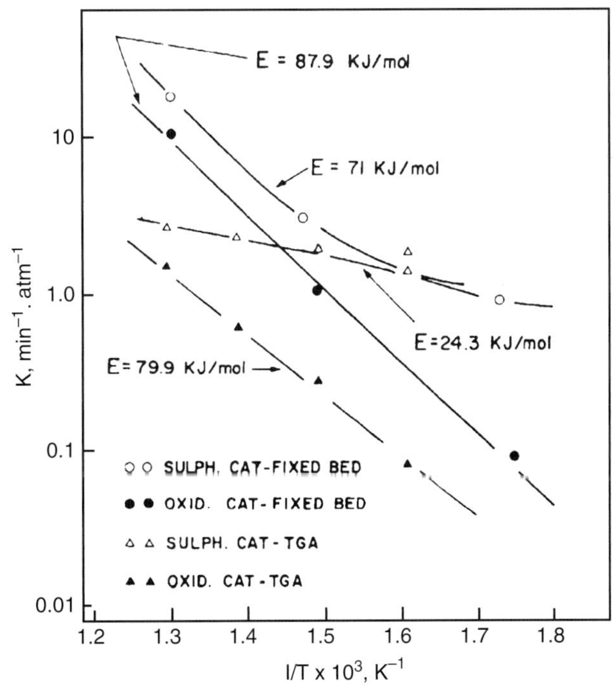

Figure 6.14: Log k versus 1/*T* correlations for burn-off performed in fixed bed and TGA reactors [From ref. 391. Reprinted with permission].

contribution from the conversion of metal sulfides to corresponding oxides. For sulfided catalyst, the plot obtained from results determined from the yields of products in the fixed bed reactor exhibited a curvature at low temperature. This resulted from a higher rate of oxidation at lower temperature. However, in view of the presence of highly reactive components in coke, the higher rate of coke oxidation should be expected. This would involve hydrogen containing entities (e.g., alkyl substituents), which can be identified in Figs 4.20 and 4.21. In other words, the structure of coke oxidized at lower temperatures was not the same as that of coke oxidized at high temperatures. Therefore, a continuous change in the structure of coke and its reactivity associated with it, is another parameter deserving attention during determination of the activation energies during the oxidative regeneration of spent catalysts. In the study of Alwarez et al. [46], the activation energy of about 157 kJ/mol was obtained. However, in this case, the rate constants were estimated at later stages of burn-off. Most likely, under such conditions, the rate constants determination was affected by diffusion.

6.2.2.2 Diffusion Controlled Kinetics

The particle size and porosity of spent catalysts are among parameters controlling kinetics in the diffusion controlled region during regeneration. However, parameters that influence chemically controlled burn-off are important as well. This is supported by the burn-off profiles in Figs 6.11 and 6.13 [374] obtained for the commercial catalyst, which was crushed to obtain particle size of 100–200 mesh. At 350 °C, the entire burn-off appeared to be chemically controlled. Thus, a low burn-off rate ensures that O_2 can diffuse into particle interior before it is consumed near the particle exterior. In this case, a uniform burn of coke across the particle is expected. To a lesser extent, the same was true for burn-off at 450 °C, although the involvement of diffusion effects becomes more evident.

According to Fig. 6.11 [374], at the latter stages of burn-off at 500 °C, the rate of chemical reactions is much higher than that of the diffusion of O_2 into the catalyst particle interior. Then, two regions, i.e., the one chemically controlled and the other diffusion controlled, are clearly present. In the former region occurring at 500 °C, the coke, which is readily accessible to O_2, undergoes rapid combustion. Gradually, the rate is slowing down before the coke combustion region, which is dominated by diffusion, is attained. Under such conditions, a coke-catalyst interface (Fig. 6.15) [13] is being developed in the interior of catalyst particles. Such interface cannot develop under chemically controlled conditions ensuring an even radial burn (e.g., at 350 °C in Fig. 6.11). A rapid reaction of coke with O_2 does not allow the latter to diffuse further into particle interior before it is consumed. Equations (6.4) and (6.7) were derived above to describe kinetics in the high rate burn-off region.

It should be noted that the events described above are applicable to spent catalysts containing a small amount of coke. For example, for fluid catalytic cracking (FCC) catalysts, an even radial distribution of coke was observed at less than 6 wt.% of the coke [397,398]. However, particle

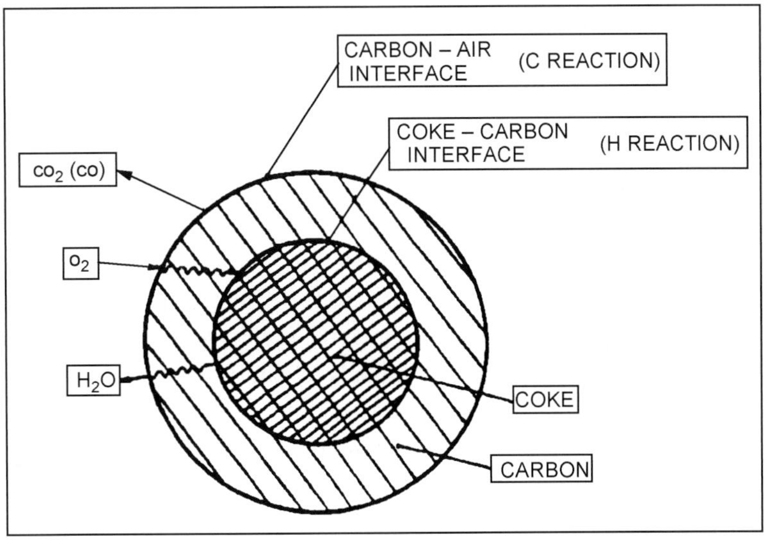

Figure 6.15: Model for simultaneous carbon and hydrogen oxidation [From ref. 13. Reprinted with permission].

size of FCC catalysts is much smaller than that of the spent hydroprocessing catalysts. It was observed that for the latter catalyst containing 7 wt.% of coke, even at 350 °C, the skin of coke had to be combusted before O_2 could diffuse into particle interior [322]. This is evident from Fig. 6.16, showing the evolution of CO_2, CO, HCN, and NH_3 during regeneration of the

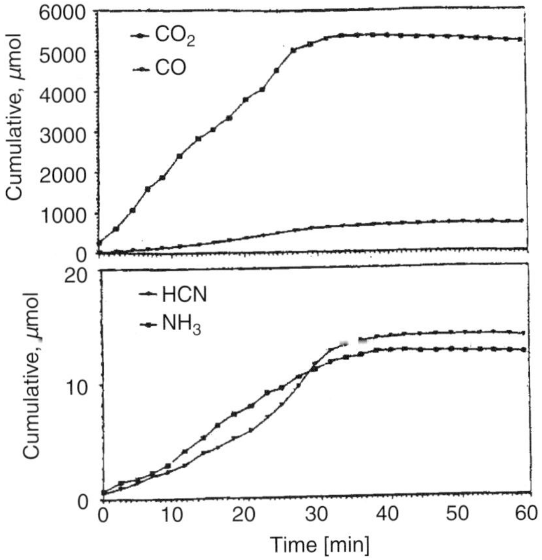

Figure 6.16: Cumulative formation of CO_2, CO, HCN, and NH_3 during oxidation of spent catalyst in 2% O_2 at 350 °C [From ref. 422. Reprinted with permission].

operating form of the spent $NiMo/Al_2O_3$ catalyst containing about 16 wt.% of coke at 350 °C using 2% O_2. In the fresh form, this catalyst had the shape of spheres of about 1 mm radius. Therefore, for such catalyst, the shape of curve at 350 °C would approach that for the crushed form of spent catalyst in Fig. 6.11 at 500 °C. This indicates the importance of spent catalyst pretreatment (e.g., de-oiling) and its particle size prior to the oxidative burn-off.

The diffusion controlled burn-off is associated with a continuous change in pore volume and size distribution. These parameters were incorporated in the model proposed by Chang [399] to describe the burn-off of spent hydroprocessing catalysts in the diffusion controlled region. According to this model, a very low burn-off rate at the later stages was attributed to the loss of reactive surface area due to the emergence of islands of catalyst surface not covered with coke. The restoring pore network of the fresh catalyst would be an ideal case of oxidative regeneration. This is however not achievable even in the case of a complete removal of carbon. Almost certainly, other factors such as catalyst recrystallization occurring during the operation and subsequent regeneration could have an effect on the pore structure of the regenerated catalysts. Also, it is unlikely that all coke is removed during the regeneration.

6.2.3 Modeling of Oxidative Regeneration

The primary objective of the models is to predict temperature profiles during regeneration. This information can be used for fine-tuning of the conditions applied during regeneration. For this purpose, the rate of reactions, which generate heat, has to be estimated. This includes the oxidation of carbon, hydrogen, and organic sulfur. In addition, the heat generated during the oxidation of metal sulfides has to be considered. Therefore, kinetic measurements form a basis for model development. A large volume of information is available on modeling of regeneration of FCC catalysts. However, compared with spent hydroprocessing catalysts, the prediction of temperature profiles is much simpler because carbon is far predominant component of coke. Also, the amount of coke in spent FCC catalyst is much smaller compared with spent hydroprocessing catalysts. Furthermore, metal sulfides in the former FCC catalysts are present in very small quantities. Then, incorporating only carbon burn-off gives reasonable information for regeneration of spent FCC catalysts.

The model for predicting ignition temperature of the spent catalyst particles during regeneration was developed by Klusacek et al. [400]. In this case, ignition temperature was defined as a sudden temperature rise compared with the temperature of oxidizing gas entering regeneration system. Both combustion of carbon and sulfur were considered in the model. The investigated catalyst was the spent $CoMo/Al_2O_3$ used during hydroprocessing a conventional distillate. This ensured the absence of V and Ni as well as an even radial distribution of coke in catalyst particles. For this case, the following heat balance equation was derived:

$$-C_{ps}(dT_s/dt) = r_c q_c + r_s q_s + 3h(T_s - T_g)/(d_p a)$$

In this equation, r_c and r_s are the rates of the oxidation of carbon and sulfur, respectively, T_s and T_g represent temperatures of catalyst pellets and that of gas between the pellets, C_{ps} and d_p the specific heat and apparent density of the catalyst, q_c and q_s heats generated by combustion of carbon and sulfur, respectively, h the heat transfer coefficient between the pellet and gas, and a the equivalent radius of spherical particle. For the heat balance of the gas in the bed voids, the following equation was derived:

$$(dT_g/dt) = (s/e)(T_{in} - T_g) + (1 - e)3h(T_s - T_g)/(eC_pd_a)$$

where T_{in} is the temperature of the oxidizing gas entering the bottom of the fixed bed, s is the air space velocity defined as the ratio of volumetric air flow rate under ambient conditions to the volume of catalyst bed, e is the bed porosity, d_a is the air density and C_p specific heat of air. Assuming that the accumulation terms were very small compared with the other terms; the above equations were simplified to obtain the following algebraic relations:

$$T_g = T_{in} + [(1 - e)d_p/(C_pd_s)](r_cq_c + r_sq_s)$$

$$T_s = T_g + [(ad_p)/3h](r_cq_c + r_sq_s)$$

The validity of the model was tested by comparison of the calculated results with those obtained during the experiments. A reasonable agreement between the ignition temperatures predicted by the model and those estimated experimentally was obtained in spite of the fact that combustion of hydrogen in coke was not included. In view of much less heat released and low content in coke, hydrogen combustion contribution to the overall heat release may not be significant. The necessity for incorporation of this source of heat may be determined on the basis of hydrogen content in coke.

6.2.4 Characterization of Regenerated Catalysts

Careful control and monitoring of operating parameters emphasized earlier suggests that the conditions employed during regeneration have decisive effects on the final properties of regenerated catalysts. In every case, the objective is to approach, as much as possible, the properties of the fresh catalyst. In this regard, a desirable effect can be achieved by an optimal combination of temperature and O_2 concentration in the oxidizing medium. Otherwise, the uncontrolled temperature runaways could not be avoided. Figure 6.17 [401] is presented to illustrate the adverse effect on the composition of the regenerated CoMo/Al_2O_3 catalyst, when temperature exceeded about 600 °C, i.e., the loss of Mo due to sublimation of MoO_3. At the same time, little change in the content of Co in the regenerated catalyst was observed. It is shown below that at such temperatures, the surface properties of regenerated catalysts are affected dramatically.

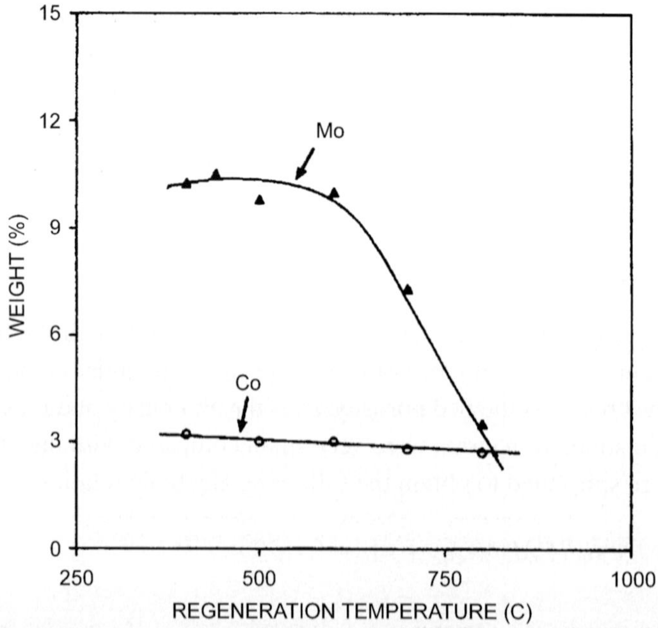

Figure 6.17: Effect of regeneration temperature on active metal content of catalyst [From ref. 401. Reprinted with permission].

6.2.4.1 Surface Properties

Surface properties such as surface area as well as pore volume and pore size distribution can be influenced by the regeneration conditions. According to Fig. 6.18 [401], the optimum of the surface area and pore volume recovery can be achieved at about 400 °C. At the optimum, a complete recovery of surface area was achieved, whereas the recovery of pore volume approached 80% of the fresh catalyst (Table 6.4) [401]. These results were obtained by a careful regeneration conducted in steps, i.e., in 5% O_2, followed by the final burn-off in air. Below 400 °C, the surface properties were not fully developed because of the burn-off of coke

Table 6.4: Properties of fresh and spent catalysts [From ref. 401. Reprinted with permission].

	Fresh	Spent
MoO_3	15.7	15.1
CoO	4.1	4.0
Carbon		10.5
Sulfur		6.4
Surface area (m^2/g)	240	108
Pore volume (mL/g)	0.54	0.22

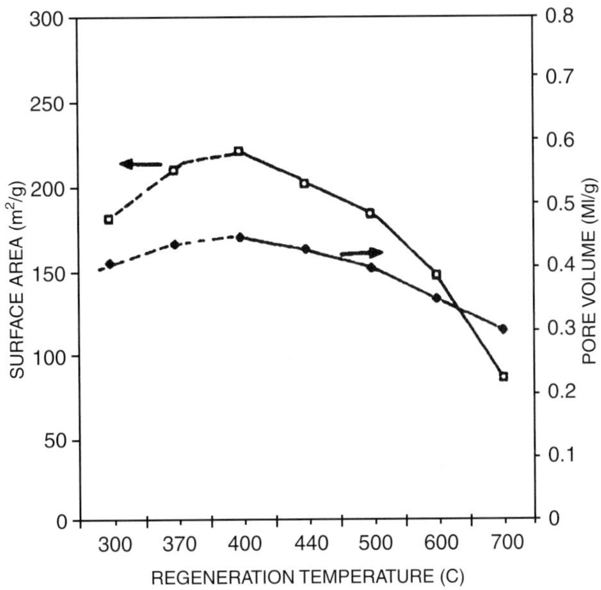

Figure 6.18: Effect of regeneration temperature on surface area and pore volume of catalyst [From ref. 401. Reprinted with permission].

was not yet completed. Above 400 °C, the loss of surface area and pore volume may be attributed to a gradual sintering of the catalyst. This is confirmed by the results in Fig. 6.19 [401] showing a gradual collapse of microporosity with increasing temperature of regeneration. According to one observation [13], only a relatively short excursion of temperature during regeneration resulted in a significant damage to catalyst material.

The O_2 concentration in the oxidizing medium is the key parameter for controlling regeneration temperature. Thus, as Fig. 6.20 [401] shows, the complete surface area recovery was achieved using 5% O_2 gas. The surface area decreased with further increase in O_2 concentration. These experiments were conducted at 400 °C. However, at higher O_2 concentrations, the actual temperature of particles was almost certainly higher than 400 °C [13]. Therefore, sintering of catalyst could not be prevented. The O_2 supply to the regenerator can also be increased by increasing the flow of oxidizing medium. Figure 6.21 [401] indicates that there is an optimal flow rate ensuring the highest recovery of surface area. An increase in the flow rate beyond the optimum decreased the surface area recovery. In practical situation, damages to catalyst properties in air can be minimized by employing an efficient excess heat withdrawing system.

The results in Table 6.5 [331] were obtained to compare two extreme regeneration conditions, i.e., the first involving 2% O_2 in N_2 and the second involving air. The spent $CoMo/Al_2O_3$ and $NiMo/Al_2O_3$ catalysts contained about 8 and 20 wt.% of coke, respectively. For both catalysts, the surface area recovery in air approached only 40% of that for the fresh catalysts. A high

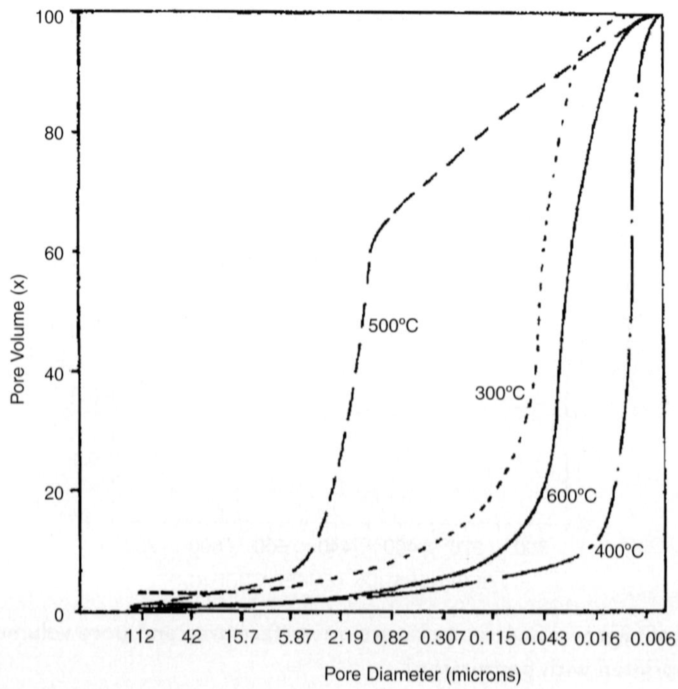

Figure 6.19: Effect of regeneration temperature on pore size distribution of catalyst [From ref. 401. Reprinted with permission].

surface area recovery for the $NiMo/Al_2O_3$ catalyst in 2% O_2 was quite surprising because this catalyst was deactivated by metals in addition to coke. A visual observation of the particles after regeneration in air indicated the presence of glassy phases, thus confirming a significant sintering.

Table 6.5: Surface area (m^2/g) loss and recovery on fresh catalyst surface area [From ref. 331. Reprinted with permission].

	$CoMo/Al_2O_3$	$NiMo/Al_2O_3$
Fresh	210	140
Spent	167	17
Treated in N_2	182	50
Surface area loss (%)	11	64
Regenerated		
In air	90	54
In 2% O_2	197	127
Surface area recovery (%)		
In air	43	39
In 2% O_2	94	91

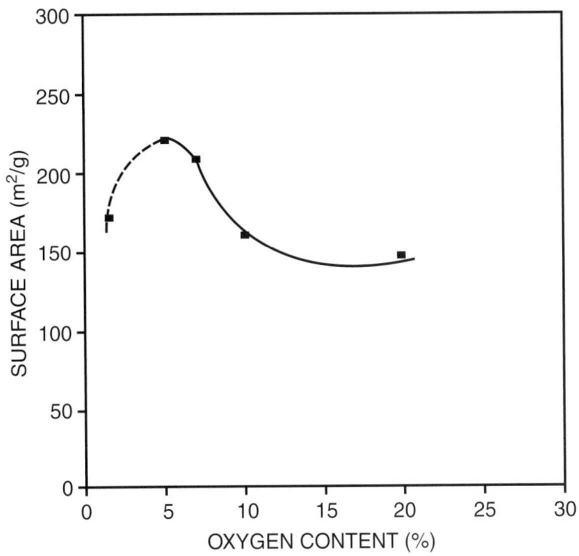

Figure 6.20: Effect of O_2 content in regeneration gas on surface area of catalyst [From ref. 472. Reprinted with permission].

The extensive database on surface properties of the fresh, spent, and regenerated catalysts (Table 6.6) established by Ammus et al. [402] is presented to indicate changes in surface properties during regeneration involving two $CoMo/Al_2O_3$ catalysts varying in particle size. In this case, particles size (L_p) was defined as the ratio of particle volume to the particle external

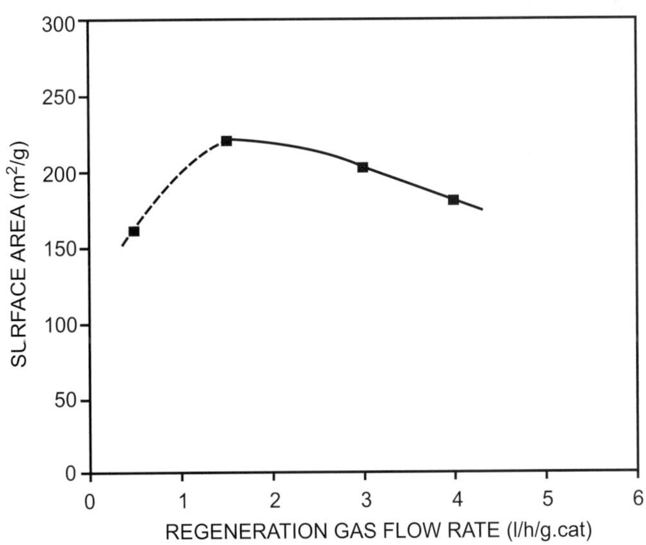

Figure 6.21: Effect of flow rate of regeneration gas on gas on surface area of catalyst [From ref. 401. Reprinted with permission].

Table 6.6: Porosity of fresh, spent, and regenerated catalysts [From ref. 402. Reprinted with permission].

Pore size and run no.		Pore volume (mL/cm^3)	Surface area		Run time (h)
			m^2/g	Loss (%)	
$L_p = 0.00567$ cm	f	0.557	253	–	
1	u	0.272	178	30	
1	r	0.537	232	8	30
2	u	0.159	107	58	
2	r	0.530	229	9	70
3	u	0.152	44	83	
3	r	0.465	211	17	140
$L_p = 0.0294$ cm					
1	u	0.229	147	42	
1	r	0.538	230	9	15
2	u	0.222	136	46	
2	r	0.509	253	0	50
3	u	0.174	93	63	
3	r	0.518	220	13	90

surface area. The catalysts were used for hydroprocessing of an atmospheric residue derived from the conventional crude containing less than 20 ppm of metals. The loss of surface area and pore volume with the duration of operation was quite evident. However, even after 140 h on stream, more than 80% recovery of surface area and pore volume could be achieved on regeneration. In the case of the NiMo/Al_2O_3 extrudate catalyst, the aged catalyst suffered a large loss in 70 to 140 A diameter pore region [403]. Most of the pore volume could be recovered on regeneration. The lower recovery of surface area and pore volume was achieved for higher metal content feeds [404].

Diffusivity measurements can provide important information, particularly when they are performed on fresh, spent, and regenerated catalysts. In the study of Johnson et al. [405], the diffusivity was determined by contacting the catalysts with the solution of coronene in cyclohexane. The diffusivity increased on regeneration of catalyst but it did not reach that of the fresh catalyst. The loss in diffusivity did not correlate with the catalyst activity recovery [404]. The former was more pronounced than the loss in activity.

6.2.4.2 Activity of Regenerated Catalysts

Figure 6.22 [367,368] shows the effect of temperature on the surface area and dynamic O_2 chemisorption (DOC) of the regenerated catalysts. The latter is usually performed on the regenerated-presulfided catalysts and is an indication of the catalyst activity recovery.

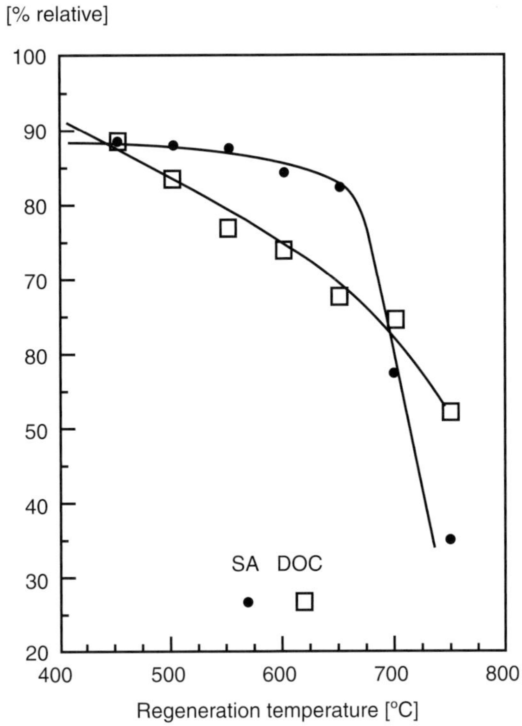

Figure 6.22: Effect of regeneration temperature on recovery of surface area (SA) and on dynamic O_2 chemisorption (DOC) [From refs 367 and 368. Reprinted with permission].

The usefulness of the O_2 chemisorption method was also confirmed by Cable et al. [406]. In agreement with Fig. 6.18 [401], the abrupt surface area decline above 600 °C was rather evident. At the same time, the decline in the DOC was more gradual. Nevertheless, only regeneration below 500 °C ensured a desirable level of the activity recovery. The DOC values were consistently lower than corresponding surface area values suggesting that a high recovery of surface area does not ensure a high recovery of activity. However, these observations cannot be generalized. For example, Inoguchi et al. [407] reported that a complete recovery of both surface area and activity was achieved for a spent catalyst deactivated only by coke, whereas in the presence of contaminant metals neither surface area nor activity of the fresh catalyst could be restored on regeneration. Another set of interesting data was published by Sakabe and Yagi [408] who studied regeneration of the spent $CoMo/Al_2O_3$ catalyst used for hydroprocessing of a residue. For this catalyst, a loss in the HDS activity on regeneration was observed. At the same time, the activity of the catalyst for hydrocracking reactions improved. Similar observation was made by Cable et al. [406]. Then, the regenerated catalyst may be used in an operation requiring a high hydrocracking service, although this observation cannot be generalized.

As it follows from Fig. 6.22 [367,368], the recovery of surface area may not be a good indication of the recovery of catalyst activity even for spent catalysts used for upgrading light feeds. It was reported that the acidity of regenerated catalysts measured by pyridine adsorption correlated well with the activity [410]. At the same time, no correlation with surface area was observed in spite of the complete surface area recovery. This may be attributed to the agglomeration and/or sintering of active phase on the support during regeneration, while little change was incurred by the latter [410]. Therefore, any additional treatment of regenerated catalysts leading to an enhanced metal distribution would almost certainly improve catalyst activity.

The regeneration of spent catalysts shown in Table 6.7 was used to illustrate the other extreme of the catalysts deactivation due to both coke and metal deposits [411]. The spent catalysts were obtained from the two different hydrodemetallization (HDM) operations using the HYVAHL process [138,412,413]. In the fresh form, the macroporous catalyst consisted of 2.7 wt.% of NiO and 14.0 wt.% of MoO_3 and had surface area of 140 m^2/g. The results of regeneration shown in Table 6.8 [411] were obtained for either fresh or regenerated catalysts that were subsequently presulfided according to the established procedures. They clearly

Table 6.7: Analysis of toluene-extracted spent catalysts [From ref. 411. Reprinted with permission].

Element (wt.%)	Catalyst A	Catalyst B
Carbon	15.0	13.6
Sulfur	5.9	13.5
Vanadium	3.3	10.1
Nickel	2.7	4.6
Iron	0	7.7

Table 6.8: Properties of fresh and regenerated catalysts [From ref. 411. Reprinted with permission].

Catalyst	Treatment	Surface area (m^2/g)	Conversion (wt.%)	
			HDS	HYD
Fresh	None	140	46.7	27.5
SpentA	None	17		
Spent B	None	11		
Spent A	2% O_2	127	15.0	10.7
Spent B	2% O_2	60	15.7	12.9
Spent A	Air	54	12.3	9.3
Spent B	Air	17	6.4	5.9

HDS: hydrodesulfurization; HYD: hydrogenation.

demonstrate that neither controlled regeneration in 2% O_2 nor that in air could ensure an acceptable activity recovery in spite of the fact that the HDS and HYD activities were determined using small molecules, such as thiophene and cyclohexene, respectively. It is believed that if a heavy feed (comprising much larger size molecules) would be used for the activity determination, the difference would be even more evident. It is, however, believed that in a pulverized form, such solids could still be suitable catalysts for hydroprocessing of heavy feeds conducted in the slurry bed reactors.

The effect of the amount of deposited metals (V + Ni) on catalyst activity was investigated by Clark et al. [414]. The coke in spent catalysts was removed by oxidative burn-off. The results in Fig. 6.23 show that the activity of the decoked catalyst containing about 8 wt.% of the

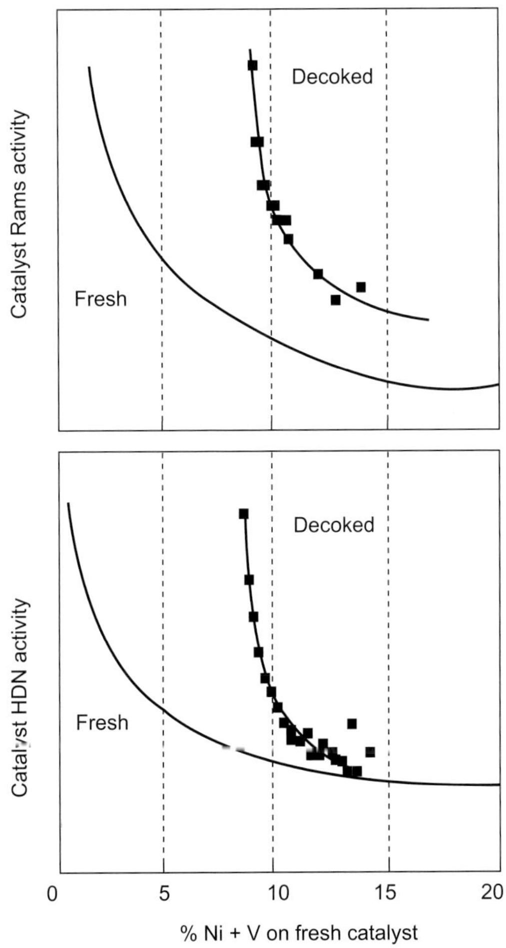

Figure 6.23: Ramsbottom carbon (RBC) conversion and hydrodenitrogenation (HDN) activities of fresh and decoked catalysts [From ref. 414. Reprinted with permission].

contaminant metals was similar as that of the fresh catalyst. This amount of metals represented the coverage of one-half of monolayer. Apparently, most of these metals were deposited on the bare support. During hydroprocessing of a heavy feed, the activity of the decoked catalyst rapidly declined with the further addition of metals and approached that of the fresh catalyst containing the same amount of metals. This suggests that the metals in excess of 8 wt.% deposited near or on the active sites and, as such, caused permanent deactivation. Then, such catalysts may be too heavily contaminated with Ni and V to be regenerated, particularly when the intention was the reuse of the regenerated catalyst for hydroprocessing of the same heavy feed. It is uncertain how this catalyst would perform during the upgrading of metals and asphaltenes free feed. It should be again kept in mind that these observations were made for a particular catalyst and they may differ for other type of spent hydroprocessing catalysts. It is again emphasized that choice of the conditions used for testing regenerated catalysts requires attention for obtaining reliable information. Thus, as it was pointed out by Dufresne et al. [307,366–368], the difference between the activity of the fresh and regenerated (contaminated with metals) catalyst was usually greater at start of the run than after several weeks/months of operation. An example of this specific behavior had been shown on the residue catalyst contaminated with the overall contaminants content as high as 21 wt.% [416]. After regeneration, the catalyst was less active for HDM and HDS than a fresh catalyst on start of run, but, after one month of the operation, the activity difference was negligible for HDM and still existing but reduced for HDS. So, in practical situation, the catalyst reuse has to be carefully examined on a case by case basis either for the same application or for cascading in less severe applications.

6.2.4.3 Chemical Structure

Spectroscopic techniques have been used to identify phases present in spent-regenerated catalysts. Also, for regenerated catalysts, the dispersion of active phase and/or potential loss of active metals due to the interaction with the support have been the focus of attention. Useful information could be obtained by using extended X-ray absorption fine spectroscopy (EXAFS), X-ray photoelectron spectroscopy (XPS), IR, X-ray diffraction spectroscopy (XRD), and other techniques.

The XRD technique was used to identify phases after regeneration of the spent $NiMo/Al_2O_3$ catalyst either in 2% O_2 or in air [331]. In 2% O_2, only γ-Al_2O_3 and MoO_3 peaks were detected. The Ni-containing species were not observed because of a high dispersion. After regeneration in air, the additional species such as $Al_2(MoO_4)_3$ was formed. The α-Al_2O_3, δ-Al_2O_3, and γ-Al_2O_3 were found in spent regenerated catalyst containing contaminant metals (e.g., V and Ni) even in 2% O_2. This signified that the contaminant metals accelerated sintering of γ-Al_2O_3. For this catalyst, crystalline phases of V_2O_5, FeO, $NiAl_2O_4$, and $NiMoO_4$ were observed after regeneration in air. Teixeira da Silva et al. [416] suggested that β-$NiMoO_4$ is the precursor to the NiMoS active phase. At higher temperature, the former tends to lose Ni because of the enhanced formation of $NiAl_2O_4$. Similarly, Mahadjev et al. [386] observed the gradual loss of Co during the repeated regeneration-utilization cycles involving the $CoMo/Al_2O_3$ catalyst. In the study of Madeley and Wanke [417], severe sintering of γ-Al_2O_3 was observed after regeneration of the spent $NiMo/Al_2O_3$ in pure O_2, while under moderate regeneration conditions a highly dispersed metal oxide phases, such as present in fresh catalyst, were found.

In the study of Yoshimura et al. [390] on the spent NiW/Al_2O_3 catalyst, the EXAFS data revealed that WS_2-like structures, which were laterally grown during hydroprocessing run, were redispersed to nearly the same level as that of the fresh catalysts when carefully controlled oxidizing conditions were used (e.g., 1.5% O_2 gas). The XPS data showed that surface compositions of Ni and W were recovered to almost the level of fresh catalysts, but the Ni/W ratio was slightly less than that of the fresh ones. Catalytic activities and selectivities were successfully recovered by low-temperature oxidation. On the contrary, for the CoMo catalyst on which MoS_2-like sulfides were laterally grown, some of the Co aggregated to Co_9S_8. In this case, it was not possible to recover the same level of structural properties as those of the fresh catalysts because of small amounts of metals such as Ni, Fe, and V being present. While the catalytic activities and selectivities were almost recovered by low-temperature oxidation, at higher regeneration temperatures, a slight loss of hydrogenation activity and a large increase in the hydrocracking activity were noted.

The study on regeneration and resulfidation of the $NiMo/Al_2O_3$ catalyst conducted by Bogdanor and Rase [392] confirmed a significant intraparticle migration of the Ni and Mo metals. This is evident in Fig. 6.24 from the radial variation in the Mo/Al, Ni/Al, and Mo/Ni ratios determined by SEM. Interestingly enough, for spent catalyst, the net result of regeneration followed by resulfidation was the initial distribution as observed for the fresh catalyst. Several qualitative observations can be made from the results in Fig. 6.24. For example, the fresh catalyst exhibited a modest maldistribution of Ni and Mo, which then changed on sulfidation followed by reaction, i.e., the Mo profile became flat, whereas the Ni profile increased in concentration towards the center. This caused the Ni/Mo ratio to increase and/or Mo/Ni ratio to decrease towards the center. In the spent catalyst, Mo was concentrated towards the pellet center and Ni towards the pellet exterior. Therefore, a high Mo/Ni ratio at the center of pellet was observed. Increasing the temperature of regeneration resulted in the migration of both Ni and Mo towards the pellet exterior in agreement with the results published by Gellerman et al. [418]. The spent catalyst and the same spent catalyst that was regenerated under different conditions and subsequently resulfided had the same Mo/Ni ratio. Generally, Ni appeared to be more mobile than Mo. During the reaction (thiophene HDS) over the fresh-sulfided catalyst, Ni migrated towards the center, whereas for the spent-regenerated-sulfided catalyst, towards the exterior. The contaminant metals, such as V and Fe, exhibited little change during the regeneration followed by resulfidation.

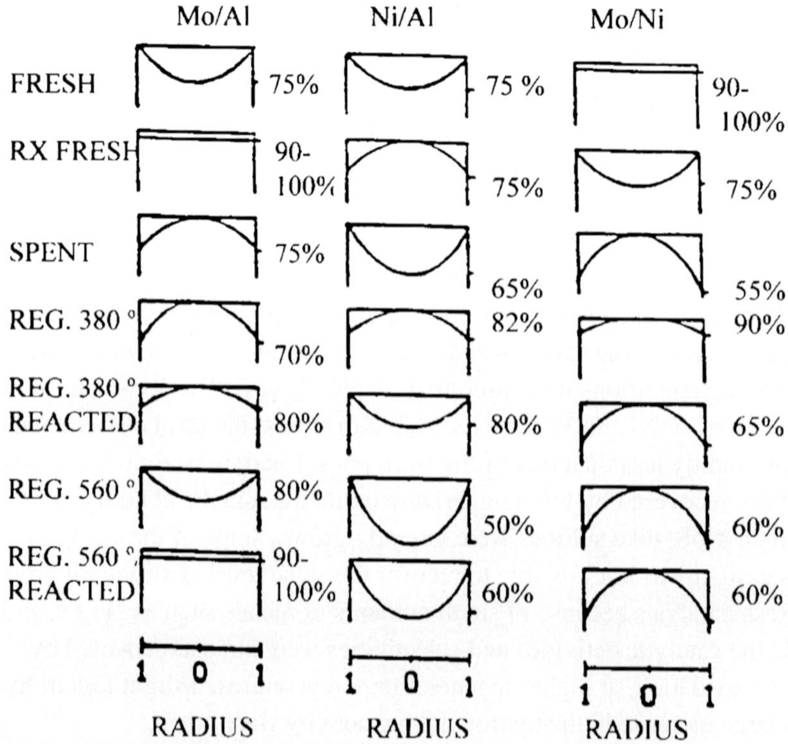

Figure 6.24: Summary of trends for intraparticle Mo/Al, Ni/Al, and Mo/Ni profiles (in Ni{Mo}/Al max.) for different catalyst treatments [From ref. 392. Reprinted with permission].

6.2.5 Safety and Environmental Aspects of Oxidative Regeneration

According to Fig. 5.4 [331], a safe contact of spent catalysts with air can only be maintained at temperatures lower than 250 °C providing that most of the flammable volatile carry-overs that were deposited on the catalyst surface were already removed. This may also involve entrapped gases, such as H_2, H_2S, and light hydrocarbons. Otherwise, an uncontrolled temperature runaway would lead to recrystallization and/or to sintering of the catalytically active phases. In the sintered form, the catalytic functionalities could not be restored in regenerated catalysts. The loss of Mo due to sublimation of the MoO_3 component caused by temperature excursions, such as shown in Fig. 6.17 [401], would make the catalyst practically unusable. From the safety point of view, a rapid surge of toxic species during temperature excursions deserves a special attention. Potential damages to the material of reactor, caused by sudden temperature increase, may be another reason for precise monitoring the regeneration temperatures.

Figure 6.25 [419] indicates the temperature runaway on the introduction of air to the bottom of the fixed bed of spent catalyst at 350 °C. A sudden temperature increase to more than 800 °C

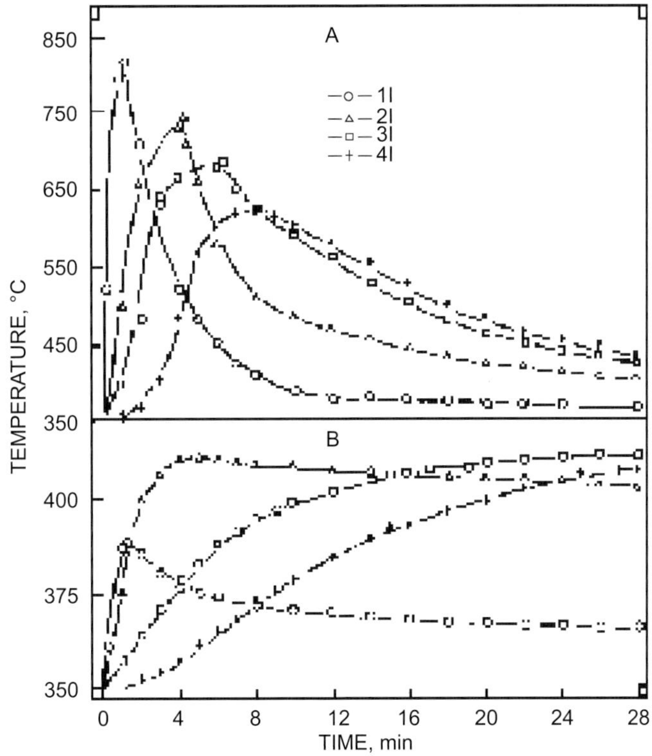

Figure 6.25: Effect of O_2 concentration in regeneration gas on temperature profile in (A) air and (B) 2% O_2 (catalyst bed at 350 °C) [From ref. 419. Reprinted with permission].

within the first few minutes of burn-off was noted. In spite of a short duration, such a rapid temperature change would have an adverse effect on catalyst properties. This was indeed supported by a glassy structure of the catalyst particles. The temperature runaways became less evident with the progress of burning zone towards the end of the fixed bed. This may be attributed to a partial consumption of O_2 at the bottom, leaving a more diluted air in contact with the catalyst in upper zone of the fixed bed. Because of a gradual removal of coke, temperature was declining with time on stream and approached the initial temperature of the fixed bed. To a much lesser extent, the temperature on the outside wall of the reactor increased as well.

The temperature runaways can be controlled by the O_2 concentration in the oxidizing gas. This is clearly indicated by the temperature increase from 350 °C to about 410 °C and almost to 850 °C in 2% O_2 and air, respectively as it is shown in Fig. 6.25 [419]. To identify a suitable O_2 concentration, the correlation between the ΔT parameter, indicating temperature rise above the temperature of reactor on the introduction of the oxidizing gas, was established. Such a correlation is shown in Fig. 6.26 [419]. In this case, the baseline temperature of the fixed bed

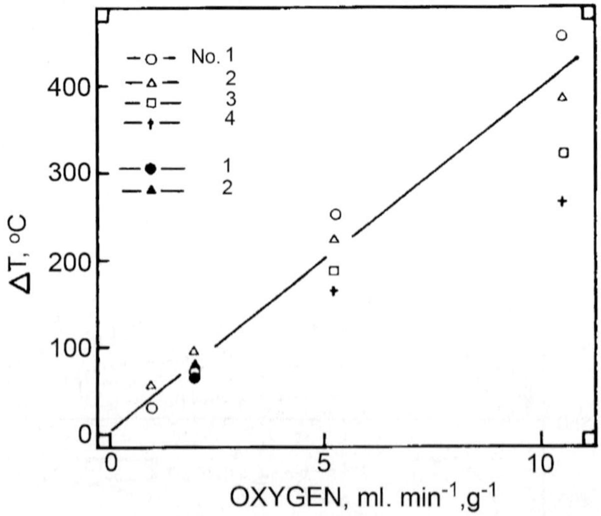

Figure 6.26: Effect of O_2 concentration in regeneration gas on temperature runway in four zones of fixed bed (solid symbols for tetrahydrofuran [THF] extracted catalyst) [From refs 419 and 420. Reprinted with permission].

of spent catalyst was 350 °C. From this correlation, it can be estimated that the O_2 concentration of about 4 vol.% can ensure the temperature excursion to less than 500 °C. Thus, according to Figs 6.17 and 6.22 [401,367,368], in order to achieve a high recovery of catalyst activity, the temperature of regeneration should not exceed 500 °C.

The correlation in Fig. 6.26 [420] is applicable to a specific spent catalyst, i.e., it cannot be applied generally. This is confirmed by the results in Fig. 6.25 [419]. The temperature profiles in Figs 6.25 and 6.27 were obtained for catalyst A and catalyst B, respectively. The properties of these catalysts are given in Table 6.9. The temperature excursion in air for the much more coke deposited catalyst B was lower compared with that for catalyst A. This can be attributed to a much smaller surface area of catalyst B, which allowed only an external chemically

Table 6.9: Composition of coke (wt.%) on spent catalysts [From ref. 419. Reprinted with permission].

	Catalyst A	Catalyst B
Carbon	7.2	26.2
Hydrogen	0.8	3.5
H/C	1.3	1.6
Nitrogen	0.40	0.27
Sulfur	7.13	5.48

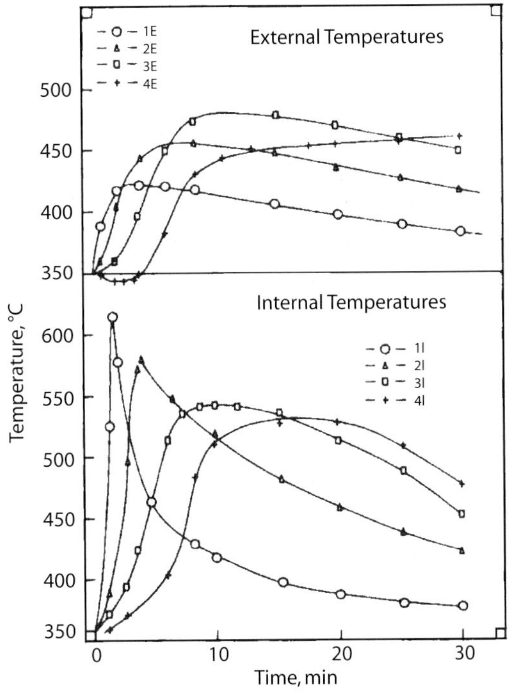

Figure 6.27: Internal and external temperature profiles during burn-off of spent catalyst pellets in 10% O_2 for four zones of fixed bed [From ref. 419. Reprinted with permission].

controlled burn-off. At the same time, much more coke surface was available for O_2 for catalyst A. Then, the diffusion of O_2 into the interior of particles played an important role during the burn-off of catalyst A. While O_2 concentration is a key parameter, there are other means of controlling temperature during regeneration. For example, the pretreatment of spent catalyst (de-oiling, thermal treatment under inert gas, addition of steam to oxidizing gas, etc.) decreased ΔT parameter [331,419].

The emissions from oxidative regeneration include regulated particulate matter and gases such as SO_X, NO_X, CO, and CO_2 as well as toxic species, e.g., NH_3, HCN, and H_2S. Theoretically, one of the most toxic species such as AsH_3 may also be formed, particularly under O_2 starving conditions. The distribution of these compounds depends on the type of spent catalyst and the conditions employed during regeneration. Beside particulate matter (dust), the solid emissions also include spent caustics. All certified regenerating companies are equipped to contain these emissions within the regulatory limits. A description of these methods is given later, i.e., in the section dealing with industrial regeneration.

Of particular importance is the potential evolution of the most toxic species, such as HCN and AsH_3, during oxidative regeneration. It is emphasized that quantities of these compounds generated during oxidative regeneration are rather small. Moreover, a large volume of oxidizing gas ensures further dilution of these toxic compounds. It is therefore believed that their concentrations in atmosphere should never approach lethal limits, particularly in a properly ventilated enclosures. Nevertheless, as the matter of awareness, a potential of HCN and AsH_3 evolution is being indicated and it should not be overlooked completely. The performance of some hydroprocessing catalysts may be improved by doping with fluoride, i.e., up to 3 wt.% on fluorine. So far, no attention has been paid to the potential release of the fluorine containing compounds (e.g., HF) during oxidative regeneration of spent catalysts. It was indicated earlier that some precautions may be necessary during all stages of handling spent catalysts due to the presence of fluoride. Although much less toxic than fluorine, phosphorus may also be present in spent catalysts. Other untypical species, which may be released during regeneration, involve Ni carbonyls [384]. It is desirable that the potential release of rather unusual species such as metal carbonyls as well as fluorine and phosphorus containing species is investigated. Continuous efforts to improve performance of hydroprocessing catalysts may result in the formation and release of other toxic compounds during oxidative regeneration.

6.2.6 Other Oxidizing Agents

Copperthwaite et al. [420,421] demonstrated that ozone is rather efficient oxidizing agent of coke at relatively low temperatures (150–190 °C). Most likely, the oxidation may involve the following set of reactions:

$$C + O_3 = CO + O_2$$

$$O_3 = O + O_2$$

$$O + C = CO$$

$$CO + O_3 = CO_2 + O_2$$

$$CO + 0.5O_2 = CO_2$$

During the reaction with coke, ozone reacted preferentially with aromatic carbon. At these temperatures, the involvement of O_2 during the oxidation of coke may be limited to the formation of surface complexes. A direct comparison of regeneration in ozone with the conventional regeneration was made by Solovetskii et al. [422]. As the results in Fig. 6.28 show, the advantage of regeneration with ozone is quite evident. At the same time, the radiation thermal treatment was the most efficient method. Presumably, ozone may be suitable for regeneration of catalysts that are unstable at higher temperatures. However, a commercial

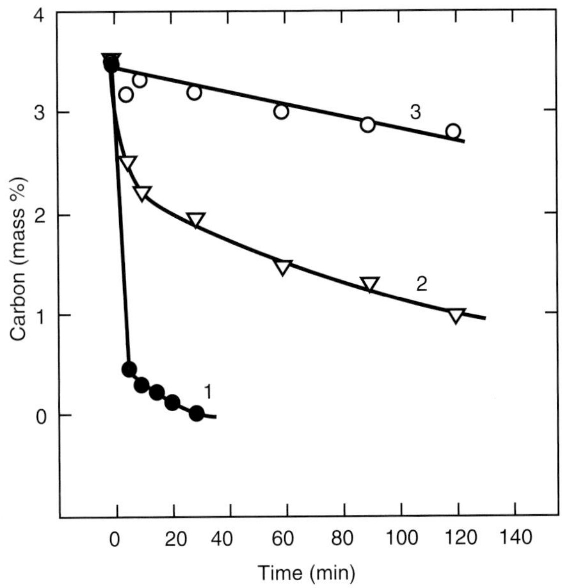

Figure 6.28: Carbon removal from spent catalyst: 1: radiation heating; 2: in ozone; 3: conventional burn-off [From ref. 422. Reprinted with permission].

use of ozone for catalyst regeneration has been prevented because of its limited availability in desirable concentrations and quantities.

6.3 Other Regeneration Methods

In this part, only the methods based on the solid-gas reactions are included. This includes steam and CO_2, as well as N_2O, NO, and H_2. The driving forces for these reactions expressed as the log K values are shown in Fig. 6.29 [423]. Among those, the reactions of carbonaceous solids with steam and CO_2 have been studied extensively as part of carbon gasification research. In this case, the reference was made to such reactions as reductive carbon conversion compared with the oxidative regeneration in O_2. Thus, while carbon was partially oxidized during these reactions, the other reactants, i.e., steam and CO_2 were reduced. Although they might exhibit an adequate reactivity, N_2O and NO have little chance to be used on an industrial scale because of the limited availability in desirable concentrations. From this point of view, the reductive regeneration of spent catalysts in H_2 yielding hydrocarbons has much more potential to be used commercially than nitrogen containing oxides.

6.3.1 Regeneration in H_2O and CO_2

The coke on catalyst surface can also be removed via its reaction with H_2O and CO_2, i.e., carbon gasification [424]. The products of these reactions include CO and H_2. This ensures the

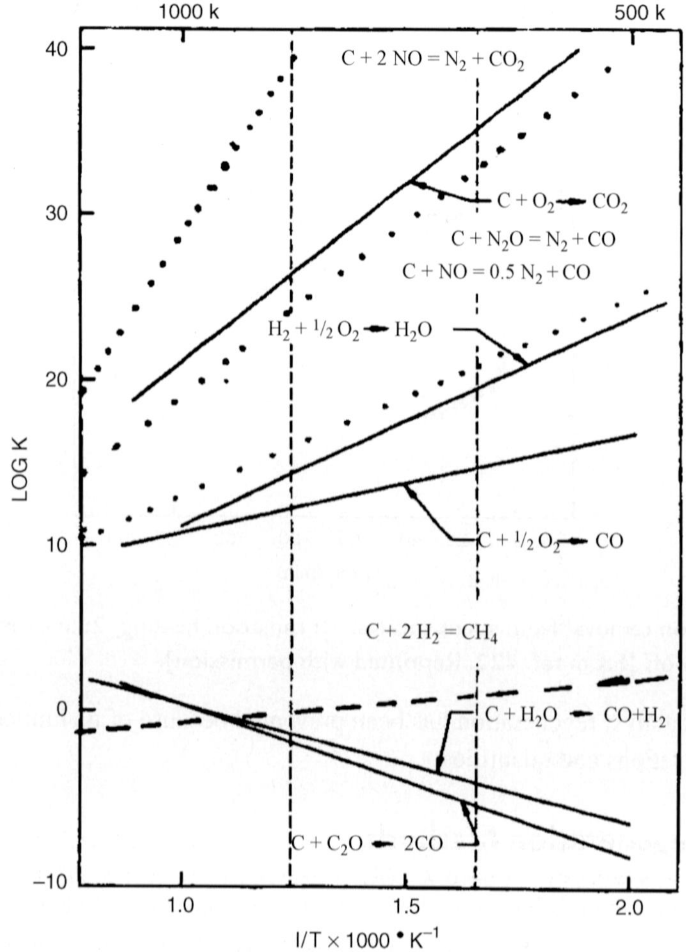

Figure 6.29: Log *K* versus 1/*T* correlations for reactions of carbon with various oxidizing and reducing agents [From ref. 423. Reprinted with permission].

presence of reducing conditions in the regeneration system. It follows from Fig. 6.29 [423] that for these reactants, much higher temperatures were necessary for achieving desirable rates of coke removal. For example, the study on the CO_2 regeneration of the catalyst used for hydroprocessing of the vacuum residue derived from Athabasca bitumen revealed that the formation of CO via Boudart reaction, i.e.:

$$CO_2 + C = 2CO$$

just began above 800 K [425]. More than 1000 K was necessary for removing most of the carbon in pure CO_2. As Fig. 6.30 shows, there was a significant effect of the CO_2 concentration on carbon removal as indicated by the increase in CO yield by increasing the

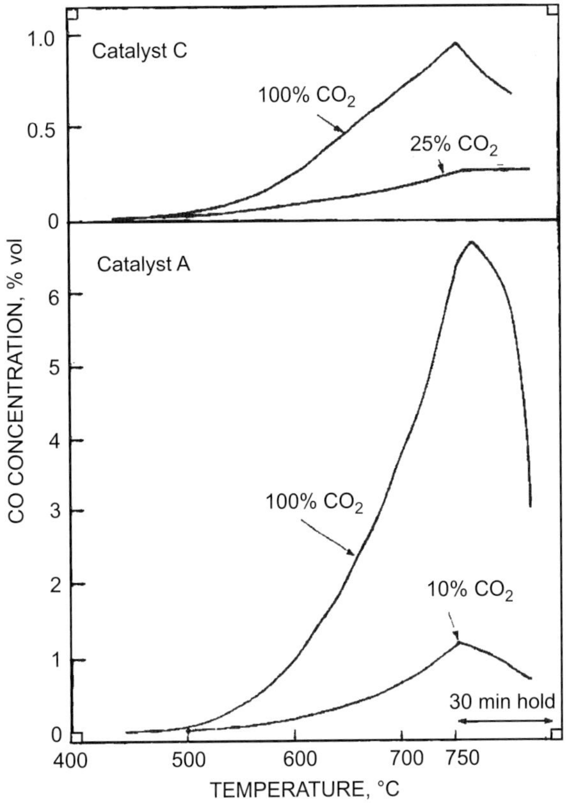

Figure 6.30: Formation of CO during temperature-programmed heating in flow of CO_2; catalyst A ($NiMo/Al_2O_3$) and catalyst C ($CoMo/Al_2O_3$) [From ref. 425. Reprinted with permission].

CO_2 concentration from 10 to 100%. In this figure, the carbon content of the catalyst A and catalyst C was about 19 and 3 wt.%, respectively.

According to Fig. 6.29, similar trends may also be observed when CO_2 was replaced by H_2O as the gasification reactant. This was indeed confirmed during the gasification of coke in steam [426]. Moreover, there is one patent [427] describing steam regeneration. According to this patent, after being washed (de-oiled), the spent catalyst was contacted with steam between 540 and 680 °C. Apparently, after 2 to 5 h contact with steam, the catalyst was ready for reuse. However, details of the composition and structure of spent catalysts could not be verified. Also, for some catalysts, steaming may enhance sintering of active phase.

6.3.2 Regeneration with Nitrogen Oxides

Thermodynamic evidence (Fig. 6.29) indicates significant driving forces for the removal of carbon using nitrogen oxides (N_2O, NO, and NO_2) via the following general reactions:

$$C + N_2O = CO + N_2$$

$$C + NO = CO + 0.5N_2$$

$$C + 2NO = CO_2 + N_2$$

Because of very weak N–O bonds in N_2O, the generation of very active O* radical and its participation in carbon removing reactions cannot be ruled out, e.g.:

$$N_2O = N_2 + O^*$$

$$C + O^* = CO$$

So far, N_2O was the only nitrogen oxide tested for regeneration of spent hydroprocessing catalysts [428,429]. In this case, a more complete carbon removal with N_2O compared with pure O_2, was achieved, although the initial rate of carbon oxidation in the presence of the latter was greater. It is believed that comparing N_2O with diluted air makes more sense than with pure O_2 because of the safety and other reasons discussed above. There is little information on the use of other nitrogen oxides in spite of their suitability predicted by thermodynamics. In this regard, a parallel may be drawn between regeneration of spent catalysts and ability of various carbons to reduce nitrogen oxides to N_2. The availability of nitrogen oxides in desirable concentrations and quantities may be an obstacle for their utilization during industrial regenerations.

6.3.3 Reactivation

The postoxidative regeneration treatment of spent catalysts with various organic agents having chelating capabilities has been also successfully applied to regenerated catalysts. Such treatment resulted in the significant enhancement of the active metals redispersion, similarly as it was observed for calcined catalysts during catalyst preparation.

Generally, a partial sintering of active metals during regeneration cannot be completely prevented. The impregnation of regenerated catalysts with various chelating agents resulted in a significant enhancement of the active metals redispersion on subsequent resulfiding. On a commercial scale, this method was introduced by Albermarble in 2003 as reactivation (REACT) technology [329,430]. This technology enables the activity of the spent regenerated STAR catalyst marketed by Albermarble to be recovered to more than 90% of the fresh catalyst activity. This was attributed to redistribution of metals and reconfiguration of the type II active phase to approach that in the original catalyst. Details of this process were given in patent applications submitted by Ginestra et al. [431] and Eijsbouts et al. [432]. The latter list

Table 6.10: Effect of reactivation on catalyst activity [From ref. 430. Reprinted with permission].

Catalyst	S in products (ppm)	N in products (ppm)	RVA-HDS	RVA-HDN
Fresh	165	83	100	100
Regenerated	290	220	68	67
Reactivated	201	105	95	93

RVA-HDS: relative volume activity-hydrodesulfurization; RVA-HYD: relative volume activity-hydrogenation.

numerous chelating agents, which may be used, e.g., citric acid, tartaric acid, oxalic acid, malonic acid, butanediolglycolic aldehyde, acetaldol, various glycols, etc. The published database confirmed that several spent hydroprocessing catalysts have been successfully reactivated using this method. Table 6.10 [430] compares the HDS and HDN activities of the fresh, regenerated, and reactivated catalyst via REACT method. The activities were measured under typical hydroprocessing conditions using VGO as the feed. A significant improvement in both HDS and HDN activities of the reactivated catalyst compared with the regenerated catalyst was quite evident. A similar revitalization process was developed by Criterion Catalyst Company for their CENTINEL catalyst [328]. Haldor Topsoe is offering a similar refresh process for reactivating regenerated catalysts by chemical treatment [433].

Although beneficial effects of reactivation with the aid of chelating agents have been confirmed experimentally, there is little information on mechanistic aspects of reactivation. In recent study, Mazoyer et al. [434] investigated the interaction of chelating agent such as diammonium salt of ethylene diaminetetraacetic acid with the calcined $CoMo/Al_2O_3$ catalyst. The study confirmed that chelating agent dissolved undesirable $CoMoO_4$ crystalline phase and redispersed Co^{2+} cation on the surface of the catalyst in a better way to facilitate the Co-Mo-S phase formation. The chelating agent was however not able to solubilize Co^{2+} from $CoAl_2O_3$ spinel structure. In a related study, Costa et al. [435] investigated the role glycol-based additives in enhancing the activity of CoMo and CoMoP catalysts. For all dried as well as calcined CoMo and CoMoP catalysts (P/Mo molar ratio less than 0.4), a redissolution phenomenon was evidenced after the additive impregnation step leading to the formation of the Anderson heteropolyanion such as $AlMo_6O_{24O}$. This redissolution phenomenon was however affected by the low solubility of this Anderson heteropolyanion. In the case of the CoMoP dried catalyst (P/Mo ratio greater than 0.4), characterization of the additive containing catalyst evidenced the $PCoMo_{11}O_{407-}$ formation. Redissolution and redispersion of the active components of the catalyst by chelating agents, thus, appeared to be responsible for the activity enhancement.

6.3.4 Regeneration Aided by Radiation Treatment

The curves 1 and 3 in Fig. 6.28 [422] compare the coke removal using the oxidative burn-off with that employing a radiation thermal treatment. A significant enhancement in the rate of coke removal should be noted, whereas the rate of oxidation with ozone (curve 2) was between that observed for the conventional burn-off method and radiation thermal treatment method. Under electron beam conditions, carbon-containing compounds are completely removed within 10–15 min. However, a careful control of the experimental conditions is necessary to avoid losses of Mo due to sublimation. At the same time, in the case of the spent $NiMo/Al_2O_3$ catalyst, the content of Ni promoter was not affected. There is little information suggesting that the radiation thermal treatment has been used for catalyst regeneration on and/or near a commercial scale.

6.3.5 Reductive Regeneration

In an ideal case, catalyst could be regenerated at the end of the operation by discontinuing the feed supply while continuing the H_2 supply. In spite of this, there is little information on the reductive regeneration of catalysts used for hydroprocessing of any petroleum feed. This may result from the fact that a desirable level of coke removal could not be achieved even at much higher temperatures than those applied during the oxidative regeneration [13]. For example, during the reductive regeneration of the $NiMo/Al_2O_3$ catalyst used for hydroprocessing of a VGO, the evolution of the most of CH_4 formed via hydrogasification of coke required temperature of more than 1000 K. As the result of this, sintering of the MoS_2 particles was observed in addition to other structural changes as indicated by Scheffer et al. [436]. Almost certainly, hydrogasification of coke catalysed by active metals was the source of the small CH_4 maximum at about 800 K.

It was noted that the reductive regeneration of Pt, Re, and Ir catalysts supported on γ-Al_2O_3 proceeded at lower temperatures and at a much greater rate [437,438] than that of the $NiMo/Al_2O_3$ catalyst [13]. For the former catalysts, the maximum of CH_4 evolution indicating coke removal occurred at about 850 K. This is not unexpected considering a high hydrogenation activity of the noble metals containing catalysts, generally observed. In refinery practice, noble metals (e.g., Pt and Pd) are part of the catalysts used for dewaxing of gas oil fractions for the preparation of lube base oil. In this case, reductive regeneration has much more potential compared with conventional hydroprocessing catalysts. In spite of its potential, the reductive regeneration of dewaxing catalysts has not attracted much attention.

A brief comparison of the reductive regeneration with the oxidative regeneration was made by Noguchi et al. [379]. The spent catalyst was from the hydroprocessing of the Arabian heavy vacuum residue conducted at 713 K and 8.5 MPa. The comparison revealed that the recovery of surface area (Fig. 6.31) and activity during the reductive regeneration was higher than that during the oxidative regeneration in spite of about 3 wt.% of coke still left behind after reductive regeneration compared with almost complete coke removal during the oxidative regeneration. However, the latter was conducted in air at 923 K, whereas the reductive regeneration at 713 K. Apparently, at 713 K, part of the coke in the vicinity of active metals

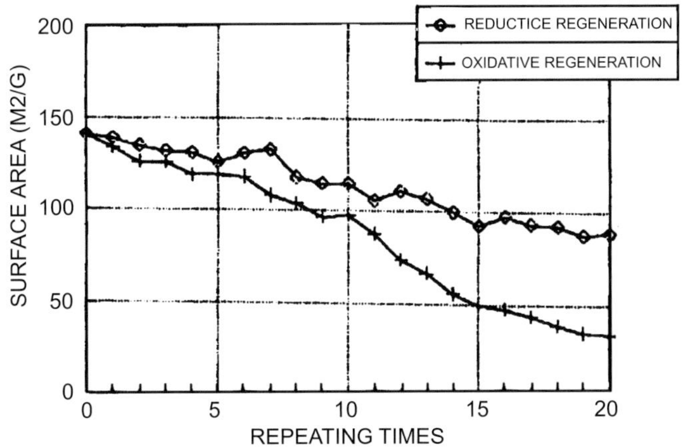

Figure 6.31: Surface area recovery during reductive and oxidative regeneration [From ref. 379. Reprinted with permission].

could be removed presumably via hydrogasification catalysed by active metals without having an adverse effect on catalyst compared with the oxidative regeneration at 923 K. In the latter case, sintering of the catalytically active phase could not be prevented. Moreover, at this temperature, transformation of the γ-Al_2O_3 support to other forms of the Al_2O_3 could become quite evident. Therefore, comparison of the reductive regeneration with the oxidative regeneration was not conducted under optimal conditions.

The coke structures in Figs 4.20 and 4.21 [238,239] may be used to illustrate difficulties, which may be encountered during the reductive regeneration in H_2. It is unlikely that such structures (e.g., tetrahydrofuran insolubles [THFIS] after 6500 h in Fig. 4.21) could be hydrogasified at a near atmospheric pressure of H_2 and temperatures at which the catalyst stability can be still maintained. The increase in temperature and H_2 pressure to increase the rate of hydrogasification of coke could have adverse effects on catalyst structure. Thus, at a temperature suitable for coke removal, the catalyst over reduction could not be avoided. This would be also accompanied by recrystallization of the catalytically active phases. Therefore, the in-situ regeneration using reductive agents may not be feasible, particularly for the catalysts used in severe hydroprocessing operations.

The spent hydroprocessing catalyst used for bitumen upgrading was reductively regenerated at 10 MPa of H_2 in the presence of hydrogen donor solvents, such as tetralin and tetralin–pyrene mixture [439]. These solvents were compared with gas oil. The following order in the effectiveness of coke removal was established: gas oil < tetralin < tetralin–pyrene. The highest effectiveness of tetralin–pyrene mixture was attributed to the ability of pyrene to form very active H-donors, such as hydropyrenes. The HDS activity of the regenerated catalyst was significantly enhanced.

6.3.6 Regeneration by Attrition/Abrasion

The objective of this method is the removal of either the "skin" of inorganic solids or this part of metals, which deposited predominantly on or near the external surface of catalyst particles. For V and Ni, the sulfides formed via the non-catalytic reaction of porphyrins with H_2S could be easily removed from spent catalysts by the attrition. Under dry conditions, such an attrition test could be conducted in a fluidized bed reactor. The spent catalyst has to be kept in a motion when the attrition test is conducted under wet conditions.

The method used by Gray et al. [243] involved rolling the mixture of the metal deposited catalyst particles with the particles of γ-Al_2O_3 in water. After drying, catalyst was separated from the γ-Al_2O_3 by sieving. Some removal of the "skin" metals (e.g., Fe and Ca) was observed, but it was less efficient than the extraction with the diluted HCl. The experiments could also be conducted under dry conditions. Nevertheless, this method can only remove metals from the external surface of catalyst. The removal of metals from pores would require a prolonged attrition, which could affect the size and mechanical properties of catalyst particles. It is believed that the attrition/abrasion methods for regeneration of hydroprocessing catalysts have limited practical applications. Apparently, such method may be more suitable for removing metals from the spent FCC catalysts.

6.3.7 Resulfidation of Regenerated Catalysts

On the industrial scale, regeneration of spent hydroprocessing catalysts has been dominated by the oxidative method employing air and/or diluted air. Thus, even industrial reactivation, which has been receiving much attention in recent years, is always preceded by oxidative regeneration. Rather than to transport the regenerated catalyst to refinery in its oxidic state, regenerating companies have been performing off-site presulfiding as an additional service to refiners. Because of the environmental and economic benefits, which result from the off-site presulfiding, many refineries have been taking advantage of this service. Handling of toxic compounds associated with on site regeneration is avoided in the case that refiners opt for off-site regeneration. In addition, faster reactor start extends production time of the overall operation.

In one off-site presulfiding method, the oxidic catalyst is first impregnated with a sulphur containing compound dissolved in an organic solution and dried in an inert atmosphere. Under these conditions, sulfur reacts with metal oxides incompletely. The catalyst is then subjected to H_2 activation at a higher temperature and pressure for complete sulfidation [440–442]. Thioglycolic acid was used as sulfiding agent in the study published by Frizi et al. [443]. This agent ensured an efficient sulfidation in addition to complexation. The latter was confirmed by a high level of active metals dispersion. The impregnation of regenerated catalyst with diethyleneglycolbutylether enhanced activity of the catalyst after subsequent sulfidation in H_2S/H_2 mixture [444]. However, below 473 K, the agent inhibited the sulfidation.

The efficiency of the conventional presulfiding method involving an H_2S containing gas can be improved by optimizing presulfiding parameters. For example, Dugulan et al. [445] showed that a more active catalyst is obtained after presulfiding at 4 MPa compared with a near atmospheric pressure. The XpresS presulfiding method introduced by TRICAT involves treatment of regenerated catalyst using H_2S in the first ebullated bed reactor [446]. This ensures uniformity of presulfidation. Subsequently, the presulfuded catalyst is transferred into the second ebullated bed reactor where the treatment with a gas renders the catalyst odourless and enables it to be handled in air.

6.4 Industrial Regeneration

Before 1980, most of the regenerations of spent hydroprocessing catalysts were conducted in-situ, using the air diluted either with steam or nitrogen. Since that time, the off-site regeneration has been gradually replacing the in-situ regeneration. According to Fig. 6.32 [447], today, the in-situ regeneration on a commercial scale was phased out almost completely. The off-site regeneration became the method of choice for several reasons, e.g., corrosion issues, safety and environment, time considerations, availability of the experienced staff, and better activity recovery. The temperature control was one of the drawbacks of the in-situ regeneration. Thus, temperature runaways damaging the catalyst structure and the reactor material could not be entirely avoided [13].

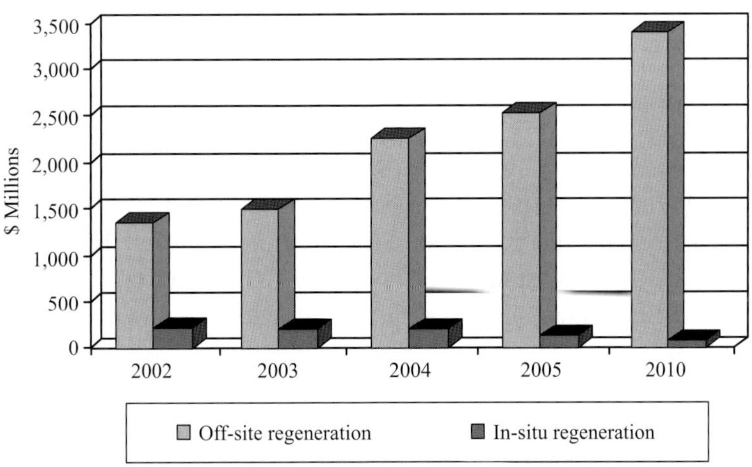

Figure 6.32: Trends in off-site versus in-situ regeneration of spent catalysts [From ref. 447. Reprinted with permission].

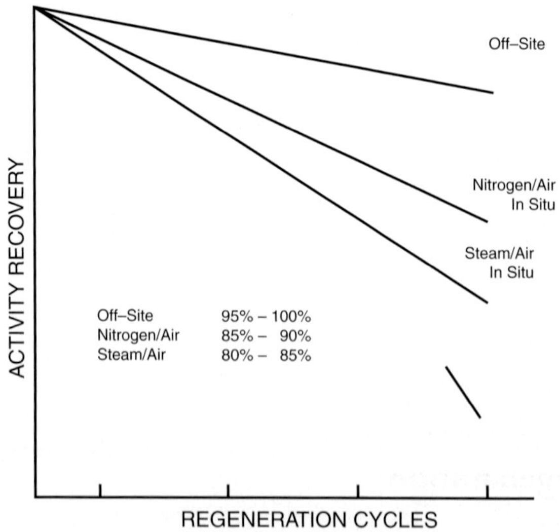

Figure 6.33: Comparison of activity recovery for off-site and in-situ regeneration [From ref. 448. Reprinted with permission].

Off-site regeneration provides the opportunity for the catalyst particle size and density grading. Thus, the presence of fines, which cannot be removed during the in-situ regeneration unless entire content of reactor is unloaded, screened, and reloaded, may lead to the unplanned shut downs of the operational cycle due to the unwanted pressure drops development. It has been established that a higher level of the catalyst activity recovery and/or a greater number of the utilization-regeneration cycles could be achieved using the off-site regeneration. This is supported by the comparison of the in-situ and off-site regeneration results shown in Fig. 6.33 [448].

6.4.1 In-Situ Regeneration

According to the procedure for in-situ regeneration conducted on a refinery site and described by Osipov et al. [449], at the end of the hydroprocessing operation, flow of the feed was replaced with the mixture of straight run diesel fraction and light catalytic gas oil in the flow of H_2. Using this mixture, the catalyst was washed during 11 h at 300 °C to remove carry-overs and a portion of coke. Subsequently, the catalyst was dried in the flow of H_2 at 380 °C for 6 h. Then, the system was flashed with an inert gas before the introduction of diluted air at reactor pressure of 1.5 MPa and temperature of 320–340 °C. The regeneration was completed at 480 °C and total pressure of 2 MPa. Under these conditions, the regeneration proceeded for 15 days. During regeneration, a circulation system involving sodium carbonate was in operation to remove sulfur oxides from the gaseous effluent. Simplified schematics used for the in-situ regeneration used in the study of Osipov et al. [449] is shown in Fig. 6.34

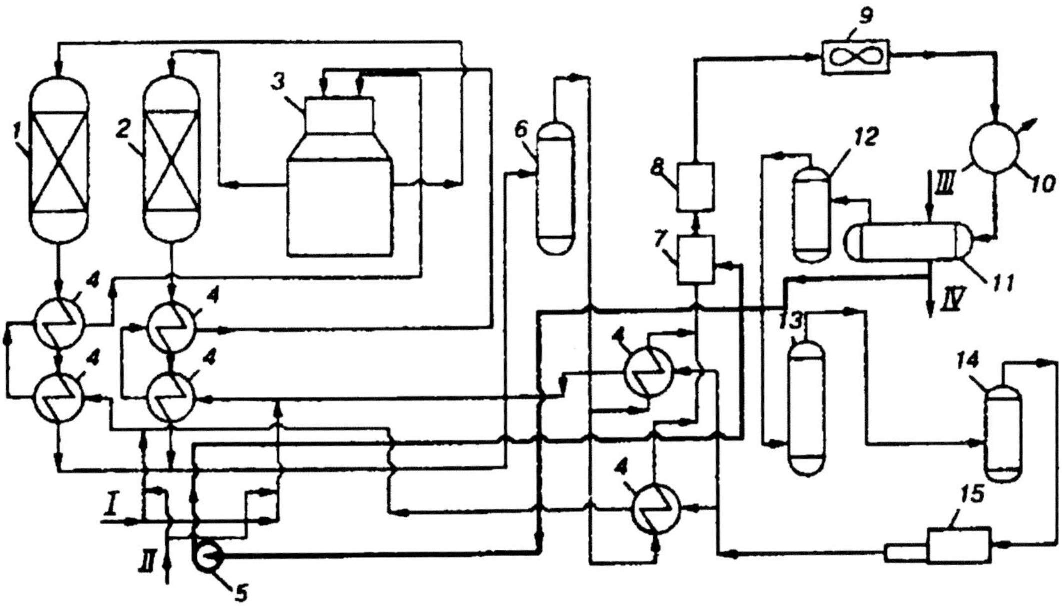

Figure 6.34: Flowsheet of in-situ regeneration plant: 1 and 2: reactors; 3: furnace; 4: heat exchanger; 5: pump; 6, 11, 12, and 14: tanks; 7 and 8: vaporizers; 9: air cooled exchanger; 10: water cooled exchanger; 13: tower; 15: piston compressor; I: air; II: inert gas; III: fresh sodium carbonate solution; IV: spent solution [From ref. 449. Reprinted with permission].

including the sodium carbonate system for scrubbing sulfur oxides from the gas exiting regenerator.

There are reports on significant problems, such as temperature runaways and mal-distribution of oxidizing gas, during the in-situ regeneration [13]. Also, fines that are usually formed during regeneration complicate the subsequent hydroprocessing operation due to the development of pressure drops and mal-distribution of liquid and gaseous streams through the bed. In some cases, this required unloading the catalyst from reactor for the removal of fines and subsequent reloading. All these problems were encountered during the in-situ regeneration of catalysts used for hydroprocessing of light feeds. Of course, even more complications could be envisaged for heavier feeds. Therefore, it is not surprising that most of the petroleum refiners prefer to have their catalysts regenerated off-site. In fact, there is little information suggesting that in recent years, any petroleum refinery would make an attempt to perform in-situ regeneration of the catalyst fixed bed after hydroprocessing of heavy feeds.

The method for the in-situ regeneration of spent hydroprocessing catalyst described by Tamayama [450] may be used to illustrate the complexity of the process. It involves de-oiling the catalyst by shutting down the feed supply lines and flushing the fixed bed with a recycle gas for about 2 h at temperature and pressure approaching those employed during the operation. A typical recycle gas has a high content of H_2 in addition to H_2S and volatile hydrocarbons. After the flushing stage, temperature and pressure are decreased and recycle gas replaced by N_2. In the flow of N_2, the reactor entrance temperature is adjusted to about 330 °C before the addition of about 0.5 vol.% of O_2. The addition of O_2 resulted in the temperature rise to about 385 °C. However, the actual temperature rise may vary from catalyst to catalyst. Subsequently, the O_2 concentration in N_2 is increased to 0.65% while the entrance temperature was increased from 330 to 345 °C to achieve the second burn-off stage. The temperature may be increased from 345 to 380 °C, if the second burning stage did not occur. However, at the same time, the O_2 concentration must be decreased from 0.65 to 0.25%. After burning at 380 °C is completed, the temperature is increased to 400 °C to attain third burn. At 400 °C, the regeneration may be completed by increasing the O_2 concentration to 1%. During this stage, all precautions must be taken to avoid temperature runaways. The progress of regeneration is monitored by analyzing the gas exiting the regenerator. Thus, regeneration is completed once the concentration of carbon oxides in the gas approached zero. After regeneration, the catalyst is cooled to 80 °C before it is exposed to air. Several days or even weeks may be required to complete the regeneration according to this method. As part of the modified method, N_2 may be replaced by steam (e.g., one ton of steam per one ton of spent catalyst per hour). To avoid wetting the catalyst, the steam temperature must be higher than 200 °C and steam pressure at the regenerator outlet lower than 7 atm.

The success of the above described regeneration method depends on the availability of skilful operators. Considering periodic regeneration requirements (e.g., once in a few years), it may not be easy to maintain such skills. It therefore appears that it may be more advantageous for petroleum refiner to have the catalyst regeneration performed by the off-site certified regenerating companies having all necessary expertise in place.

6.4.2 Off-Site Regeneration

The off-site regeneration allows the size or density grading of spent catalyst prior to regeneration. Then, the unusable and/or non-regenerable portion of spent catalyst, i.e., either due to unsuitable particle size or excessive deactivation, can be eliminated from the regeneration process. In this way, the efficiency of regeneration is improved. This is an incomparable advantage of the off-site regeneration compared with the in-situ regeneration. The off-site regeneration processes began to emerge in 1970s [451,452]. During the early stages, the off-site regeneration was performed in muffle furnaces. However, the lacks of uniform distribution of oxidizing medium and insufficient temperature control were the major disadvantages of this method. Under such conditions, occasional temperature excursions could not be prevented. The next stage of the off-site regeneration development involved a fixed bed reactor. In this case, a more uniform distribution of oxidizing media and the control of regeneration process associated with this was achieved particularly when the spent catalyst was prescreened prior to regeneration. Apparently, rotary kilns, which have been used in various industrial operations, have also been used for oxidative regeneration of spent hydroprocessing catalysts [446].

Significant improvements in the control of temperature and overall regeneration were achieved by introducing processes in which the regeneration is conducted with the spent catalyst in a continuous motion. The improved control of the process parameters which resulted in significant enhancement in the catalyst activity recovery, as it is demonstrated in Fig. 6.33 [448]. Thus, at least four utilization-regeneration cycles could be achieved using the off-site regeneration. Worldwide, there are three major companies namely, Porocel (previously CRI), EURECAT, and TRICAT that offer off-site catalyst regeneration services [7,453]. Each of these companies employs different regeneration technologies. Porocel uses fluidized bed combined with moving belt [454], Eurecat uses rotatory kiln [455], and TRICAT uses a fluidized bed [456] in their regenerators.

Presulfiding services after regeneration are also offered by these companies. In the case of some new generation diesel hydrotreating catalysts with high metals (Mo and Co or Ni) loading, additional chemical treatments of decoked catalysts with oxygen containing chelating agents are done to redisperse the metals [457,458]. Recently, Albemarle has introduced REACT technology that enables the activity of regenerated STARS catalyst to be recovered to over 90% when compared with fresh catalyst. It is claimed that Albemarle's REACT technology redistributes the metals and reconfigures the actives sites to the type II formation of the original STARS catalysts [458,113]. Criterion Catalyst Company has developed a similar revitalization technique for regeneration of their CENTINEL catalysts. Recovery of the activity levels very close to that of fresh catalyst activity has been reported in the revitalization technology [459,460]. In this regard, the best-known and most established commercial processes include the belt regeneration process developed by the Catalyst Recovery International (currently Porocel) [459] and a cylindrical drum arranged to rotate slowly on a horizontal axis, which has been developed and used commercially by EUROCAT [452,453,461].

A brief description of the strategy adopted by EUROCAT is given to illustrate the necessity of a preliminary spent catalyst evaluation prior to regeneration [462]. This strategy relies on the extensive database of the laboratory results and the results from commercial operation. This database may be used to predict the regeneration difficulty after the spent catalyst evaluation on a laboratory scale. For this purpose, a composite sample is split into three sub-samples for laboratory regeneration under different conditions. This is followed by the analysis of carbon and sulfur content as well as surface area. The regenerated samples are sulfided before performing a dynamic oxygen chemisorption, which is an indication of the level of catalyst recovery. A typical information obtained from these evaluations is shown in Fig. 6.22 [366–368] showing the effect of temperature on the carbon and sulfur content as well as that on the dynamic oxygen chemisorption and surface area of regenerated catalyst, respectively.

With aim to predict regeneration difficulty on a scale 1 to 10, attempts have been made to correlate various laboratory and pilot plant results [379]. In this regard, the best correlation was established when the regeneration difficulty was related to the total amount of metals in the catalyst. On the basis of this information, the final regeneration temperature for achieving an optimal activity recovery can be identified. Murff [461] reported that TGA and differential thermal analysis (DTA) techniques can also be used for predicting temperatures in a commercial regeneration plant. It is believed that a similar set of tests prior to commercial regeneration is conducted by other regeneration companies, although their methodology could not be verified in the literature. In the case of these companies, similar information may be considered proprietary.

6.4.3 Mechanical Separation of Spent Catalyst

The efficiency of regeneration may be improved if the catalyst particles of an unsuitable size are removed. Also, for some spent catalysts (e.g., from ebullated bed reactors), a variable level of deactivation has been noted. In this case, the spent catalyst represents a mixture of particles comprising those still possessing a high activity (e.g., nearly fresh) and heavily fouled fraction of particles that cannot be regenerated anymore. Because of its non-regenerability, it would make little sense to subject the latter fraction to regeneration. Figure 6.35 [462] shows the mass balance of spent catalyst particles separated to three fractions, i.e., heavy catalyst, medium catalyst and light catalyst fractions, using a density grading method. This information was the basis for a preliminary design of a rejuvenation process having a capacity of about 6000 tons of spent catalyst annually. According to this scheme, only light catalyst would be subjected to regeneration (decoking). The medium and heavy fractions of the spent catalyst required rejuvenation (metal leaching) prior to decoking. Thus, the preliminary design represents the integration of the oxidative regeneration with rejuvenation involving metal leaching. Without such integration, only a relatively small portion of the spent catalyst could be regenerated to achieve a sufficient activity recovery unless the contaminant metals were leached out via rejuvenation.

Density grading method can separate particles of a similar size but with the density differences smaller than 10% [463]. The density increase may be caused by deposition metals, mainly vanadium. For example, in the fixed bed reactor used for upgrading a heavy feed, the front bed is contaminated by vanadium, whereas the end of the bed is not. After unloading, a mixture of both contaminated and uncontaminated fractions of catalyst is obtained. The former fraction can be easily separated from the uncontaminated fraction using density grading. The density grading may be applied either to spent catalyst or to regenerated catalyst. The former case may translate into a smaller amount of spent catalyst requiring regeneration by eliminating non-regenerable catalyst from the mixture.

The fines that are part of the spent catalyst mixture can be simply removed by screening providing that the spent catalyst was adequately de-oiled and dried on the refinery site.

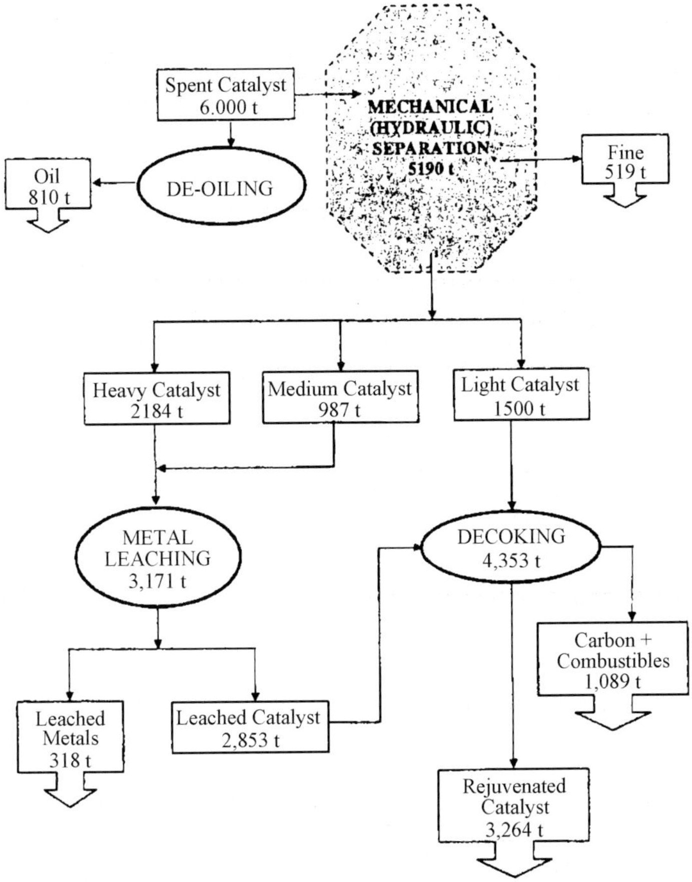

Figure 6.35: Flowsheet and mass balance involving mechanical separation, decoking, and rejuvenation steps [From ref. 462. Reprinted with permission].

Otherwise, such pretreatments have to be performed by regenerating company. In this case, de-oiling and mechanical separation processes may be integrated with regeneration process as it is shown in Fig. 6.35 [462]. The phase separator employing fluid mechanics, which is suitable for density grading spent catalyst particles, is shown in Fig. 6.36 [462]. Separation depends on the particle terminal velocity, which is a function of the physical and transport properties of the fluid and catalyst particles. By adjusting a fluid flow for each compartment, the catalyst particles with different densities are separated. For the steady flow of fluid past the moving solid, boundary layers are established. The force exerted on the solid is a combination of the boundary layer drag, form drag, external force, and buoyant force. The rate of separation of phases settling in a gravitational field is usually limited by the rate of fall of the smallest particles with residence time, which is important in determining the size of unit. In an ideal case, the solvent, which was used for de-oiling, would also be used for separation. This would avoid necessity for drying the de-oiled catalyst. The density grading can be successfully

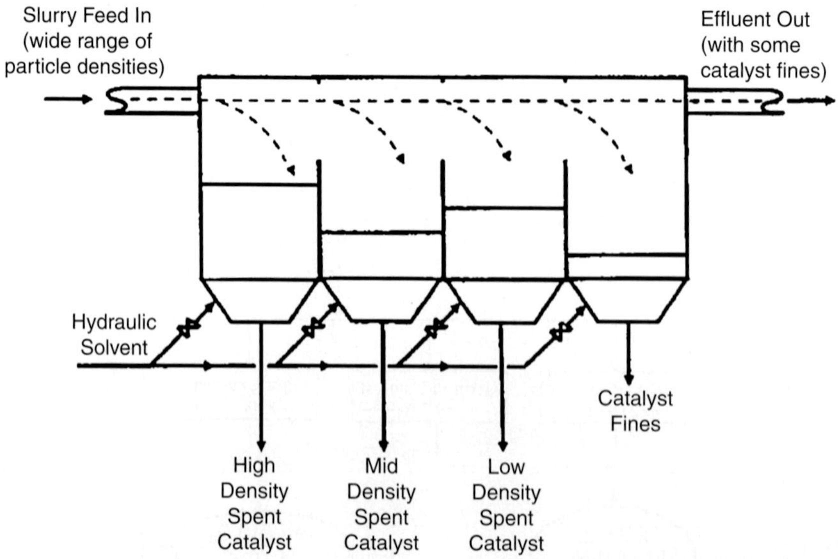

Figure 6.36: Schematic of gravity settling unit for catalyst separation [From ref. 462. Reprinted with permission].

applied to hydroprocessing catalysts of varying shapes and size particles as well as to the FCC catalysts [464].

The effectiveness of density grading is enhanced if the particles are of a uniform size. For extrudate shape of particles, the length grading may be necessary prior to density grading. The length–density grading according to vanadium content has been successfully applied [465]. This was based on the observations that the compacted bulk density and surface area of spent catalyst correlated with the level of vanadium contamination. The results of length grading of the spent $NiMo/Al_2O_3$ catalyst are shown in Table 6.11, whereas the particle distribution of spent catalyst as received and the fraction of 1.5–3.5 mm in Fig. 6.37 [465]. The 1.5–3.5 mm fraction accounted for more than 70 wt.% of the spent catalyst. The fraction of particles shorter than 1.5 mm had to be rejected before regeneration. The remaining spent catalyst was subjected to density grading to obtain four fractions, i.e., light, medium-light, medium-heavy,

Table 6.11: Yield and average length of particles after length grading [From ref. 465. Reprinted with permission].

Catalyst	Yield (wt.%)	Average length (mm)
Fresh	0	2.3
<1.5 mm	16	1, 2
1.5-3.5	71	2.3
>3.5	13	3.6

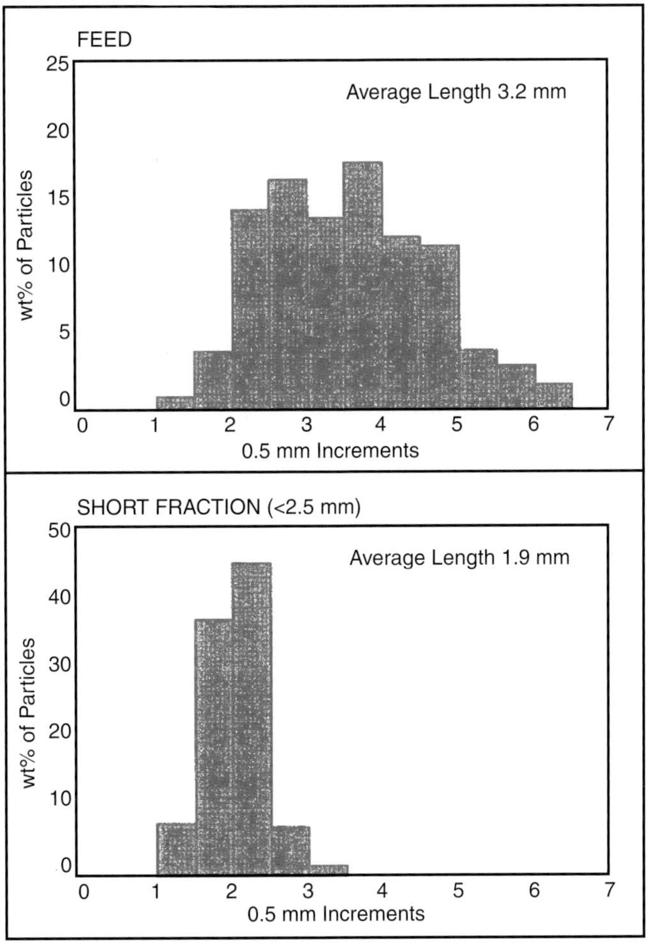

Figure 6.37: Length distribution of spent catalyst (feed) and its short fraction [From ref. 465. Reprinted with permission].

and heavy fractions. The properties of these fractions and relative HDS activity after their regeneration are shown in Table 6.12 [465]. It is evident that only about 28 wt.% of the spent catalyst could be regenerated to an acceptable level of catalyst activity. Therefore, the yield of the regenerable catalyst may be a factor to be considered when justifying regeneration after the length–density grading. Otherwise, all catalyst may be considered non-regenerable.

As it was indicated, screening is the simplest way of the size grading. In the case of one refinery, screening of regenerated catalyst was routinely performed after receiving the catalyst from regenerator [466]. Thus, an undesirable pressure drop development forced discontinuation of the operation unless regenerated catalyst was screened to remove fines.

Table 6.12: Results of density grading of NiMo/Al_2O_3 catalyst [From ref. 465. Reprinted with permission].

Catalyst	Yield (wt.%)	Compacted bulk density (lbs/cu ft)	Surface area (m^2/g)	Vanadium (wt.%)	RVA[a] (%)
Fresh	-	36	330	0	100
Composite	-	53	232	9.0	74
Light	28	44	293	4.7	84
Med-light	36	48	269	7.2	76
Med-heavy	16	52	233	10.0	69
Heavy	20	66	125	14.7	51

[a] *Relative volume activity (hydrodesulfurization [HDS]) compared with fresh catalyst.*

6.4.4 Commercial Regeneration Processes

A brief description of the best-known commercial regeneration processes as well as the emerging processes is the intention of this section. Thus, this is by no means an exhaustive review of the regeneration processes. There might be other processes operating on a commercial scale and/or near commercial scale, although their details are not available in an open literature. Apparently, some companies are reluctant to provide even brief information on their process. Therefore, a more detailed account of the industrial regeneration processes cannot be given.

6.4.4.1 Belt Regeneration Process

This process has been operated by Porocel (previously CRI). The schematics of the belt regeneration process are shown in Fig. 6.38 [467]. The process comprises of the de-oiling and regeneration sections. The former has a provision for sieving the spent catalyst to remove fines. The heat required for this step is supplied by the combustion of a fuel gas. The details of the process such as radiant tube burner, bed thermocouple and gas flow are shown in Fig. 6.39 [468]. To achieve an efficient oil removal, the entrance temperature of de-oiling section is maintained at 150–300 °C, whereas the exit at 300–330 °C.

The regeneration section of the belt regeneration process consists of a rectangular, stationary tunnel furnace, about 3 m high, 3 m wide, and almost 25 m long. A fine mesh stainless steel belt can continuously move inside of the vessel. The regeneration section is divided into four distinct zones. Each zone has a number of inlets for oxidation medium located above and below the belt. The spent catalyst is placed on the moving belt at a bed depth varying from 0.5 to 5 cm. As the catalyst moves through each zone, it is subjected to different O_2 concentrations and different temperatures. Usually, in the last zone of regeneration section, the catalyst is contacted with air. The heat to regeneration section is provided by the combustion of fuel gas

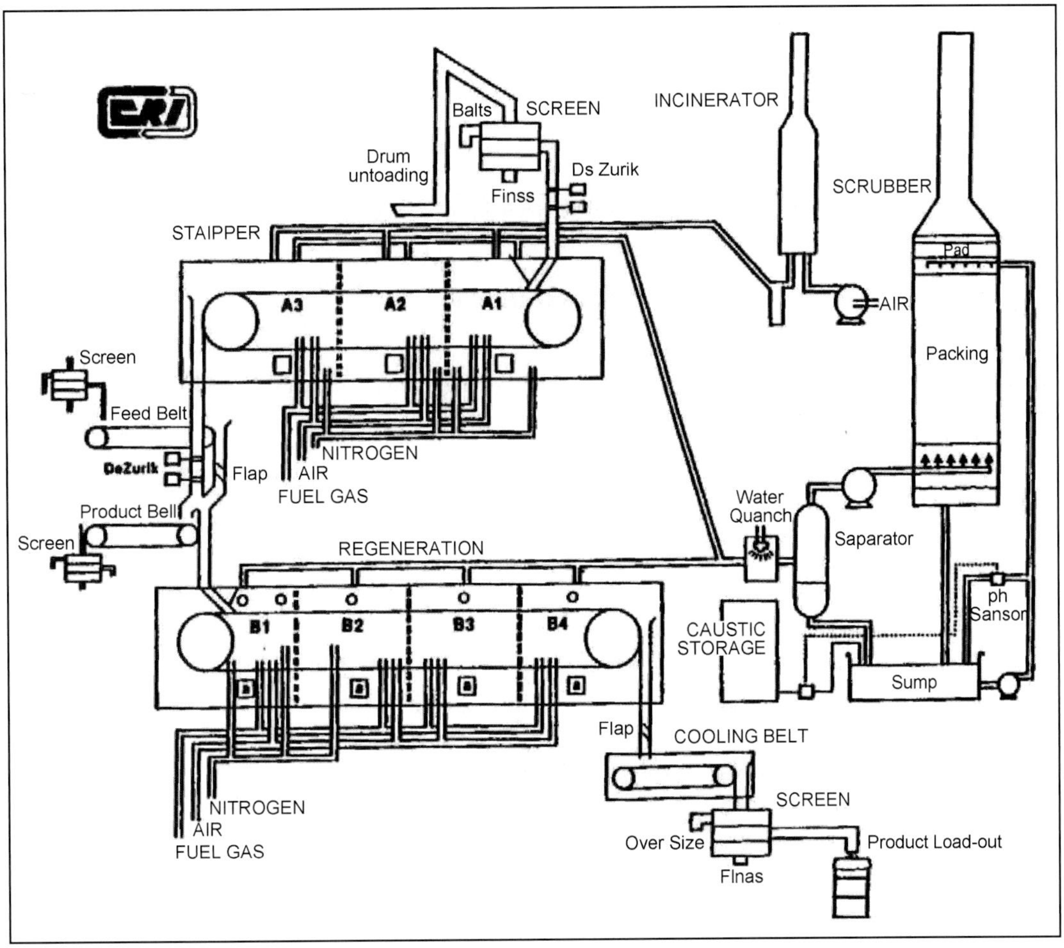

Figure 6.38: Schematics of CRI-MET regeneration process [From ref. 467. Reprinted with permission].

using four separate burners. This provides a flexibility required for temperature control. Catalyst bed temperature is continuously monitored at 30 different points. The thickness of catalyst bed, the speed of conveyor, air supply, and fuel input into the burners are all important parameters for maintaining a precise temperature control. If necessary, the nitrogen system is available for additional temperature control.

The movement of catalyst through the stripping and regeneration zones is more clearly viewed in the simplified flowsheet of the belt process shown in Fig. 6.40 [468,469]. Figure 6.40 also shows the integration of the moving-belt process with the fluidized bed pretreater, i.e., OptiCat-Plus reactor. This integration resulted in the enhancement in the overall efficiency of regeneration. The process consists of the fluidized bed reactor that strips hydrocarbons using

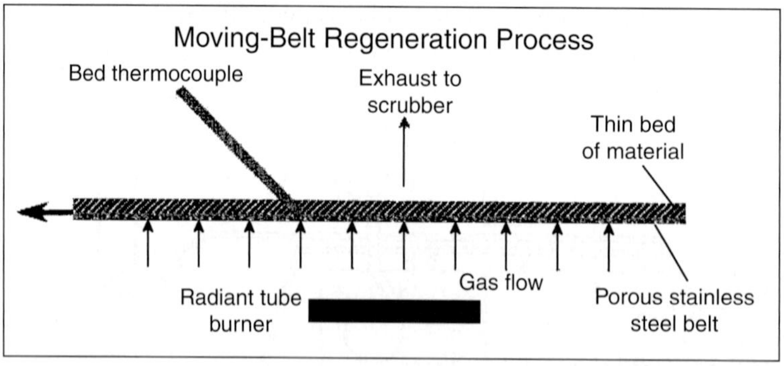

Figure 6.39: Internal details of belt-moving regeneration process [From ref. 468. Reprinted with permission].

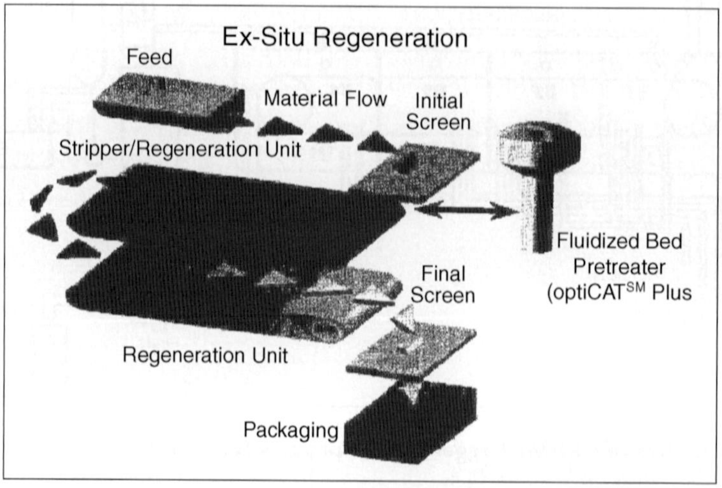

Figure 6.40: Movement of spent catalyst in belt-moving regeneration process [From ref. 468. Reprinted with permission].

air rather than an inert gas. Under these conditions, besides being stripped, the catalyst is also partially regenerated. At the same time, stripping under inert conditions results in the formation of additional coke on catalyst. The OptiCAT stage is in fact regeneration stage conducted in the chemically controlled regeneration region. A long heat soak removal of the remaining coke and sulfur is conducted in the regenerator. For most part, this stage of regeneration is in the diffusion controlled region requiring a longer residence time than that in the pretreater.

The emission control system must be an integrated part of any commercial regeneration process. This is evident from Fig. 6.38 [467] for both the stripper and regeneration sections.

The incinerator attached to the former section converts volatile matter removed during stripping. If necessary, the flue gas from incineration may be introduced to the scrubbing tower attached to regeneration section. The off gas from regeneration is water quenched before entering absorption tower (Packing) used for SO_X and NO_X emission control. The quenched water is neutralized in the sump by caustic supplied from the adjacent storage tank. To maintain pH necessary for scrubbing, the caustic can bypass sump and introduced directly to the top of tower. Spent caustic is considered a hazardous waste because of corrosive nature caused by pH exceeding 12. Also, some caustics may be reactive and on acidification can release H_2S [470]. Therefore, the spent caustic must be handled in accordance with environmental and safety regulations.

6.4.4.2 TRICAT Regeneration Process

Simplified flowsheet of the TRICAT regeneration process (TRP) is shown in Fig. 6.41 [471,472]. In this process, screened spent catalyst enters the first of two ebullated bed reactors shown in Fig. 6.42. The regeneration air entering at the bottom through distribution plate is used as the fluidization medium. The design of ebullated bed ensures that regeneration temperature is maintained below that observed in other commercial regenerators. This is attributed to the fluidization hydrodynamics of the bed. Air is supplied to each reactor independently. Temperature control is maintained through the number of independent variables such as catalyst feed rate, air temperature, cooling coils and catalyst level in the

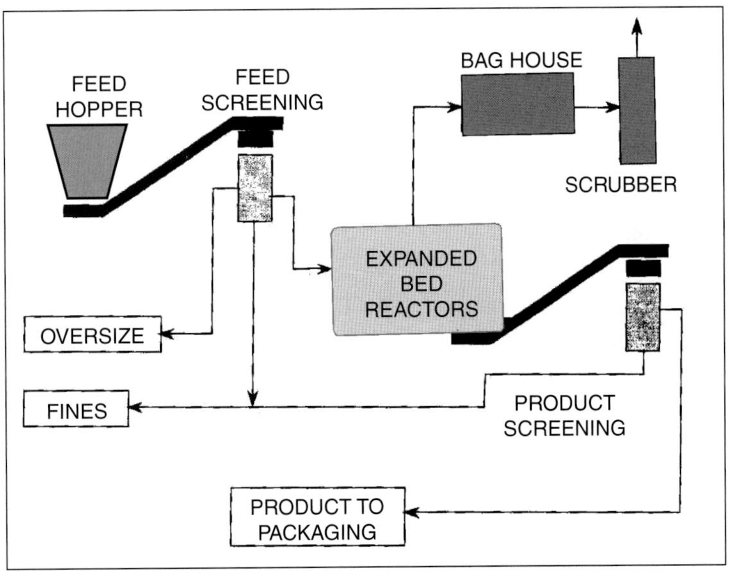

Figure 6.41: Simplified flowsheet of TRICAT regeneration process [From refs 471 and 472. Reprinted with permission].

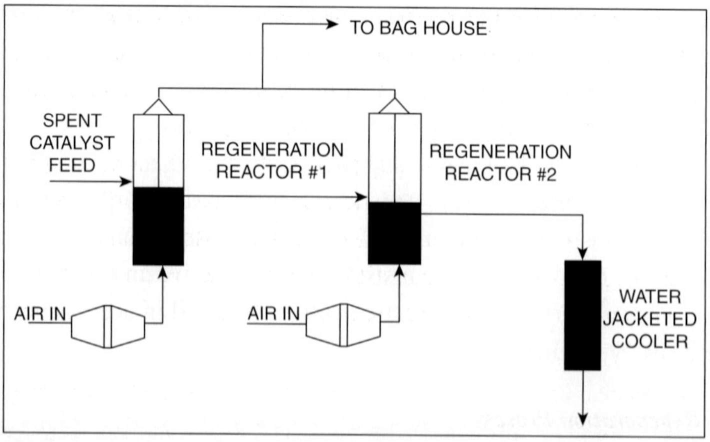

Figure 6.42: Ebullated bed reactors of TRICAT regeneration process [From refs 471 and 472. Reprinted with permission].

reactor. The regeneration temperature is maintained in the range of 450–510 °C. Regenerated catalyst passes through the water-jacketed cooler before screening and packaging.

The stripper (Fig. 6.43) [471], which is part of the TRP process, is usually engaged to achieve a uniformity of regeneration. Without stripping, an erratic performance and poor temperature control could be encountered. The aim is to remove the excess water and hydrocarbons. This is

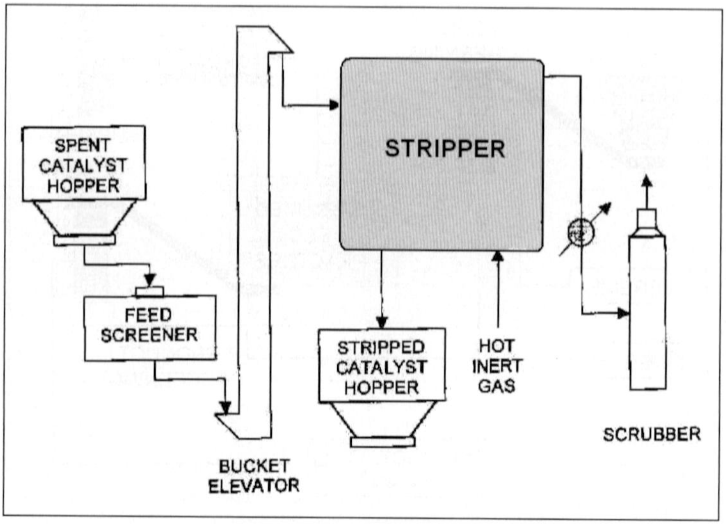

Figure 6.43: Catalyst stripping system of TRICAT regeneration process [From ref. 471. Reprinted with permission].

achieved by passing hot inert gas (e.g., N_2 or CO_2) across the downflowing stream of spent catalyst. This permits higher regeneration rates and reduced residence time. Stripping prior regeneration improves temperature control and fluid properties of the catalyst.

A bag house for removing dust during regeneration is part of the TRP process (Fig. 6.41) [471]. To further ensure environmental acceptance, the dust free gas is then scrubbed to remove SO_x generated during the oxidation of sulfur in spent catalyst. Only after scrubbing, the gas is released to the atmosphere. The scrubber solution is neutralized and then discharged. The composition of the sludge removed from the scrubbing solution approaches that of gypsum. Therefore, it is classified as a non-hazardous solid.

To provide complete service to petroleum refineries, the new XpresS presulfiding process was added to the TRP process [473]. The catalyst presulfided using the former process exhibits a high stability in air and its activity approaches that of the in-site presulfided catalyst. With this catalyst, the start-up of the reactor operation was greatly accelerated without an isotherm being observed.

6.4.4.3 EUROCAT Process

The central part of the EUROCAT process shown in Fig. 6.44 [448] includes a continuous rotolouvre, which is a sophisticated cylindrical drum rotating slowly on a horizontal axis and enclosing a series of overlapping louvres. The louvres extend the full length of the drum (~11 m) to form a conical inner shell, which increases in diameter toward the discharge end

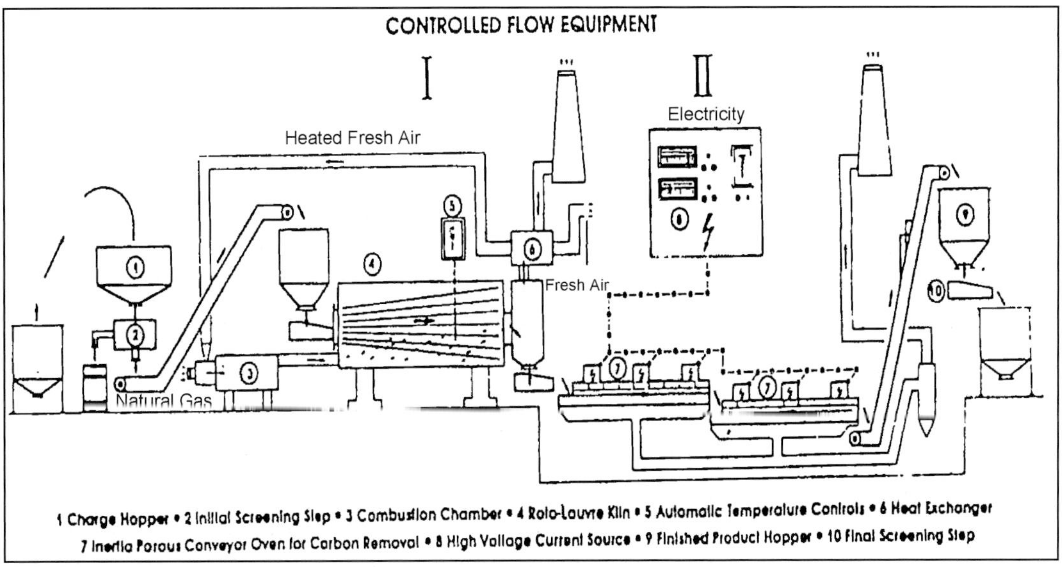

Figure 6.44: Schematics of EUROCAT continuous off-site regeneration process [From ref. 448. Reprinted with permission].

(2 to 2.5 m). The louvres supports are continuous radial partitions that form longitudinal passages extending the entire length of the drum. The catalyst to be regenerated is contained on the louvres and moves toward the discharge end by rotation of the drum. Hot air produced in an external, independent combustion chamber passes through the spaces between the louvres and permeates a thin layer of catalyst bed. A composition and high velocity of gases combined with the rotation ensuring the movement of catalyst prevent the development of hot spots. In addition, the countercurrent of the catalyst and oxidation medium are favourable for a steady monitoring the temperature. Other important operating parameters include depth of catalyst bed, feed rate of catalyst residence time. All these parameters can be varied to contain temperature with acceptable limits. Under these conditions, a high recovery of catalyst activity is ensured. Similarly as for the CRI (belt) process, all units required for the control of gaseous and solid emissions are also integrated with the EUROCAT process.

6.4.4.4 REACT Process

This process was developed by Albemarle Catalysts for activity recovery of their STARS portfolio catalysts with type II active sites. Some features of this process were discussed above in the section on reactivation of spent hydroprocessing catalysts [430–432]. An Albemarble catalyst used for hydroprocessing VGO was regenerated using conventional oxidative regeneration method as well as reactivated using the REACT process [474]. In the former case, the HDN activity of 60 to 70% was achieved compared with 92 to 97% using the STAR process and/or about 10 and 2 °C activity loss for conventional regeneration and that followed by REACT process, respectively.

6.4.4.5 ReFRESH Process

This process was developed and commercialized by Haldor Topsoe [433]. The details of this process are not available although the simplified schematic in Fig. 6.45 indicates that it may involve an oxidative regeneration in the first step followed by a reactivation presumably involving chelating agents. In the first step, the activity recovery reached 80–85 % of the fresh catalyst activity. The further enhancement in activity to more that 95% achieved in the second stage suggests that the refiner can use the same catalyst at least in two cycles.

6.4.4.6 Rotary Kilns

Rotary kilns have been used in various industrial applications (e.g., oil shale retorting, tar sands coking, incineration, cement production, etc.). The rotation of a cylinder-shaped vessel positioned longitudinally approximately 30° of the horizontal position ensures a continuous motion of catalyst between the entrance and exit of the kiln. With regard to the spent catalyst regeneration, the description of rotary kilns was given by Ellingham and Garrett [451]. There are two types of rotary kilns, i.e., direct fire and indirect fire.

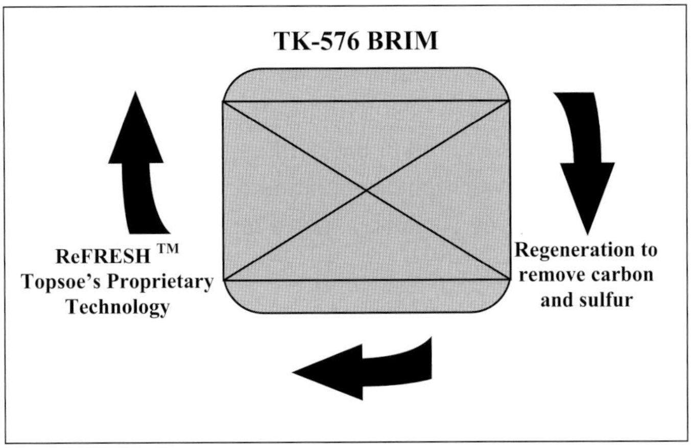

Figure 6.45: Simplified flowsheet of ReFRESH regeneration process [From ref. 433. Reprinted with permission].

The direct fire is a single shell vessel with rings added inside to slow the catalyst as it tumbles from the inlet (elevated part) towards outlet (lower part). The oxidation medium flows countercurrent to catalyst movement. The O_2 concentration in the medium will decrease in the same direction because of its consumption. Therefore, the zone in the vessel located near the inlet may function as a stripper of volatile components of coke. The kiln is fired by gas burners directly against the outer shell of the vessel. The temperature inside the kiln is controlled by adjusting the burner heat, varying concentration of O_2 in the oxidizing medium and its flow. The indirect fire kiln comprises a double-shell cylinder vessel. The inner shell is similar as that of the direct fire kiln. The space between the shells is heated either by combustion gas or steam. In some cases, the inner cylinder shell is ebullated allowing hot gases or steam to enter and contact the tumbled catalyst. The catalyst temperatures are controlled by monitoring the temperatures of the inlet and outlet gases. It is believed that Eurocat process evolved from a rotary kiln process by be improving the control of operating parameters such as temperature, gas flow, speed of rotation, etc.

6.4.5 Comparison of Regeneration Processes

Obviously, at this point, it would be desirable to provide some comparison of commercial processes that have been used for regeneration of spent hydroprocessing catalysts. Unfortunately, this is prevented because of the limited information on details of the individual processes available in open literature. It is believed that differences in the operating costs may be more evident than that between the efficiency of regeneration processes. The information on quantities of consumables, i.e., gases, electricity, chemicals, etc. as well as emissions such as sludges, dust, waste water, and gaseous emissions would be necessary to make estimates of regeneration efficiency. Such information is generally considered as proprietary.

With respect to the choice by petroleum refineries, the location of Regeneration Company may be a decisive factor. Thus, refiner has to deal with fewer environmental and safety issues in the case, that regeneration company is located in the refinery proximity. Accordingly, some regeneration companies either relocated their operation or constructed a new plant in the regions with large petroleum refining capacities, i.e., a large volume of spent catalysts requiring regeneration, available.

CHAPTER 7

Rejuvenation

It is again emphasized that the removal of both contaminant metals and coke with the aim to recover activity of the catalyst is the principal objective of rejuvenation. This differs from regeneration in which case only coke removal has been the objective. Most of the methods for catalyst rejuvenation are based on selective leaching of contaminant metals from spent catalysts. At the same time, all efforts are made to ensure that the leaching of active metals is kept at a minimum. The non-leaching methods have been receiving attention, although to a much lesser extent.

Figure 7.1 [462] describes the strategy for rejuvenation of spent catalysts developed by Kuwait Institute for Scientific Research (KISR). It should be noted that it is there where the only of its kind database on various aspects of spent hydroprocessing catalysts has been established with primary focus on rejuvenation. It is noted that some steps shown in Fig. 7.1 are common for regeneration, e.g., starting with de-oiling up to mechanical separation of particles. The proposed strategy took into consideration three options described briefly as:

- Option I, where spent catalyst is de-oiled by washing with kerosene in the first step. Then the oil-free catalyst is mechanically separated into catalyst fines (<0.5 mm particles), as well as lightly, medium, and heavily fouled portions. The medium dense and heavy portions are subjected to chemical treatment to remove metals by leaching. Subsequently, the leached catalysts are mixed with the lightly fouled catalyst portion and decoked by controlled burn-off to produce rejuvenated catalyst.
- Option II
  After de-oiling and separating the spent catalyst into three groups as in Option I, the heavily fouled portion is sold to metal reclamation companies, while the medium fouled portion is subjected to leaching. The leached catalyst and lightly fouled catalyst are decoked to produce rejuvenated catalyst.
- Option III
  De-oiling and grouping is performed as in Options I and II. The heavy and medium fouled portions are sold to metal reclaimer, whereas lightly fouled catalyst is decoked under conditions approaching those employed during the oxidative regeneration.

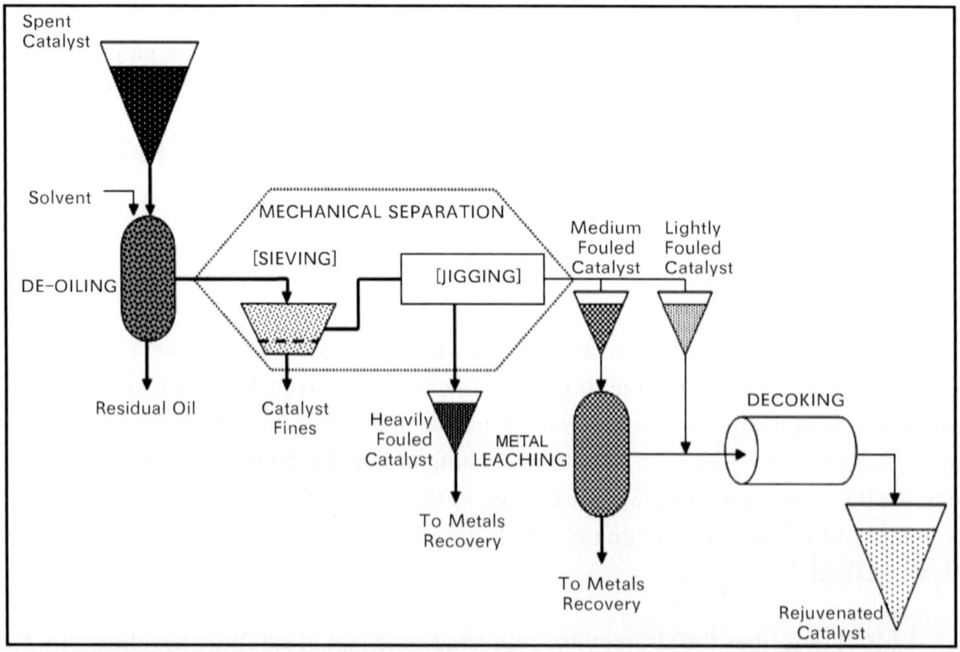

Figure 7.1: Options and process for spent catalyst handling and utilization [From ref. 462. Reprinted with permission].

It has been indicated above that the primary objective of rejuvenation is the removal of metals deposited on the catalyst surface during the operation, while leaving the active metals intact. Ideally, if contaminant metals could be removed selectively by chemical treatments without significantly altering the chemical and physical characteristics of the original catalyst, then spent catalysts could be rejuvenated and reused in the operation. It is also more advantageous to conduct rejuvenation using spent catalyst particles in their original size and shape, rather than using an altered form (e.g., a pulverized form). In the latter case, a reprocessing of the solid material would be necessary to obtain an operating size and form of catalyst. It was indicated earlier that in the conventional hydroprocessing catalysts the active metals exist in the Co(Ni)-S-Mo(W) phase, although under hydroprocessing conditions, the presence of other phase cannot be excluded. At the early stages of operation, the active phase accounts for most of the metals on the support. With time on stream, inorganic solids in the feed as well as the V and Ni produced during hydrodemetallization (HDM) reactions are deposited on catalyst surface. The active phase metals together with these deposits represent rather complex mixture, particularly in the later stages of the operation. This may include sulfides and mixed sulfides of V, Ni, Fe, and other metals. To certain extent, Co and Ni promoters in the active phase may be replaced by V and Fe and, as such, decrease the catalyst activity. The objective of rejuvenation is to liberate the active phase from the unwanted deposits in the mixture.

Moreover, because of the deposition in pores, the metals have to be removed to approach, as much as possible, the original porosity of catalyst. This appears to be rather challenging task. It should be noted that several methods, employing a wide range of leaching agents, have been tested for the removal of contaminant metals with a varying degree of success.

Besides removal of the unwanted metals, catalyst has to be decoked by oxidative burn-off. The latter may be performed either prior to decoking or on the decoked catalyst. General observations confirm that the removal of metals by various solvents from spent catalysts can be accomplished more readily on decoked catalyst. However, the selectivity of leaching out contaminant metals may be more favorable in the presence of coke. Therefore, for rejuvenation purposes, the spent catalyst still deposited by coke has been usually employed. Nevertheless, it is desirable that during decoking the temperature is carefully controlled to prevent sintering, which would affect the availability of active metals on catalyst surface during subsequent utilization cycle. Some studies indicate that the extraction affected mechanical properties of the rejuvenated catalyst [414]. Contrary with this observation, numerous studies confirmed that desirable mechanical properties of rejuvenated catalysts could be established on rejuvenation [475–478].

A wide range of solvents has been used for the metal extraction from spent hydroprocessing catalysts. Both organic and inorganic solvents have been evaluated. The former ensure the non-corrosive environment, requiring much less safety precautions than that which has been present during the leaching using inorganic agents. Moreover, their leaching efficiency can be increased and/or modified by various additives with the aim to enhance the selectivity for removal contaminant metals.

At KISR, the world leader in research on various aspects of rejuvenation, considerable effort has been made as part of the research program on handling and utilization of spent hydroprocessing catalysts [462]. Factors influencing the selectivity for removal of major foulant metals (e.g., V and Ni) from spent catalysts were investigated [475–478]. Improvements in the essential surface properties such as surface area and pore volume as well as the activity for hydrodesulfurization (HDS) and HDM of atmospheric residues after metal leaching were achieved. Extensive database established in the bench scale unit was used for a preliminary design and economic assessment of the commercial rejuvenation process.

7.1 Organic Agents

This method of metal extraction is based on the pioneering work of Beuther and Flynn [479] who recognized that some organic agents were capable of forming complexes with the transition metals, which are part of the hydroprocessing catalysts. Rather small concentration of acids, i.e., ~1 wt.%, was sufficient for removing substantial amount of the metals from spent catalysts. The tested organic agents included acids such as oxalic, lactic, citric, glycolic, phthalic, malonic, succinic, and salicylic as well as acetylacetone, ethylenediamine, *o*-aminophenol, and slicylaldehyde. Among these agents, acids were much more efficient leaching agents than non-acid compounds.

The leaching with organic agents has been significantly advanced by the researchers at KISR, who established the following order of the leaching efficiency of the most efficient agents: oxalic > malonic > acetic acid. In addition, combination of these acids with various oxidizing agents such as H_2O_2, HNO_3, $Al(NO_3)_3$, $Fe(NO_3)_3$, etc. were evaluated as well [475,480–482]. The effect of oxidizing agent differed from acid to acid. For example, when added to oxalic acid (0.66 M), $Al(NO_3)_3$ (1 M) had a dramatic effect on leachability of V and Ni, whereas for malonic and acetic acids the effect was much less pronounced [483].

As the most efficient agent, the oxalic acid (0.66 M) in the mixture with H_2O_2 (0.66 M) was used for rejuvenation of the spent catalyst obtained from the first stage of the atmospheric residue desulfurization (ARDS) process used for hydroprocessing of the Kuwait atmospheric residue [476,483]. The composition and surface properties of the fresh and corresponding spent catalysts are shown in Table 7.1 [483]. Leaching experiments were performed in the continuous upflow reactor at 298 K using both decoked and the spent as received catalysts. The results of the experiments are shown in Table 7.2 [475]. The selectivity for the V removal was much higher for non-decoked catalyst. The trends in the recovery of surface area and pore volume with the amount of V leached out are shown in Fig. 7.2 [484]. They indicate that these parameters, surface area in particular, improve significantly before about 35% of V was removed. As it is shown in Fig. 7.3 [476], the HDS activity recovery followed the same trends as that of the surface area. The HDS activity data of the rejuvenated catalysts were obtained in the continuous fixed bed reactor system at 623 K and 4.0 MPa. For the activity estimate, the AGO containing about 2 wt.% of sulfur was used as the feed. After removing ~30 wt.% of V,

Table 7.1: Chemical composition and physical properties of the fresh and spent catalysts [From ref. 483. Reprinted with permission].

Property	Fresh catalyst	Spent catalyst
Surface area (m^2/g)	240	52
Pore volume (mL/g)	0.48	0.12
Bulk density (kg/L)	0.73	1.18
Mo (wt.%)	8.80	5.40
Co (wt.%)	3.20	1.90
Ni (wt.%)	0	3.07
V (wt.%)	0	14.90
Fe (wt.%)	0	8.00
Carbon (wt.%)	0	15.60
Sulfur (wt.%)	0	5.30

Table 7.2: Surface area, pore volume, and HDS activity of fresh, spent, and rejuvenated catalysts [From ref. 475. Reprinted with permission].

Catalyst	Agent	Surface area (m^2/g)	Pore volume (mL/g)	HDS activity	
				%	% Recovery
Fresh		240	0.40	61	100
Spent		52	0.12	17	28
Method A	Oxalic	63	0.15	22	34
Method A	Oxalic + H_2O_2	141	0.27	52	85
Method B	Oxalic + H_2O_2	181	0.43	59	97
Method C	Oxalic + H_2O_2	197	0.46	36	59
Method C	Oxalic	180	0.42	37	61

Method A: leached but not decoked; method B: leached and decoked; method C: decoked and leached.
HDS: hydrodesulfurization.

the HDS activity approached 85% of the activity of the fresh catalyst in spite of the coke still being present on the catalyst. On the basis of the extensive testing program, which has been conducted at KISR, one may conclude that the catalyst rejuvenation using organic agents is approaching a commercial stage [475,476,480–485]. As the consequence, the rejuvenated catalyst may be cascaded for less severe applications and/or at least used as the solid for guard beds. In both cases, economics are a decisive factor.

Apparently, leaching with oxalic acid can be further optimized by varying concentration of the acid and temperature. Figure 7.4 [480] shows that at 25 °C, increasing the concentration of oxalic acid from 0.33 to 1.0 M had little incremental effect on leaching of V. The 0.33 M concentration represents about 3 wt.% solution, whereas in the study of Beuther and Flinn [479] 1 wt.% solution was successfully tested. The wide range of concentration suitable for leaching offers flexibility when optimizing other leaching parameters. For example, at concentration of 0.66 M, increasing temperature from 25 to 50 °C about doubled the amount of V leached out (Fig. 7.5) [480].

McVicker et al. [486] reported that buffering of oxalic acid solution at varying concentrations enhanced significantly selectivity for leaching contaminants metals. At the same time, with buffered solution, little removal of catalytically active metals from catalyst was observed in spite of long contact times of oxalic acid solution with contaminated catalyst even at elevated extraction temperatures.

Several combinations of the organic acids with different agents were also evaluated. In agreement with results in Table 7.2, the results in Table 7.3 [475] show that the leaching efficiency can be enhanced in the presence of H_2O_2. This was attributed to the oxidation of the

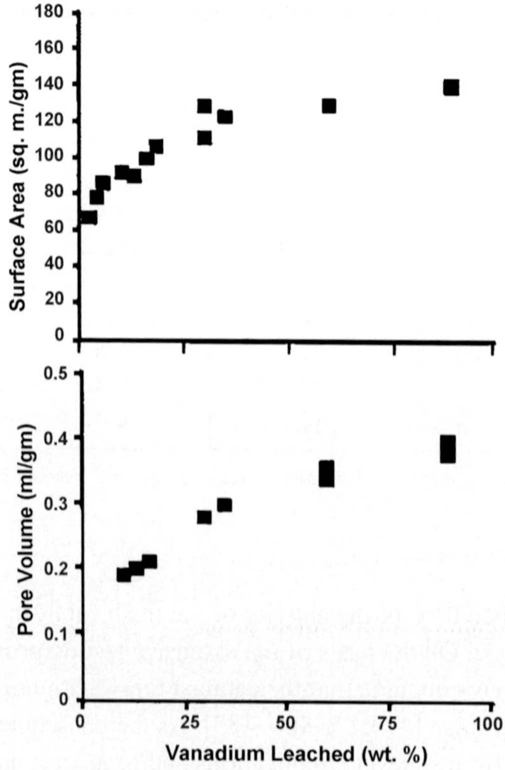

Figure 7.2: Effect of vanadium leached out on surface area and pore volume [From ref. 484. Reprinted with permission].

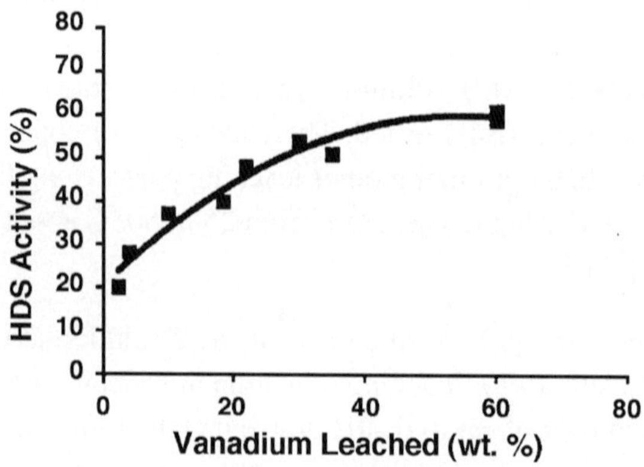

Figure 7.3: Effect of vanadium leached out on hydrodesulfurization (HDS) activity [From ref. 476. Reprinted with permission].

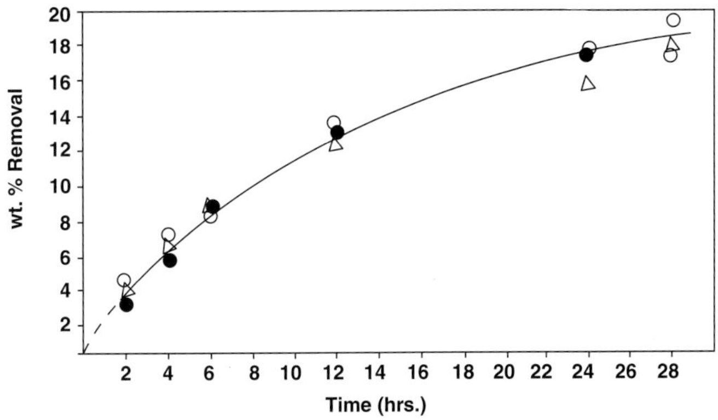

Figure 7.4: Effect of oxalic acid concentration on vanadium removal at 25 °C. Δ: 0.33 M; ○: 0.66 M; ●: 1.0 M [From ref. 480. Reprinted with permission].

metal sulfides in spent catalysts to corresponding oxides [484]. The presence of latter is more favorable for the formation of the water-soluble complexes with organic agents. Furthermore, the surface area and pore volume improvement as well as the increased recovery of activity were observed in the presence of $Al(NO_3)_3$ [476]. Similar observation was made by Farrell and Ward [487] using the mixture of oxalic acid with HNO_3 and $Al(NO_3)_3$. Compared with H_2O_2, $Al(NO_3)_3$ was more efficient additive for the removal of Ni, whereas for V removal

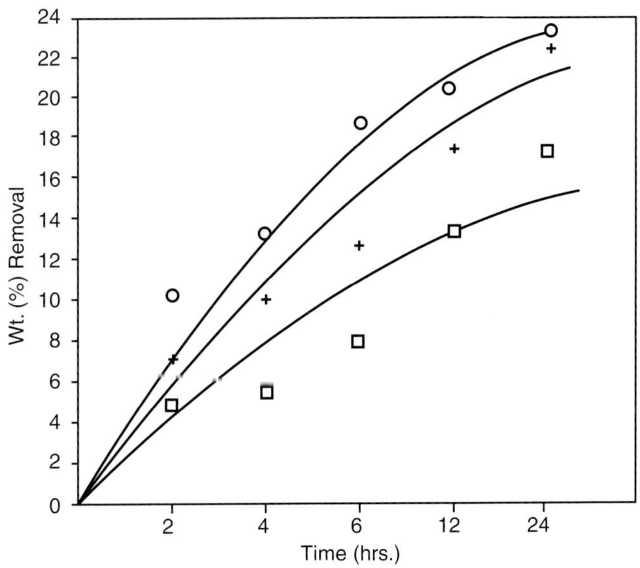

Figure 7.5: Effect of temperature on vanadium removal using oxalic acid (0.66 M). □: 25 °C; +: 50 °C; ○: 75 °C [From ref. 480. Reprinted with permission].

Table 7.3: Effect of H_2O_2 on leaching of Ni [From ref. 475. Reprinted with permission].

	Amount of Ni leached out (wt.%)	
	Coked catalyst	Decoked catalyst
Oxalic acid	3.6	80.3
Oxalic acid + H_2O_2	47.0	76.5

both additives had a similar effect [485]. However, as Fig. 7.6 shows, for both V and Ni, the most efficient system comprised the mixture of $Fe(NO_3)_3$ + oxalic acid. However, with this system, the unwanted removal of Mo occurred as well. The $Fe(NO_3)_3$ was more efficient than $Fe_3(SO_4)_2$ presumably because of the higher oxidation strength of the nitrate group [488]. It

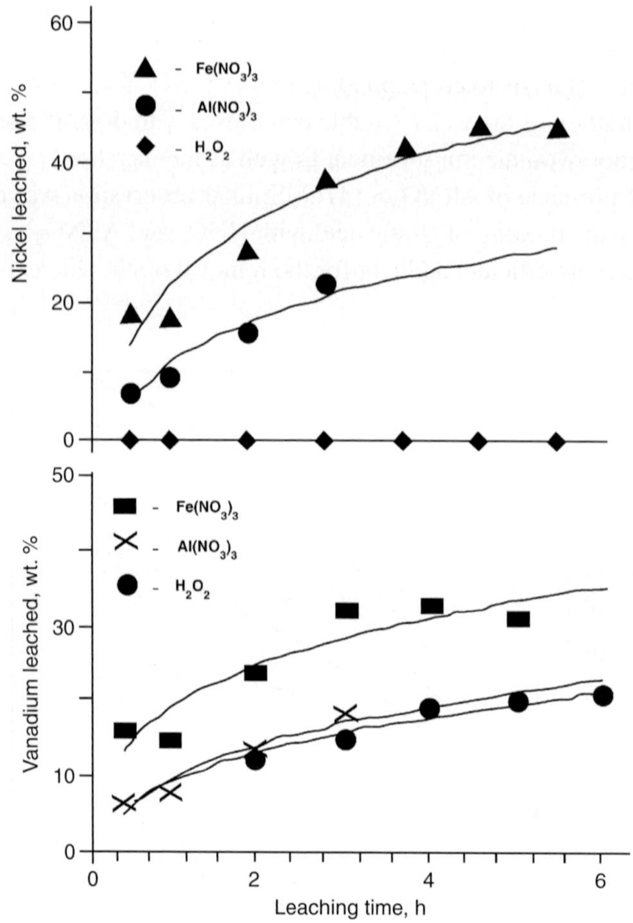

Figure 7.6: Effect of additives on leaching efficiency of oxalic acid [From ref. 488. Reprinted with permission].

was postulated that HNO_3 participated in leaching reactions when metal nitrates were used as oxidation agents with oxalic acid [476]. Apparently, the acid was released according to the following reaction:

$$2Fe(NO_3)_3 + 3(COOH)_2 = Fe_2\{(COO)_2\}_3 + 6HNO_3$$

In similar manner, HNO_3 can also be released from $Al(NO_3)_3$. This suggests that the control of pH of the leaching solution may be necessary to avoid excessive release of HNO_3. Otherwise, a desirable selectivity for leaching contaminant metals could not be maintained.

The $Fe(NO_3)_3$ + oxalic acid system was evaluated in the fixed bed and ebullated bed pilot plants [485]. The latter system was specially designed for leaching experiments. The amount of V leached out was significantly greater in the ebullated bed. Moreover, three modes of the $Fe(NO_3)_3$ addition to oxalic acid, i.e., continuous, successive, and batch additions, were tested [477]. During these experiments, the mixture of oxalic acid (0.66 M) and $Fe(NO_3)_3$ (0.66 M) was pumped continuously through the fixed bed of spent catalyst (Table 7.1) upwards. The liquid was collected in the reagent vessel and recirculated through the bed. The results in Fig. 7.7 show that with progress of leaching, continuous addition gained in leaching efficiency relative to batch and successive additions. This was complemented by the largest recovery in pore volume and surface area as well as the removal of V. Figure 7.8 compares the HDS activity of the catalysts rejuvenated via three different modes. Almost 85% recovery in HDS

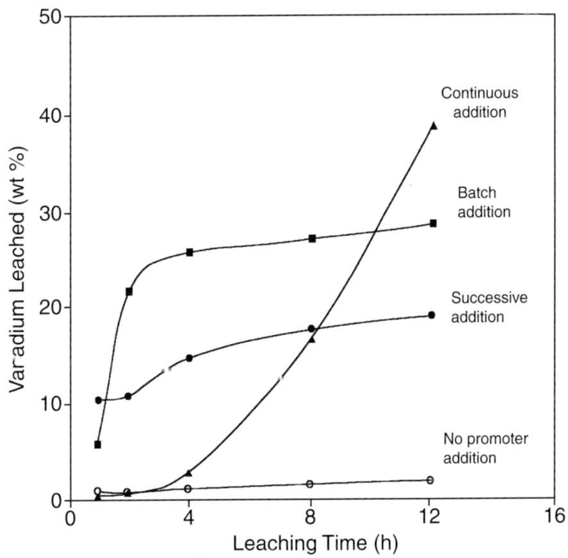

Figure 7.7: Vanadium leached versus leaching time for different modes of promoter addition [From ref. 477. Reprinted with permission].

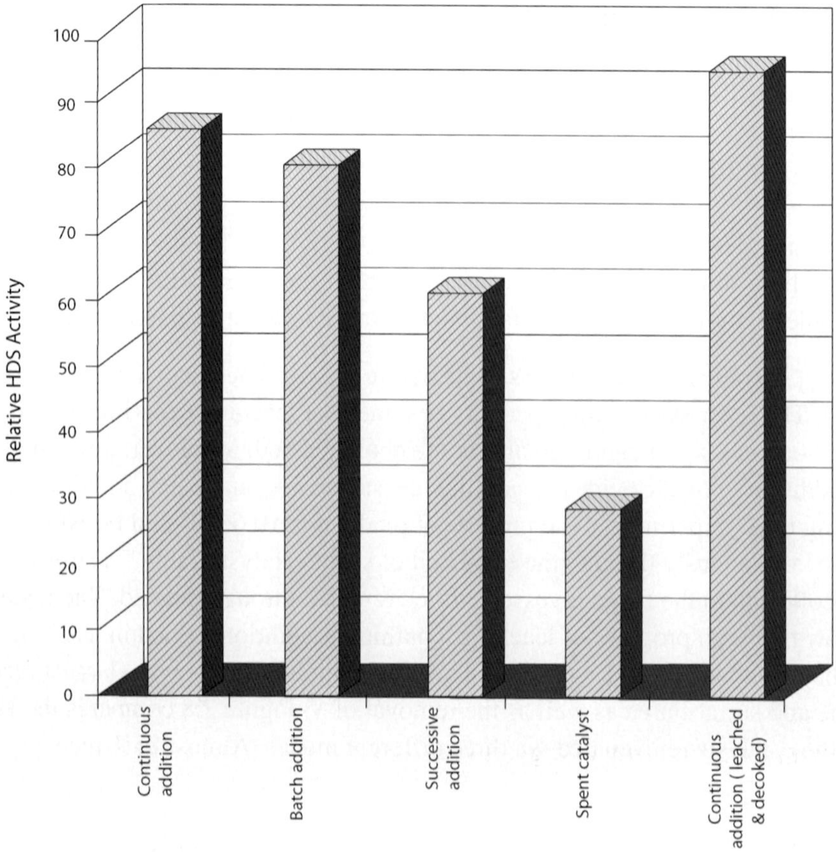

Figure 7.8: Effect of different modes of promoter addition on hydrodesulfurization (HDS) activity [From ref. 477. Reprinted with permission].

activity was achieved using the continuous mode. The activity further increased to more than 90% after decoking. In contrast to the report by Clark et al. [414], the rejuvenated catalyst only lost about 5% of the side crushing strength compared with the fresh catalyst [476].

Some evidence suggests that there are other Fe ions containing systems, i.e., the redox couple such as Fe^{2+}/Fe^{3+} which when added to an organic acid, may enhance both leaching efficiency and particularly the selectivity for V [489]. This was evident for oxalic acid in particular. For example, without redox cycle in 1% oxalic acid, the removal of Mo and V was 34 and 48%, respectively, whereas with redox cycle (10^{-2} M), it was 5 and 34%, respectively.

There might be some relation between rejuvenation and reactivation of spent catalysts. The details of the latter were given in the patent disclosed by Eijsbouts et al. [432]. In this case, after oxidative decoking under controlled conditions, the spent hydroprocessing catalyst is impregnated with organic agents (e.g., glycols) dissolved in a solvent. Beneficial effects of

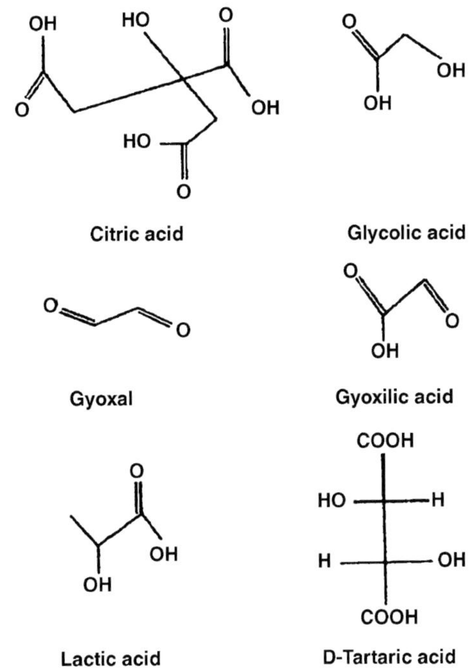

Figure 7.9: Structures of organic agents used for catalyst rejuvenation.

chelating agents on the activity of regenerated catalysts were evident after sulfidation. This method is discussed in the section on Reactivation.

7.1.1 Mechanism of Rejuvenation by Organic Agents

Ability of an organic agent to form a complex and/or a compound with contaminant metal ions either in solution or in contact with a solid is an essential requirement to be an efficient rejuvenation agent. The overwhelming information confirmed that this requirement is best fulfilled by di-carboxylic acids (e.g., oxalic, citric, tartaric, etc.), while monocarboxylic acid are less efficient, though more efficient than aldehydes and ketones. The structures of some organic agents are shown in Fig. 7.9. The dissociation of an acid to anion and hydronium ion precedes the formation of a complex with a metal in solution. For oxalic acid, this can be depicted as follow:

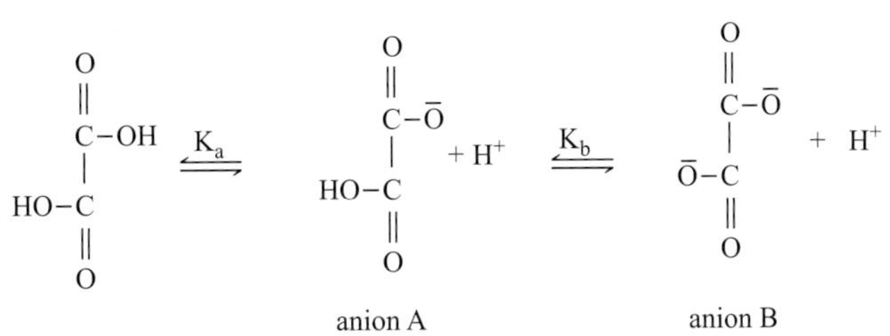

{7.1}

Table 7.4: Equilibrium constants of organic acids

Acid	K_a	K_b
Acetic	1.8×10^{-5}	
Butyric	1.5×10^{-5}	
Glycolic	1.5×10^{-4}	
Lactic	1.4×10^{-4}	
Tartaric	9.6×10^{-4}	2.9×10^{-5}
Citric	8.7×10^{-4}	1.8×10^{-5}
Malonic	1.4×10^{-3}	8.0×10^{-7}
Oxalic	6.5×10^{-2}	6.1×10^{-5}

Initially, the complex formation with metallic cation in the solution will involve anion A. In this case, it is assumed that metallic ions entered the solution by partial dissolution of solid. Thus, according to the K_a value (Table 7.4), the concentration of anion A will be in significant access compared with anion B. In the case of Ni^{2+}, the complex formation may involve the following tentative scheme:

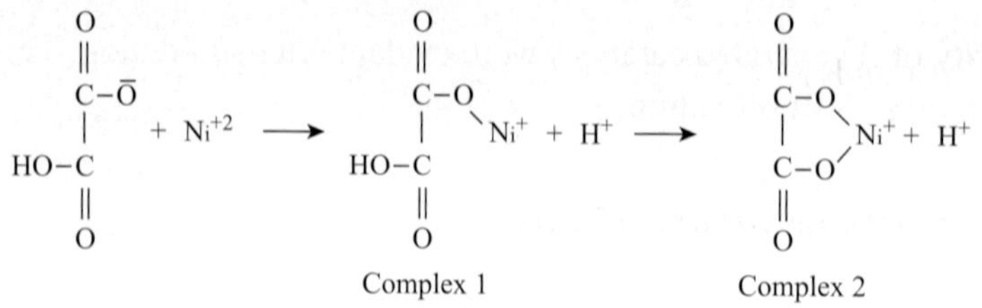

{7.2}

Complex 1 in this scheme represents a transition state before the final product (complex 2) is formed. It is believed that in the transition state the O–H bond of the non-ionized carboxylic group is weakened considerably. This results in the enhancement in the acidity of the group, i.e., H^+ is released from this group more readily. Therefore, as the K_b value (Table 7.4) indicates, the initial contribution of the completely ionized acid to the complex formation will be minor, although it may enhanced via formation of complex 2 from complex 1. Two anions A (Reaction {7.1}) can also be involved during complexation with metallic cations in solution, e.g.,

$$2(HOOCCOO^-) + Ni^+ \Rightarrow Ni(^-OOCCOOH)_2 \qquad \{7.3\}$$

The removal of anions via Reaction {7.2} shifts the equilibrium {7.1} to the right. Consequently, additional acid is ionized. Table 7.4 indicates the differences in the equilibrium constants of several acids. According to these results, the availability of anions A required for

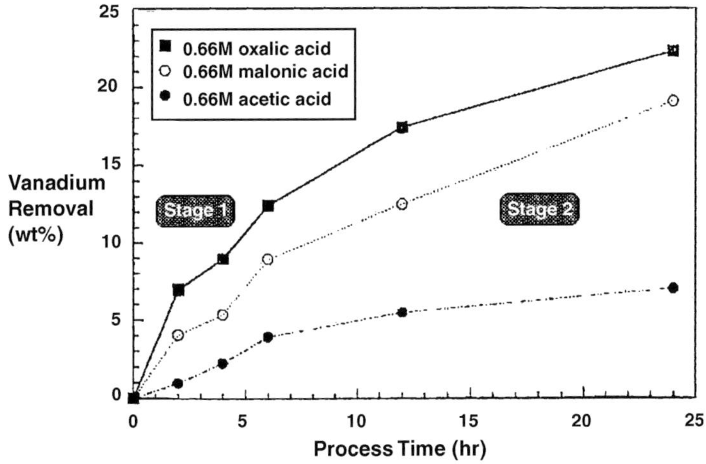

Figure 7.10: Effect of type of organic acid (0.66 M) on vanadium removal at 50 °C. ■: oxalic acid; ○: malonic acid; ●: acetic acid [From ref. 490. Reprinted with permission].

complex formation in solution will increase in the following order:

$$\text{Acetic} \sim \text{butyric} < \text{glycolic} \sim \text{lactic} < \text{tartaric} \sim \text{citric} < \text{malonic} \ll \text{oxalic}$$

The results on V removal in Fig. 7.10 [490] are in the agreement with these trends. The study of Reda [489] confirmed similar trends as well. However, different trends were observed for Ni removal [480–482]. For example, for Ni, malonic acid was much more efficient leaching agent than oxalic acid. According to Reaction {7.2}, two carboxylic groups are required to form a complex. Then, a complex can also be formed using two molecules of a partially ionized acid as shown in Reaction {7.3}.

It has been generally observed that the solubility of metal oxides in water is significantly greater than that of the corresponding metal sulfides. To a certain degree, beneficial effects of the oxidizing agents on leachability may be attributed to the conversion of a sulfidic form of metals to an oxidic form according to the following tentative reaction in which H_2O_2 is used as the model oxidant:

$$MeS_X + 2H_2O_2 \Rightarrow MeS_{X-1}O + SO_2 + H_2O \qquad \{7.4\}$$

$$H_2O_2 + SO_2 \Rightarrow SO_3 + H_2O \Rightarrow H_2SO_4 \qquad \{7.5\}$$

The oxidation process may continue until MeS_X is completely oxidized to MeO_X. The release of SO_2 in Reaction {7.4} should be noted. If sufficient concentration of H_2O_2 is maintained, SO_2 can be oxidized to SO_3 and subsequently converted to H_2SO_4 via Reaction {7.5}. The participation of the in-situ produced H_2SO_4 during leaching cannot be ruled out. However, when a flow reactor is employed, the H_2SO_4 build-up is prevented by its continuous removal from the system. Contrary to this, in a recirculation system, the participation of the in-situ produced H_2SO_4 can quite important. The involvement of the in-situ produced HNO_3 was reported by Marafi et al. [480] in the study using the mixture of oxalic acid with $Al(NO_3)_3$.

Besides reacting with anions in solution, the metal dissolution may proceed via liquid–solid reaction, which may involve a direct interaction of the non-ionized acid with metal oxide on the surface of catalyst, yielding a complex 2 shown in Reaction {7.2}. This reaction is always accompanied by the formation of H_2O. Such a dissolution may also involve two non-ionized molecules of acid. To depict these events, again oxalic acid and NiO are used as an example.

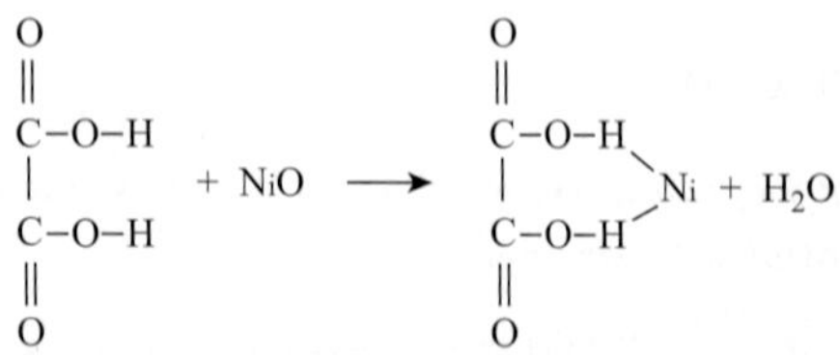

{7.6}

Compared with the other acids (Fig. 7.9), oxalic acid would be the least sterically hindered during the metal oxide dissolution. Moreover, an intramolecular rearrangement such as required during the formation of complex 2 from complex 1 in Reaction {7.2} is more easily attainable compared with more complex acids. Therefore, besides the strength of acid, the structure of organic acids may be an important factor while selecting the most efficient leaching agents. Of course, Reaction {7.6} may not proceed without the ionization of the acid in true sense, although this process is not depicted. Most likely, the final complex in Reaction {7.6} arose from a transition state in which the NiO aided weakening of O–H bonds resulted in the H^+ release while Ni^{2+} partially coordinated with the oxygen in OH groups. As the last step, H^+ coordinated with the O in NiO to form H_2O while the final complex in Reaction {7.6} was liberated from the transition state.

Reactions {7.1} to {7.3} assumed the presence of metallic species in a dissolved form in solution, whereas Reaction {7.6} in a solid oxidic form. This may approach the conditions for either decoked catalyst or a sulfided catalyst leached in the presence of an oxidizing agent. Leaching of a coked catalyst in the absence of an oxidizing agent may involve a similar set of reactions as proposed for the oxidic form of metals with MeS_X replacing MeO_X species [483]. Because of a low solubility of the former, the dissolution near or on the surface such as shown in Reaction {7.6} for the oxidic catalyst is expected to be much more pronounced and may proceed according to the following general equations:

$$4(COOH)_2 \dagger 4H^+ + 4(HOOC\text{-}COO^-) + MeS_2 \rightarrow Me(HOOC\text{-}COO^-)_4 + H_2S \qquad \{7.7\}$$

$$2(COOH)_2 \dagger 4H^+ + 2(^-OOC\text{-}COO^-) + MeS_2 \rightarrow Me(^-OOC\text{-}COO^-)_2 + 2H_2S \qquad \{7.8\}$$

Table 7.5: Stability constants of oxalic acid complexes [From ref. 480. Reprinted with permission].

Metal ion	Temperature (°C)	Stability constant (mol/L)
Co^{2+}	25	2×10^{-3}
Ni^{2+}	20	6.3×10^{3}
$MoO_2{}^{2+}$	25	39.8
VO^{2+}	25	3.2×10^{6}

The release of H_2S in Reactions {7.7} and {7.8} should be noted. This suggests that precautions have to be taken to ensure safety and environmental acceptance of the rejuvenation process.

The stability of metal complexes may influence leaching process. Generally observed, the reactions proceed towards the most stable product. Table 7.5 [480] shows the stability constants for metal complexes with oxalic acid. According to the values in Table 7.5, in the case of oxalic acid, it is much more easy to remove V than Ni, whereas the stability of the $Co(COO^-)_2$ complex is very low. Compared with the VO^{2+} and Ni^{2+} complexes, the stability of the $MoO_2(COO^-)_2$ complex is rather low. Low stability constants for $MoO_2(COO^-)_2$ and $Co(COO^-)_2$ complexes compared with the constants of VO^{2+} and Ni^{2+} are favorable for an efficient removal of contaminant metals from spent $CoMo/Al_2O_3$ catalyst, while leaving active metals intact. Unfortunately, the database of the stability constants is rather limited. This prevents the use of stability constants as a tool to predict leaching process.

Overwhelming information confirmed that it is easier to leach out metals from spent catalysts in an oxidic form than in a sulfided form because of a greater solubility of the former in water. This was clearly confirmed by the increased leaching efficiency of the aqueous solutions of organic compounds in the presence of oxidizing agents (e.g., H_2O_2, HNO_3, $Fe(NO_3)_3$, etc.). Comparison of the coked catalysts with the corresponding decoked catalysts revealed that for the latter, the rate of leaching was greater. This would indicate on the interference of coke layer with the leaching process. Thus, the coke layer represents a barrier through which the molecules of leaching agent have to diffuse in and the metal–agent complexes and/or metal compounds to diffuse out. However, according to the results in Table 7.3 [475], this barrier is significantly diminished in the presence of oxidizing agent. Thus, for oxalic acid alone, the amount of Ni leached out from decoked catalyst was greater by a factor of 17 compared with the coked catalyst, whereas for the oxalic acid + H_2O_2 mixture this factor decreased to less than two. In the latter case, the difference cannot be entirely attributed to the coke barrier. However, selectivity for contaminant metals is key factor in comparing the leaching efficiency using coked catalyst versus decoked catalyst. It has been demonstrated that the selectivity for leaching contaminant metals from coked catalyst is significantly greater than that from decoked catalysts. This was confirmed by the results in Fig. 7.11 [476] showing the radial distribution of contaminant metals and active metals before and after leaching of spent catalyst

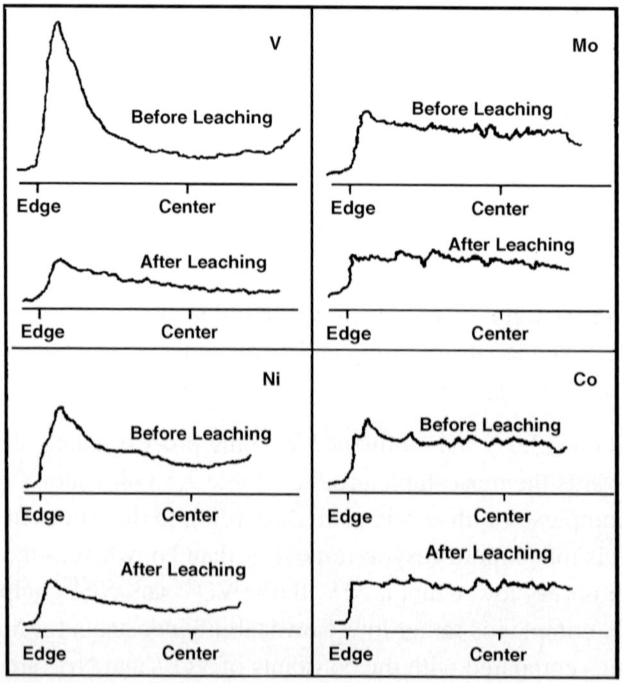

Figure 7.11: Effect of leaching on radial distribution of metals [From ref. 476. Reprinted with permission].

with the solution of oxalic acid + H_2O_2. Thus, after leaching, the distribution of the V and Ni was significantly altered compared with little change in that of Co and Mo.

For coked catalyst, the metal oxidation began during the first contact of the catalyst surface with the oxalic acid + H_2O_2 mixture. At the same time, in the decoked catalyst, most of the metals were already converted to an oxidic state during decoking. For coked catalysts, both molecules of the acid and H_2O_2 have to diffuse from the solution through the layer of coke to reach the surface of particles. Among the metals present in spent catalysts, V will be oxidized and leached out to the greatest extent compared with the other metals. Thus, the extensive information in the literature confirmed that in spent catalysts, V is concentrated on the external surfaces of particles as it is confirmed in Fig. 7.11 [476]. Under such conditions, a high selectivity for leaching V is ensured. This is clearly confirmed in Fig. 7.12 [484] indicating significantly enhanced selectivity for leaching V from coked catalyst compared with decoked catalyst, i.e., for coked catalyst at about 80% V removal, the removal of Mo approached about 20% compared with more than 60% for decoked catalyst.

It is believed that there is an optimal ratio of the acid/H_2O_2 giving the highest selectivity. The optimum may be established experimentally. Moreover, the optimum depends on the origin of organic acid and oxidizing agent. For example, a much greater enhancement in the selectivity

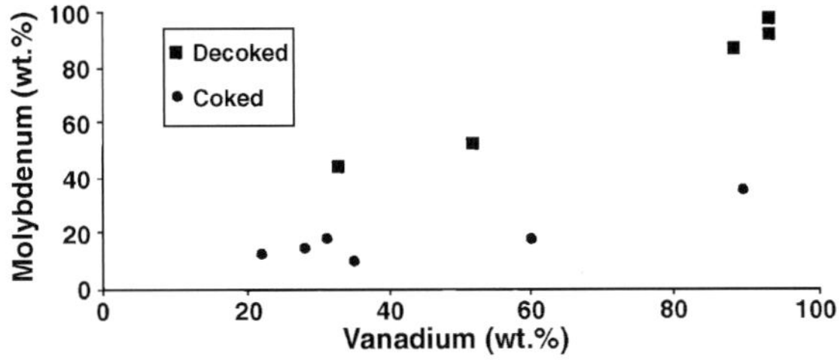

Figure 7.12: Vanadium leached versus molybdenum leached for coked and decoked catalysts with oxalic acid + hydrogen peroxide reagent [From ref. 484. Reprinted with permission].

for V removal was achieved when malonic acid was combined with $Al(NO_3)_3$ than when the same oxidizing agent was added to oxalic acid [483].

7.1.2 Kinetics of Rejuvenation

It is postulated that in the case of organic acids, leaching process comprises both complexation of metals in solution released by catalyst dissolution and liquid-solid reaction. As part of the latter reaction, metals are released from the solid to the solution with the aid of organic agents such as depicted in Reactions {7.6}, {7.7}, and {7.8}. It is believed that the complexation in the solution proceeds at a much higher rate than the liquid–solid reaction. Also, the rate of complexation in solution is much greater than the rate of metal dissolution in water. It is however believed that the liquid–solid reactions in which metals are released from the solid to the solution with the aid of organic agents dominate rejuvenation process. In this case, diffusion phenomena are expected to play an important role. This is supported by the shape of leaching curves (Fig. 7.10) [490] indicating a faster leaching during the initial stages, followed by a gradual slow down indicating a growing involvement of diffusion in overall reaction. Thus, the early stages of leaching involve the chemically controlled removal of metals from the external surfaces of particles with only a limited involvement of diffusion.

Kinetic treatment of different leaching systems developed by Levenspiel [491] was adapted by Marafi et al. [480] to study kinetics of the rejuvenation process. Among several models, progressive conversion or shrinking particle model and shrinking core (or constant particle size) model are relevant to leaching processes employed during rejuvenation. Rate equations for two cases, i.e., particle constant size and particle changing size are summarized below.

7.1.2.1 Particle Constant Size

Liquid phase diffusion controls

$$\frac{t}{\tau} X_B \tag{7.1}$$

where t is time of leaching; X_B is fractional conversion; τ is time for a complete conversion of a single solid particle.

Diffusion through reaction rim controls

$$\frac{t}{\tau} = 1 - 3(1 - X_B)^{2/3} + 2(1 - X_B) \tag{7.2}$$

Surface chemical reaction controls

$$\frac{t}{\tau} = 1 - (1 - X_B)^{1/3} \tag{7.3}$$

7.1.2.2 Particle Changing Size (Shrinking Particle) Model

Chemical reaction controls

$$\frac{t}{\tau} = 1 - (1 - X_B)^{1/3} \tag{7.4}$$

Liquid phase diffusion through the pores controls

$$\frac{t}{\tau} = 1 - (1 - X_B)^{2/3} \tag{7.5}$$

Comparison of the data predicted by these models with the experimental data should identify the rate-controlling process involved during leaching. It is obvious that diffusion in liquid phase cannot be rate-controlling process. The data from leaching of pure metal sulfides (Table 7.6) [480] gave the best-fit using the Eqn (7.4) (Fig. 7.13), suggesting that chemical reactions were the rate-controlling process. This is not surprising considering very low porosity and surface area of the metal sulfides used for leaching.

For coked catalyst, the presence of the layer of coke has to be considered. Thus, in order for leaching to take place, the acid molecules have to diffuse through the layer to contact catalyst surface. Therefore, the mass transfer through the layer of coke deposited on catalyst surface is the rate-controlling process. This is confirmed by the near-perfect fit of the experimental data using Eqn (7.2) shown in Fig. 7.14 [480]. However, it may not be easy to describe the leaching of metals from the pores of spent catalysts. According to Fig. 7.10 [490] leaching process

Table 7.6: Surface properties of metal sulfides [From ref. 480. Reprinted with permission].

Compound	Surface area (m^2/g)	Pore volume (mL/g)
V_2S_3	7	0.11
MoS_2	4.3	$<$ 0.01
CoS	1.5	$<$ 0.01
NiS	1.2	$<$ 0.01

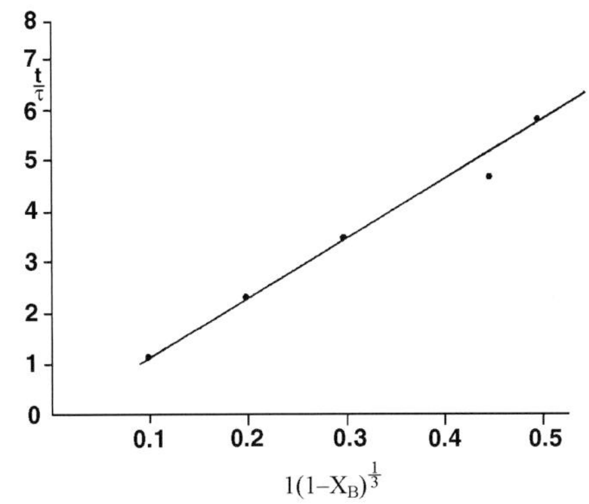

Figure 7.13: Chemically controlled vanadium removal from vanadium sulfide at 25 °C (0.66 M oxalic acid) [From ref. 480. Reprinted with permission].

proceeds in two stages. The change from stage 1 to stage 2 appears to be at a particular time (e.g., 3 to 5 h). This depends on the type of leaching agent, which has an impact on both kinetics and mass transfer coefficients. Therefore, a parameter such as effective diffusivity has to be taken into consideration to describe leaching phenomena in more realistic terms. The effective diffusivity (D_e) of a leaching agent can be determined using the following approximation [493]:

$$D_e = \varepsilon^2 D \quad (7.6)$$

where D is the diffusion coefficient and ε is the catalyst porosity. It is evident from Table 7.7 [490] that effective diffusivities for acetic acid, malonic acid, and oxalic acid change with

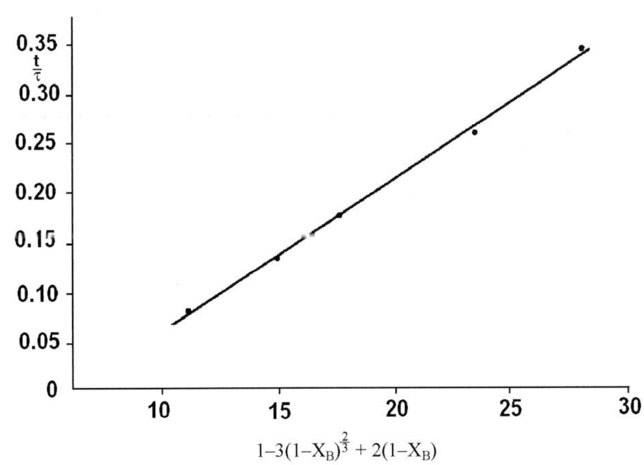

Figure 7.14: Diffusion controlled vanadium removal from spent catalyst at 25 °C (0.66 M oxalic acid) [From ref. 480. Reprinted with permission].

Table 7.7: Mass transfer and kinetic data for leaching V (0.66 M; 323 K) [From ref. 490. Reprinted with permission].

	Acetic acid	Malonic acid	Oxalic acid
Diffusion coefficient, D (cm^2/s) × 10^5	1.908	1.881	2.088
Effective diffusivity, D_e(cm/s) × 10^7			
Spent catalyst	2.75	2.70	3.01
After 1st stage leaching	23.4	23.0	25.6
After 2nd stage leaching	33.6	33.2	36.8
Fresh catalyst	43.9	43.3	48.1
Leaching rate (wt.%/s) × 10^5	13.19	31.25	41.48
Thiele modulus			
1st stage leaching	1.73	4.17	4.96
2nd stage leaching	0.21	0.49	0.58

catalyst porosity in the course of leaching. As expected, the D_e values are higher in the stage 2 than in stage 1. The values also confirm the established efficiency of the acids for leaching, i.e.:

Oxalic acid > malonic acid > acetic acid

The combined effects of reaction rate and diffusion on leaching process can be examined using the Thiele modulus defined as the ratio of the reaction rate to diffusional rate, i.e.:

$$\varphi^2 = \frac{R_{c2}k}{D_e} = \frac{R_{c2}}{D_e} \times \frac{\text{Reaction rate}}{(C_{Ao})^n} \tag{7.7}$$

where R_c is the catalyst pellet radius, k is the rate constant, n is the order of reaction, and C_{Ao} is the initial concentration of leaching agent. The values of Thiele modulus in Table 7.7 indicate improvement in intraparticle mass transfer due to metal removal. Thus, as leaching progresses, the intraparticle transfer is gradually improved in the pore network and chemical reaction becomes rate-controlling. This is confirmed by much higher values of Thiele modulus for the stage 1 (Table 7.7), whereas significantly diminished restriction indicated by low values of Thiele modulus confirm that in the stage 2, the reaction is chemically controlled. This situation can be represented by "A" progressing to "B" as it is illustrated in Fig. 7.15 [491] and the mass balance can be expressed as:

$$\frac{t}{\tau_D} = -2\frac{r^2}{R^2}[\ln(R) - (r)] + \left[1 - \frac{r^2}{R^2}\right] \tag{7.8}$$

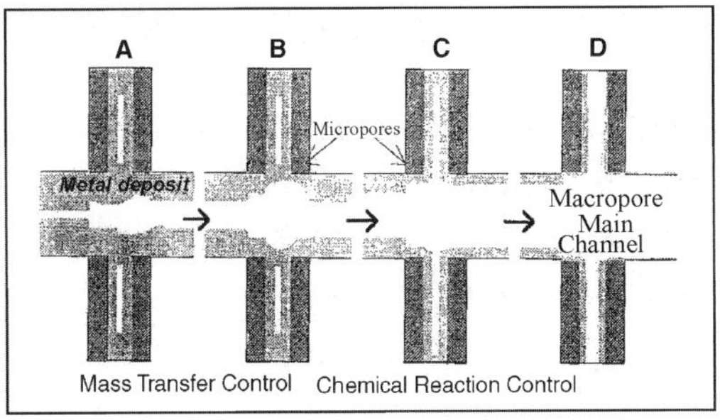

Figure 7.15: Schematic diagram of two leaching stages in catalyst pore network [From ref. 491. Reprinted with permission].

and

$$\tau_{\mathrm{D}} = \frac{R^2 \rho_{\mathrm{B}}}{4C_{\mathrm{A}} D_{\mathrm{e}}} \tag{7.9}$$

By defining fractional (χ) conversion as:

$$\chi = \frac{\text{Amount leached}}{\text{Total foulants}} = 1 - \frac{r^2}{R^2} \tag{7.10}$$

the mass balance equations can be reorganized to:

$$\frac{t}{\tau_{\mathrm{D}}} = 2(1-\chi)\ln(1-\chi)^{1/2} + \chi \tag{7.11}$$

The chemical kinetics of the leaching process, assuming first-order reaction can be represented schematically in the situation progressing from "B" to "C" and then to "D" as shown in Fig. 7.15. According to Levenspiel [491], the chemical kinetics of the leaching process on a cylindrical pore can be represented by:

$$\frac{t}{\tau_{\mathrm{K}}} = 1 - \frac{r}{R} \tag{7.12}$$

where

$$\tau_{\mathrm{K}} = \frac{R\rho_{\mathrm{B}}}{kC_{\mathrm{A}}} \tag{7.13}$$

and finally

$$\frac{t}{\tau_{\mathrm{K}}} = 1 - (1-\chi)^{1/2} \tag{7.14}$$

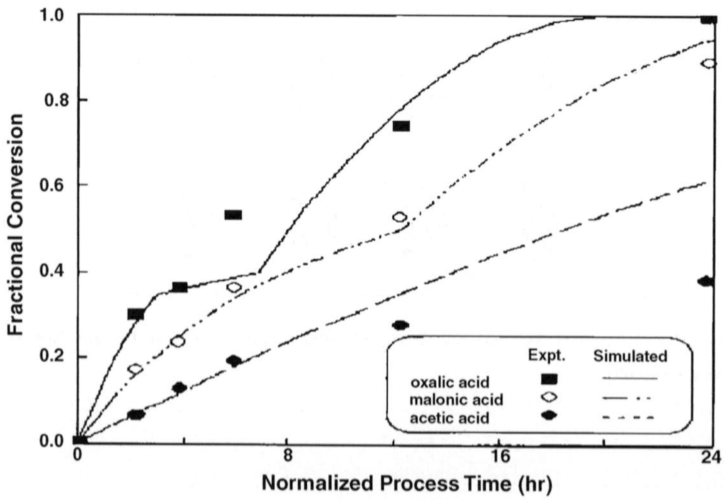

Figure 7.16: Comparison of experimental data with simulated data [From ref. 490. Reprinted with permission].

When the reaction rate is much faster than mass transfer, i.e.:

$$k = D_e$$

the rate is under diffusion control. Otherwise, it is under kinetic control.

In this set of equations, r and R represent radial coordinate of pores and pore radius, respectively, τ_D and τ_K time in diffusion control and kinetic control, respectively, ρ_B density and C_A concentration of leaching agent.

Normalized conversions determined experimentally during leaching V using 0.66 M concentration of oxalic acid, malonic acid, and acetic acid were compared with the values predicted by the model. The results are shown in Fig. 7.16 [490]. The simulated values are represented by lines, while experimental data are indicated by symbols.

7.1.3 Emissions from Rejuvenation by Organic Agents

The compliance with environmental regulations is a requirement before any process can be commercialized. Apparently, for a rejuvenation process employing aqueous solutions of organic agents, the control of emissions can be maintained by the established methods. Thus, there is no need for development of any special techniques to suit the rejuvenation process. Gaseous, liquid, and solid emissions are released to various degrees. Because of the non-corrosive environment, the safety requirements during the rejuvenation using aqueous solution of organic acids can be fulfilled without any difficulties. Moreover, most of the units of rejuvenation process can be constructed from the readily available materials.

7.1.3.1 Gaseous Emissions

Most of the gaseous emissions are released during decoking step, which has to be an integral part of the rejuvenation process. These emissions include SO_X, NO_X, CO, CO_2 and particulate matter. The same methods that have been used for treating the off-gases from oxidative regeneration can also be applied here. This may include a particulate removing system and caustic scrub. Some concerns about release of HCN and NH_3 during decoking of catalyst have been noted [19]. However, the conditions of oxidative decoking, i.e., very low concentrations of O_2 combined with low temperatures and large volume of diluted air used for decoking ensure that the amount of these toxics are at near and/or below detection limits.

Reactions {7.4} as well as {7.7} and {7.8}, which were discussed as part of the rejuvenation mechanism, indicate a potential formation of SO_2 and H_2S, respectively. The formation of the former requires the presence of an oxidative agent, suggesting that the oxidation of SO_2 may continue to produce SO_3 followed by H_2SO_4 as shown in Reaction {7.5}. Almost certainly, metal sulfates are the final products of this process. Therefore, SO_2 emissions during the leaching should not be of a concern. However, for coked catalysts in the absence of an oxidative agent, the formation of H_2S cannot be avoided. Because of toxic character of H_2S, its potential release has to be taken into consideration during the design of commercial process.

7.1.3.2 Liquid Emissions

Depending on the extent of de-oiling of spent catalyst before decoking, various amounts of the condensible volatiles may be released during decoking. In their properties, the volatiles approach properties of the feed being upgraded and those of products. The volatiles may be consumed on the site as a fuel. Before reuse, rejuvenated catalysts must be dried to remove water from the catalyst pores. This step should be benign, although some mild odor could be present.

The leachate from rejuvenation is the main liquid phase requiring attention. If isolated in a pure form, the contaminant metals removed from catalyst (e.g., mainly V and Ni) are marketable products. This may proceed by the adjustment of pH to liberate metals from complexes with organic acids. Subsequently, the metals can be separated either by selective precipitation or by the methods that have been part of the commercial hydrometallurgical operations. However, an increase in the metal concentration in leachates may be needed to improve the efficiency of these operations (e.g., by evaporation). An additional treatment of the liquor, i.e., a treatment with activated carbon and/or with a caustic, before the water can be either disposed or reused, may be necessary. If disposal option is considered, the concentration of regulated metals in disposed water must not exceed limits prescribed by environmental standards.

7.1.3.3 Solid Emissions

A breaking and crushing of the catalyst particles during operation cannot be avoided. This may be evident after decoking of spent catalyst. It might be necessary to remove the crushed particles from spent catalyst before rejuvenation. These aspects of spent catalysts were discussed earlier. Fine particles formed during rejuvenation process end up in the leachate. If removed from the leachates, the fine particles of catalyst have to be either stored or disposed using methods ensuring environmental acceptance unless some other safe utilization options were identified.

7.1.4 Rejuvenation Process Design

Important information to be considered during process design includes the process capacity, operating conditions, and special precautions needed due to corrosion and environmental regulations. All stages of rejuvenation process, i.e., spent catalyst de-oiling, mechanical separation of catalyst particles, selective metal leaching, and decoking, have to be carefully assessed. In addition, there might be requirements for water systems, fuel systems, and power generation. Also, facilities for treatment of gaseous and liquid effluents have to be installed to prevent release of harmful species into the environment. If integrated with petroleum refinery, some facilities required for rejuvenation may be integrated with those used during refining. Therefore, it might be more attractive to construct rejuvenation plant on or near the site of petroleum refinery than a stand-alone rejuvenation operation.

7.1.4.1 De-Oiling

The aspects of de-oiling which are common for regeneration process were discussed in the Chapter 6. It was indicated that it is preferable that de-oiling is performed by spent catalyst generator (e.g., petroleum refinery) unless rejuvenation process is integrated with refinery. It is believed that an efficient mechanical separation cannot be achieved without prior de-oiling of spent catalyst particles.

7.1.4.2 Mechanical Separation

It is desirable that particles that are non-regenerable and/or non-reusable are removed from process streams before rejuvenation step. Two methods are employed for mechanical separation of spent de-oiled catalyst particles, i.e., fines sieving and particle grouping according to density or jigging. With efficient removal of fines, the emissions of particulate matter are significantly reduced. Apparently, fines removal is much more critical for spent catalysts removed from ebullated bed reactor than that from the fixed bed reactor. This results from the small inside diameter (ID) of catalyst particles (1 mm or less) used in the former reactor compared with fixed bed reactors. Also, because of a continuous motion, a cracking of catalyst particles may occur more readily in ebullated bed than in fixed bed. Table 7.8 [12]

Table 7.8: Properties of spent catalyst particles from ebullated bed reactor [From ref. 12. Reprinted with permission].

Property	Spent mix	Lightly fouled	Heavily fouled
Vanadium (wt.%)	10.6	4.4	13.8
Nickel (wt.%)	4.0	3.5	5.2
Carbon (wt.%)	16.2	15.8	16.3
Surface area (m^2/g)	68	122	55
Bulk density (kg/L)	1.09	0.97	1.21
Side crushing strength (lb/mm)	1.8	2.1	1.2
Pore volume (mL/g)	0.17	0.21	0.11
Particle length distribution (wt.%)			
<1.5 mm	25.2	14.4	40.0
1.5–3.0	42.3	23.5	37.0
3.0–6.0	32.5	61.3	23.0
>6.0	0	0.8	0

shows the properties of spent mix obtained from ebullated bed reactor as well those of lightly fouled and heavily fouled portions obtained from the spent mix by jigging technique using a mineral jigg. Lightly fouled particles accounted for about 30 wt.% of the mix. According to the classification discussed earlier, lightly fouled particles could be still regenerated; however, the other particle fractions would almost certainly require rejuvenation step before decoking. Particle size distribution in Table 7.8 is of particular interest. Thus, for the fresh catalyst, particle size was dominated by 3.0–6.0 mm particles. For the heavily fouled particles, almost 70% of these particles were broken to less that 3.0 mm length. This suggests that the size separation of the heavily fouled particles would be desirable prior to their use in rejuvenation process.

The rejuvenation strategy discussed in the introductory part of Chapter 6 included three options. Each of these options requires de-oiling of spent catalyst followed by particle size and/or density separation. Figure 7.17 [462] shows how these two steps can be integrated into a single process that could be applicable for each option. In this case, the dotted lines show the gas flow, thin solid lines show the liquid routes, and thicker solid lines shown the solids streams. Details of the gravity settling unit, termed also as a surface velocity classifier shown in Fig. 7.17 as unit "01V03" are given in Fig. 6.36 [462]. In this unit, separation depends on the particle terminal velocity, which is a function of the physical and transport properties of the solvents and catalyst particles. By adjusting the liquid flow from each compartment, the catalyst particles with different densities are separated. After removing solvent in the drier, particles are transferred to metals leaching process.

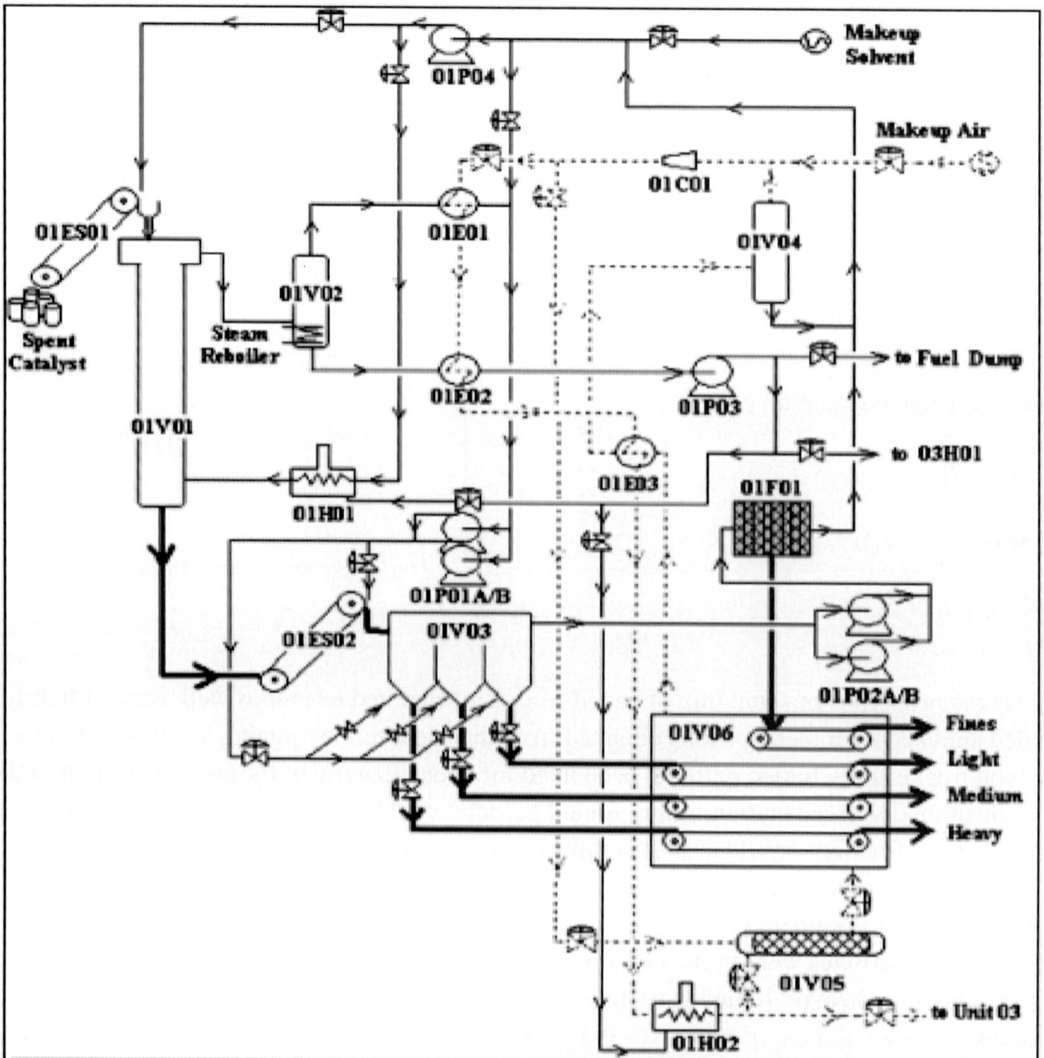

01C01	Air Compressor	01P01A/B	Filter Press Feed Pump
01E01	Solvent Distillate Heat Exchanger	01P03	Heavy Oil Fuel Pump
01E02	Residual Oil Raffinate Heat Exchanger	01P04	Selvent Recycle Pump
01E03	Dryer Flue Gas Heat Exchanger	01V01	Deoiling Column
01ES01	Spent Catalyst Escalater	01V02	Distillation Column
01ES02	Deoiled Catalyst Escalater	01V03	Gravity Setting Unit
01F01	Catalyst Fines Filter	01V04	Solvent Stripper
01H01	Solvent Feed Heater	01V05	Drying Air Mixer
01H02	Drying Air Feed Heater	01V06	Dryer
01P01A/B	Gravity Setiler Unit Feed Pump		
03V01	Decoker	03H01	Furance

Figure 7.17: Integrated de-oiling and mechanical separation process flow diagram [From ref. 462. Reprinted with permission].

7.1.4.3 Metals Leaching Process

Either fixed bed reactor or ebullated bed reactor require careful evaluation before being selected for the design of leaching process. The design of the process shown in Fig. 7.18 [462] incorporates two ebullated bed reactors operating in a semi continuous mode. Thus, while one of the reactors is in an operating mode, the other is on standby. Leaching reaction proceeds while catalyst particles are ebullated. After leaching is complete, catalyst is transported to the leached catalyst holding tank by fluidization using leaching solution as the fluidizing media. The leaching solution continues to be sent to the leaching agent holding tank until all of the catalyst was unloaded. The regeneration of the spent leaching solution takes place in one of the three regenerators. The leached catalyst is washed to remove any traces of the leaching solution, while contaminated water from washing is sent to the water treatment plant.

7.1.4.4 Decoking of Leached Catalyst

The oxidative regeneration used for decoking spent catalysts has been practised commercially for decades. It involves controlled burn-off of coke using air and/or diluted air. Details of commercial process were given in Chapter 6 dealing with catalyst regeneration. It is believed that the same processes can be adapted to suit decoking of the leached catalysts.

7.1.4.5 Other Auxiliary Processes

Additional processing units are necessary, if the rejuvenation process is not integrated with a refinery operation. Details of the additional process requirements for stand-alone rejuvenation process were given by Marafi et al. [462]. They include fuel system, general water system, power and steam generation and distribution, gas treatment plant, liquid effluent treatment plant, and water treatment plant. Chemical analysis laboratory and process control room may be necessary as well. Obviously, the requirement for these processes is minimized in the case that the operation is conducted on or near the site of refinery. Because of their impact on capital and operating costs, the need for the additional processes has to be carefully evaluated during the design of rejuvenation plant.

7.1.4.6 Design Basis

For preliminary process design Marafi et al. [462] assumed a plant processing 6000 tons of spent hydroprocessing catalyst annually. Table 7.9 shows details of mass balance. In Option I, 3171 tons of the heavy and medium portions of de-oiled catalysts are processed to yield 318 tons of metals and 2853 tons of leached catalysts. However, in Option II, only 987 tons of medium fouled catalysts are processed to produce 60 and 927 tons of metals and catalyst, respectively. In this case, the leaching solution contained 6% oxalic acid and 8% $Fe(NO_3)_3$ in water. The optimal leaching temperature has been found to be 313 K. In Option III, only the portion of lightly fouled catalyst was decoked without being subjected to leaching.

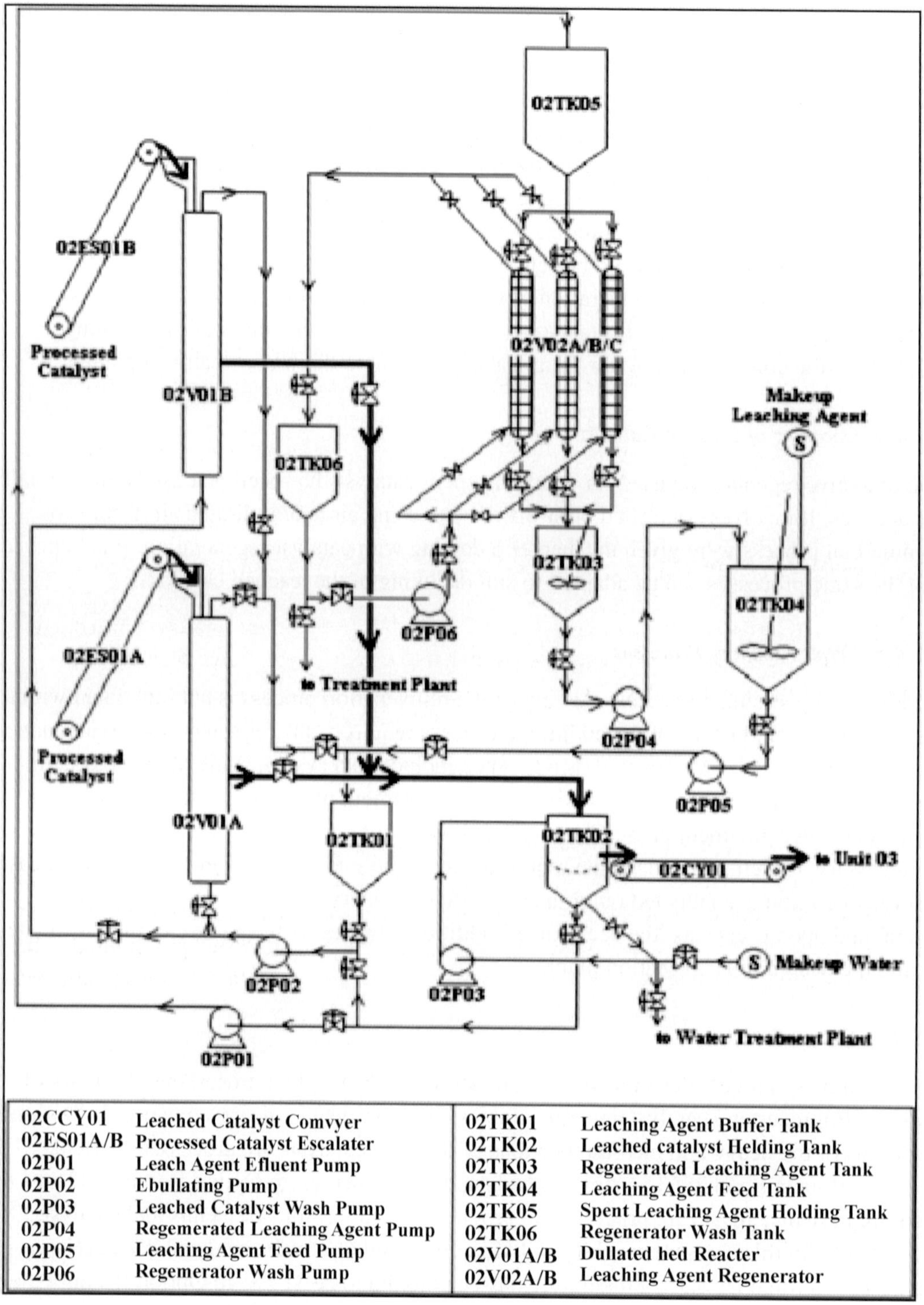

02CCY01	Leached Catalyst Comvyer	02TK01	Leaching Agent Buffer Tank
02ES01A/B	Processed Catalyst Escalater	02TK02	Leached catalyst Helding Tank
02P01	Leach Agent Effluent Pump	02TK03	Regenerated Leaching Agent Tank
02P02	Ebullating Pump	02TK04	Leaching Agent Feed Tank
02P03	Leached Catalyst Wash Pump	02TK05	Spent Leaching Agent Holding Tank
02P04	Regemerated Leaching Agent Pump	02TK06	Regenerator Wash Tank
02P05	Leaching Agent Feed Pump	02V01A/B	Dullated hed Reacter
02P06	Regemerator Wash Pump	02V02A/B	Leaching Agent Regenerator

Figure 7.18: Flow diagram of rejuvenation process [From ref. 462. Reprinted with permission].

Table 7.9: Mass balance for rejuvenated spent catalyst process (6000 tons/year) [From ref. 462. Reprinted with permission].

Option	Products			By-products		
	Fines	De-oiled catalyst	Rejuvenated catalyst	Residual oil	Leached metals	Combustibles
I	519	0	3264	810	318	1089
II	519	2184	1820	810	60	607
III	519	3171	1125	810	0	375

The estimated economic indicators for all three options revealed that it is much more attractive to construct rejuvenation plant on and/or near refinery site than a stand-alone independent plant. The integration of rejuvenation process with petroleum refinery ensures that the latter needs not to pay third party to dispose of spent catalyst and to risk potential future liabilities. Among three options considered, the Option II was the most profitable.

7.2 Inorganic Agents

The solutions of strong acids (e.g., HCl and H_2SO_4) alone and/or their mixtures with metal ions (e.g., HCl + $FeCl_3$) as well as and bases (NaOH, ammonia, etc.) have been used for rejuvenation of the spent catalysts as well. Corrosive nature of these solutions suggests that special precautions have to be taken to ensure safety of the operation and environment, although many years experience with similar methods gained in hydrometallurgy can be applied during rejuvenation as well.

In a summary of general observations one may say that during the rejuvenation of spent catalysts using inorganic agents, a high selectivity for contaminant metals (V, Ni, and Fe) may require a delicate control of pH [493]. At high pH of basic solutions an unwanted removal of Mo/W simultaneously with V may occur. In acidic solution, Ni is removed preferentially, but removal of Mo/W cannot be avoided unless the pH is carefully controlled. It is also anticipated that the adverse effects of inorganic solutions on mechanical properties of catalyst particles after rejuvenation may be more evident that those during rejuvenation using organic acids solutions. It is therefore believed that rejuvenation using inorganic agents has some limitations caused by complexity of leaching conditions. Nevertheless, some studies in literature indicate interests in this route of catalyst reactivation.

7.2.1 Acidic Agents

The removal of metals in both oxidic and sulfided form from spent catalysts using inorganic acids involves the dissolution based on the general reactions, e.g.:

$$H_2SO_4\ (2HCl) + MeO = MeSO_4\ (MeCl_2) + H_2O$$

$$H_2SO_4\ (2HCl) + MeS = MeSO_4\ (MeCl_2) + H_2S$$

These reactions suggest that for sulfided catalyst, attention has to be paid to the H_2S emissions. For oxidic catalyst, sulfur emissions were kept under control during decoking step. Jocker [311] reported that during leaching of the oxidic catalyst more use was made of H_2SO_4 than of HCl. For the former, a greater flexibility and recirculation possibilities could be achieved.

To various degrees, all metals in spent catalysts are soluble in strong acids. The solubility can be monitored by the acid concentration and various additives added to the former. It is however believed that it is more difficult to attain a desirable selectivity compared with the aqueous solutions of organic agent.

Perhaps, the simplest case of the catalyst rejuvenation using inorganic agents is the removal of the "skin" of inorganic solids, which is deposited on the exterior of catalyst particles. The "skin" may comprise clays, alkali and alkali earth metals compounds, and V and Ni sulfides formed via the reaction of porphyrins with H_2S. The study of Gray et al. [243] showed that such "skin" could be easily removed by the treatment with diluted HCl. Much more severe conditions are necessary to remove inorganic solids deposited in pores. It is believed that little contaminant metals would be removed in diluted HCl.

According to method disclosed by Sherwood et al. [494–496], coked spent catalyst was steam treated (550–680 °C) before being sent to a rejuvenation unit where it was first washed with water to fill the catalyst pores. Next, the catalyst particles were passed to an acid treating column in which a solution of 5–50 wt.% H_2SO_4 and 0–10 wt.% ammonium ion was introduced into the lower section of the column. The acid solution was recovered from the top of the acid treating column and recycled to the bottom of the column together with additional make-up acid. Usually the acid treatment is conducted at temperatures ranging from room temperature to 120 °C. In a final step the catalyst particles are passed to a washing unit where the acid-treated particles are washed with water to remove the acid following which the recovered, washed catalyst particles are dried in air or in air diluted with nitrogen. Other examples of treating acids include acetic acid, HCl, and HNO_3. The acid treatment solution, temperature and time duration are optimized to ensure that the active metal removal does not exceed about 4 wt.%. Because the polar solvent and acid liquids used for the washing and treatment steps are soluble in both oil and water, this catalyst rejuvenation process can utilize a two-step water washing procedure for the solvent washed and the acid-treated catalyst, after which the solvent and the acid liquids may be recovered by distillation for reuse in the rejuvenation process.

The rejuvenation process described by Duddy et al. [497] used the same acids but experimental system was different. Thus, it comprised a pressurizable vertically oriented vessel with inlet and outlet openings for the catalyst and washing liquids. With this arrangement, solvent liquid washing, water washings, and acid treatment could be carried out in successive steps. After rejuvenation, catalyst is withdrawn from the vessel conical-shaped lower portion downwardly for further processing. The water-soluble solvent and acid treatment liquids can be usually recovered by distillation for reuse in the catalyst rejuvenation process. The solvent-washed and acid-treated catalyst is dried and decoked to obtain completely rejuvenated catalyst. A similar rejuvenation process was disclosed by Tesker and Milligen [498]. In this process, the vessel assembly is arranged to permit solvent washing, vacuum drying and acid treatment, and gas drying of the used catalyst in a bed supported above the conical grid, by upward flow and recycle of the washing liquids and fluidization of the catalyst. Following rejuvenation step, the catalyst is withdrawn from the vessel downwardly through the conical shaped grid for further processing, i.e., decoking.

Several studies were selected to illustrate the use of inorganic agents for selectively leaching the contaminant metals from spent hydroprocessing catalysts [499–501]. For example, the spent $CoMo/Al_2O_3$ catalyst used for hydroprocessing of a vacuum residue was decoked by the oxidative burn-off before being extracted with either $Fe^{3+} + HCl$ or $Fe^{3+} + H_2SO_4$ solutions [499]. The spent catalyst contained about 15 wt.% of coke, 14 wt.% of V, and 3.5 wt.% of Ni. The solutions removed contaminant metals as well as the active metals. However, after presulfiding of the decoked catalyst, selectivity for the removal of the former metals significantly improved. The activity of the rejuvenated catalyst was measured in an autoclave using a residual feed. In this case, the ratios of rate constants for the rejuvenated catalyst to that of the fresh catalyst were determined. The values of 0.7, 0.7, 3.9, and 2.5 were obtained for HDS, hydrodenitrogenation (HDN), hydrodevanadization (HDV), and hydrodenickelization (HDNi), respectively. Although only 70% of the original HDS and HDN activity was recovered, the HDM activity of the rejuvenated catalyst was superior to that of the fresh catalyst, suggesting that the former may be suitable for cascading, e.g., in a guard reactor. However, this would have to be confirmed during long-run experiments. In another treatment, the spent catalyst was presulfided at 813 K in 5 vol.% H_2S in He and then extracted with the boiling solution of $Fe^{3+} + H_2SO_4$ for 20 min before being decoked [500]. This treatment resulted in the significant removal of both V and Ni. This is confirmed in Fig. 7.19 [500] comparing the spent and rejuvenated catalysts. Significantly greater amount of V than that of Ni in spent catalysts should be noted. In spite of small amount of Mo and Co removed, the HDM activity of the rejuvenated catalysts was not affected. Physical properties of the fresh, spent, and rejuvenated catalysts are shown in Table 7.10 [500]. Most likely, the improvement in HDM activity resulted from the increase in the mean pore diameter.

The efficient leaching of the metals (V + Ni) from spent catalyst with H_2SO_4 could be achieved by optimizing the concentration of the latter [501]. For example, for 15% H_2SO_4, the

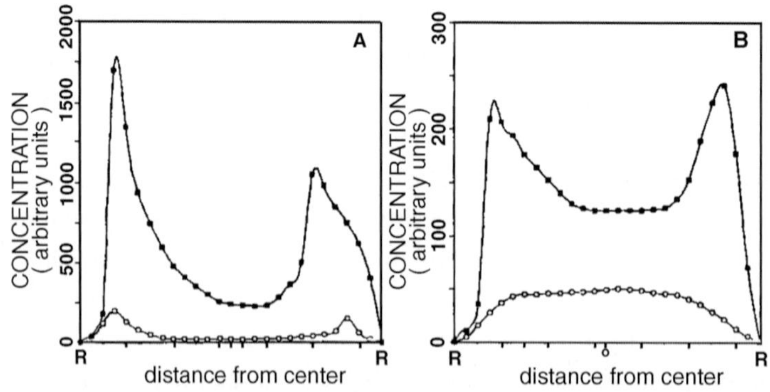

Figure 7.19: Electron microprobe analysis (■: spent catalyst; ○: treated catalyst). A. Vanadium. B. Nickel [From ref. 500. Reprinted with permission].

removal of V and Ni was significantly greater than that of Mo and Al. Prior to leaching, catalysts were de-oiled but not decoked. After removing contaminant metals, catalysts were decoked by oxidative burn-off. A good selectivity for the removal contaminant metals could not be achieved for decoked catalysts because decoking increased the removal of Mo relative to that of V and Ni.

7.2.2 Basic Solutions

The decoking spent catalyst prior to leaching using basic solutions appears to be a requirement. Tentative reactions occurring during leaching in basic solution are:

$$2NaOH + MoO_3 = Na_2MoO_4 + H_2O$$

$$2NaOH + V_2O_3 = 2NaVO_2 + H_2O$$

$$2NaOH + Al_2O_3 = 2NaAlO_2 + H_2O$$

With respect to catalyst rejuvenation, the reaction of NaOH with V_2O_3 is of primary importance. The rate of this reaction relative to the reactions involving MoO_3 and Al_2O_3 can

Table 7.10: Physical properties of trilobe catalysts [From ref. 500. Reprinted with permission].

Catalyst	Density (kg/L)	Pore volume (mL/g)	Mean pore diameter (MPD) (A)	Surface area, (m^2/g)
Fresh	1.26	0.50	70	243
Spent	1.67	0.12	60	85
Rejuvenated	1.46	0.40	110	178

be maximized by carefully controlled leaching parameters such as the concentration of base, pH, and temperature. Little Ni and/or Co can be leached out under optimally selected conditions. However, the residual alkali metal remaining in the catalyst after leaching may have an adverse effect on catalyst activity. Therefore, rejuvenation of spent hydroprocessing catalysts using basic solutions may have some limits.

The readiness of V oxides to form water-soluble alkali vanadates can be explored as the potential method for rejuvenation of spent catalyst using alkaline solutions. Apparently, this route of catalyst rejuvenation has been receiving little attention in spite of the potentially high selectivity for leaching V compared with Ni and Co. However, simultaneous removal of Mo/W may occur as well. The predominance of V deposits on the external surfaces of spent catalyst particles, as indicated by the high V/Mo ratio shown in Figs 7.11 and 7.19 [476,500], may alleviate the problem with simultaneous removal of Mo. Olazabal et al. [502–504] observed very low selectivity for leaching Ni and Co using a NaOH solution, while selectivity for leaching Mo and W was relatively high. However, a high V/Mo ratio near the surface of particles suggests that much more of V than Mo can be leached out under optimally selected conditions. In this regard, the pH of solution and contact time between the catalyst and leaching solution may be key parameters. It is believed that this method of rejuvenation has not yet been fully explored. For example, potential of the NH_4OH solutions in comparison with various alkaline solutions during catalyst rejuvenation should be investigated.

Apparently, such solutions were tested for metal reclamations by Gryglewicz and Rutkowski [505], who attempted to selectively leach out V from the catalyst used during the HDM of a petroleum residue. The catalyst was a macroporous $Fe/SiO_2–Al_2O_3$ containing neither Mo nor W. Table 7.11 shows the results of leaching lasting 1 h using 5% solutions of the agents. The solution of K_2CO_3 exhibited a high selectivity for leaching V. The results confirmed negligible leachability of Ni and Fe in basic solutions. Obviously, if present, Mo and/or W would be leached out in basic solutions. This results from the amphoteric nature of Mo and W. However, a high concentration of V on the external surfaces of catalyst particles would be favorable for an increased rate of leaching of V relative to Mo. It is therefore believed that an efficient

Table 7.11: Selective metal leaching (%) from spent catalyst[a] (5% concentration, 1 h) [From ref. 514. Reprinted with permission].

Reagent	V	Ni	Fe	Al
KOH	73	0	0	10
K_2CO_3	65	0	0	3
CH_3COOK	21	0	0	0
H_2O	6	5	0	0
CH_3COOH	6	5	0	0

[a] *Reagent concentration of 5%, 1 h duration.*

rejuvenation of spent catalysts using basic solutions can be achieved by optimizing leaching parameters such as concentration of leaching agent, temperature, and contact time of leaching solution with catalyst. A high selectivity for V leaching was also confirmed using a NaOH solution [506]. The leaching was the most extensive during the initial contact of spent catalyst with the solution, i.e., during the first 30 min more than 80% of V was leached out. The leaching efficiency was further enhanced by applying electric current.

7.2.3 Environmental and Safety Aspects

Corrosive and hazardous nature of both acidic and basic solutions used for the rejuvenation of spent catalysts indicates on a much higher level of precautions and control required to ensure the acceptable environment and safety of operation, compared with the rejuvenation using aqueous solutions of organic acids. Chemical methods must be used to trap and neutralize vapours released during leaching. Similarly, after extraction of metals from the leachate, a chemical treatment of the latter, followed by removal of the remaining contaminants from the water, must be performed before the reuse and/or disposal of water. For health protection, use of special clothes and protective means by operators must be obligatory. The selection of material for construction of the equipment used for rejuvenation is much more demanding compared with the processes employing aqueous solutions of organic acids.

7.3 Solvent Extraction

A high solubility of coke with the incorporated contaminant metals in a solvent is an essential requirement for this method to be attractive. In an ideal case, this technique may be applied at the end of operation by replacing the flow of feed with a solvent. A removal of spent catalyst would be necessary in the case that solvent extraction is carried out under supercritical conditions. Some studies on solvent extraction appeared in the literature [495,507–509], although commercial viability of this method is rather remote.

Potential applications of the solvent extraction for catalyst activity recovery were discussed by Seapan and Guohui [507], with focus on the solvents such as CO_2, SO_2, and pyridine. Under supercritical conditions, these agents exhibit lower viscosity and surface tension and higher diffusivity compared with the normal conditions. Under such conditions even highly aromatic materials can be displaced from catalyst surface and solubilized. The level of coke removal was influenced by temperature, pressure, and duration of extraction. However, even for the best case, only about 50% of the coke remaining on the catalyst surface after toluene extraction could be removed. This increased pore volume from 0.17 to 0.22 mL/g and surface area from 106 to 137 m^2/g for the spent and extracted catalysts, respectively. There is little information available on the removal of metals under similar conditions, although one study indicated that less than 50% of Co could be removed from the spent $CoMo/Al_2O_3$ catalyst using the supercritical and sub-critical solution of ammonia [508]. However, with respect to catalyst rejuvenation, the removal of Co is an unwanted reaction.

After a solvent wash to remove process oils, a substantial removal of coke was achieved by treating the spent catalyst with an organic solvent, such as N-methyl-2-pyrrolidone, at a temperature of 100 to about 250 °C for a period of about 1 to about 12 h [495]. After this treatment the catalyst was ready for reuse providing that the content of the contaminant metals in spent catalyst was low. Optionally, the solvent-treated catalyst can be oxidatively regenerated. In another version, the solvent-treated catalyst can be acid leached using diluted H_2SO_4 and/or HCl to remove contaminant metals and then decoked.

The spent hydroprocessing catalyst used for bitumen upgrading was extracted with hydrogen donor solvents such as tetralin and tetralin-pyrene mixture [439]. These solvents were compared with gas oil. The following order in the effectiveness of coke removal was established: gas oil < tetralin < tetralin–pyrene. The highest effectiveness of tetralin–pyrene mixture was attributed to the ability of pyrene to form very active H-donors such as hydropyrenes. The HDS activity of the regenerated catalyst was significantly enhanced. Because this extraction was conducted under H_2 pressure, this mode of spent catalyst treatment was discussed in the section dealing with reductive regeneration.

Strong solvents such as pyridine and quinoline were used for the extraction of coke from the spent $CoMo/Al_2O_3$ catalyst [509]. In this case, the amount of coke was decreased from 27 wt.% to 14 and 12 wt.%, respectively. The isolated extract contained V, Ni, and Fe, whereas Co and Mo were not present. The drawback of this method was that a special treatment was necessary to remove the solvent adsorbed on the catalyst surface, unless catalyst was subjected to decoking.

7.4 Biorejuvenation

From time to time, biorejuvenation as the subject of discussion has been noted. In this case, a microorganism-aided removal of V and Ni from spent catalyst while leaving active metals intact would be an ideal method for rejuvenation. Bioleaching of metals from various solids including spent catalysts has been attracting attention and is the subject of the later chapter. It will be shown that potential of this method for metal recovery is limited. It is believed that biorejuvenation of spent hydroprocessing catalysts is still in its infancy. Thus, no experimental studies dealing with this issue could be found in literature. This sub-chapter was added only with the aim to clarify the issue of biorejuvenation, which so far only attracted very limited interest. However, there is good scope for research in this area, particularly in the development of microorganisms, which can selectively remove the contaminant metals (e.g., V) leaving active metals (Co and Mo) in the catalyst.

7.5 Non-Leaching Methods for Contaminant Metals Removal

Hettinger [510] concluded that there is a potential of the magnetic separation method for removing contaminant metals, such as V, Ni, and Fe from spent cracking catalysts. Similarly, this technique may be applied to hydroprocessing catalysts, although so far no experimental results could be found in the literature.

A well-established fact that contaminant metals accumulate predominantly on the external surfaces of catalyst particles (Figs 7.11 and 7.19) suggests that abrasion/attrition method may be suitable for rejuvenation of spent catalysts. A reference to this method was made by Trimm [345] in his review article. Assuming that abrasion/attrition removes the outermost layers of the catalyst particles, the attrited material should have a high content of V and Fe. However, the removal of metals from pores would require a prolonged attrition. It is not certain whether the strength of catalyst particles can be maintained under such conditions. Apparently, this method may have only limited applications in rejuvenation of spent hydroprocessing catalysts. It is believed that this method is more suitable for rejuvenation of the spent fluid catalytic cracking (FCC) catalysts.

CHAPTER 8

Cascading

While the objective of regeneration and rejuvenation is the reuse of spent catalysts at the point of their origin as many times as possible, cascading offers an additional option for the reuse of spent catalysts in less and more severe operations. Generally, severity of hydroprocessing increases with increasing temperature and H_2 pressure is employed during the operations. This is usually accompanied by the increased hydrogen consumption. The cases in which spent-regenerated catalysts are used either in more or less severe operation can be encountered in the refinery practice. Thus, with increasing size of refinery, the number hydroprocessing reactor increases. Consequently, the number of applications for regenerated/rejuvenated catalysts increases as well. According to the literature, the following options for catalyst cascading may be identified [511–513]:

(1) Cascading to less severe operation
After regeneration of catalyst used in a severe service (e.g., gas oil) the catalyst is returned to a less severe operation (e.g., naphtha or straight run distillate).

(2) Cascading to high metal service
After regeneration, catalyst is used to treat high contaminant metals content feed. The catalyst can be used in a guard bed as metal trap or a guard reactor to protect catalysts in the downstream reactors of a multistage process.

(3) Cascading to a high silicon service
In this case, the catalyst contamination due to silicon from the anti-foaming agent such as used during delayed coking may be more serious problem than deactivation by coke. For this application, using a relatively cheap catalyst than the fresh catalyst may be more advantageous.

(4) Make-up catalyst after reactor skim
In some cases, catalytic reactors, i.e., front of the fixed-bed require skim to alleviate pressure drop build-up at the front of the bed. Such service can be adequately provided by spent regenerated/rejuvenated catalysts.

8.1 Cascading of Spent Catalyst

A variable level of deactivation of spent catalyst removed from reactor has been noted. If separated from the mixture, the least deactivated portion of the spent catalyst may be returned either directly to the same operation or cascaded to the less severe service. In the case of ebullated-bed reactors, the spent catalyst may comprise a portion of the fresh and/or very lightly deactivated catalyst which was admixed with spent catalyst during ebullation. The cost of catalyst inventory could be decreased if this portion of the catalyst can be separated from spent catalyst mixture and reused.

It was indicated earlier that density grading of spent catalyst after de-oiling might separate the least deactivated portion of spent catalyst. Sherwood et al. [514] disclosed the process in which spent catalyst is subjected to elutriation by air. The flow of air directed upwardly was maintained at the velocity sufficient to expand the bed at least 25% over its settled height. At a sufficient time, a substantial segregation of catalyst particles occurred. The least deactivated potion withdrawn from the top of fluidized bed was returned directly to the operation. In the case of several ebullated-bed reactors operating in series, the catalyst is used in a preceding reactor, whereas for fixed-bed systems, the elutraited catalyst may be placed on the front of the bed in the same reactor. In large refineries, additional options for the reuse of the least deactivated catalyst separated from the spent catalyst mixtures may be identified.

8.2 Cascading of Regenerated Catalysts

Several examples from literature are presented to illustrate practical cases of cascading. In the study of Mittal et al. [515], the spent-regenerated catalyst after more than four years of operation during the hydrodesulfurization (HDS) of naphtha was used for mild hydrocracking of a vacuum residue. The catalyst exhibited a good activity and selectivity to middle distillates. A marginal cost of catalyst, low hydrogen consumption and a high yield of the desirable product made this option economically attractive. In another study, the spent catalyst from residue upgrading was used for the upgrading of naphtha [516]. In this case, the purpose was to selectively remove sulfur, while avoiding the extensive hydrogenation of olefins with the aim of minimizing the loss of octane number of the gasoline product.

In the case of the multistage processes employing either fixed-bed reactors (e.g., atmospheric residue desulfurization [ARDS] and HYVAHL processes) or ebullated-bed reactors (e.g., LC-Fining and H-Oil processes) in a series used for upgrading heavy petroleum feeds, the severity decreases from the first reactor towards the last reactor. The extent of catalyst deactivation by metals decreases in the same order. This suggests that cascading of the regenerated catalyst from fourth reactor to third reactor makes more sense than vice versa. In fact, according to Suchanek [517], regenerated catalyst from the downstream reactors could be successfully used in the first reactor. In the same reactor, regenerated catalyst can also be used as part of the layered bed with fresh catalyst. In the first two hydrodemetallization (HDM) reactors, the regenerated catalysts from the third and fourth reactors may be used as part of a layered bed to provide HDS/hydrodenitrogenation (HDN) activity. In this case, the layer of the regenerated catalyst would be placed at the end/bottom of fixed-bed. The HDS is usually the main function of the last reactor in series. It has been established that the HDS activity of catalyst is severely inhibited by the metals deposited from the feed, particularly V [197]. This suggests that the cascading of the spent-regenerated catalyst from preceding reactors may not be feasible. However, in one case, even a heavily metal-fouled catalyst after regeneration exhibited an adequate activity during mild HDS of gas oil as the lowest activity user option [518]. However, this gas oil was derived from sweet crude; therefore, this option may not be representative of the current situation in refineries. Another example of the strategy used for catalyst cascading in a multireactor system was presented by Rockwell [519]. In this case, a portion of spent catalyst withdrawn from third reactor would be cascaded to the front of the second reactor.

Apparently, it makes more sense to use the spent-regenerated catalyst in a preceding reactor though it operates at a higher severity. In this regard, the level of catalyst deactivation may be a decisive factor. For example, in the last reactor of the ARDS and/or HYVAHL process, the catalyst deactivation is dominated by coke deposition. After decoking, such catalyst may still pose an adequate activity to be used in one of the preceding reactors. As part of the mixed bed or layered bed with an HDM catalyst possessing a high metal storage capacity, spent-decoked catalysts may exhibit desirable hydrogenation and hydrocracking activities required for asphaltenes conversion in the upstream reactors with dedicated HDM functions. One report indicates on the use of spent-regenerated catalyst from the most severe operation to the next most severe service in the multistage system [520].

Fixed-bed guard reactors are sometimes placed upstream of the HDM and HDS/HDN reactors to prolong the operation of the latter. For example, the advanced version of the HYVAHL process incorporates two guard reactors operating in a perturbating mode. A composite bed comprising a typical guard reactor material and spent-decoked catalyst may be another option for the reversed cascading of spent catalysts. Primary function of guard chamber is the removal of the solids dispersed in heavy feed, whereas in guard reactor removal of these solids and a partial HDM is anticipated. It is believed that after decoking, heavily deactivated non-regenerable catalysts can still be utilized either in guard chamber or added to the front of the fixed-bed in the first reactor to prevent the crust formation.

Practical cases of cascading of spent hydroprocessing catalysts from low severity to high severity service can be found in the literature. For example, Sakabe and Yagi [408] used a tubular type reactor to evaluate the spent hydroprocessing catalyst for upgrading atmospheric and vacuum residues. Rather high activity for hydrocracking of the residues was observed. Refinery practice indicates the practical case of cascading from lower to higher severity service during the operation of LC-Fining process employing three ebullated-bed reactors in the series. Thus, the least deactivated catalyst from the third reactor is transferred to the second reactor before it must be replaced. In this way, the consumption of the fresh catalyst in the second reactor can be decreased. Perhaps, the simplest case of cascading of spent-regenerated catalyst was reported by Duddy et al. [521]. These authors used the regenerated catalysts from the end of the fixed-bed and placed it on the front of the same bed before the subsequent utilization cycle.

8.3 Cascading of Rejuvenated Catalysts

According to the established refinery practice, the desirable recovery of activity via oxidative regeneration cannot be achieved in the case that spent catalysts contain more than 5 wt.% of contaminant metals [517,518]. Such catalysts can be successfully reactivated by combining the removal of contaminant metals with decoking as applied during rejuvenation. The activity of catalyst obtained by oxidative regeneration of the spent catalyst lightly deactivated by metals may approach that of a catalyst obtained by rejuvenation of the spent catalyst severely deactivated by metals. This suggests that cascading options for rejuvenated catalysts may be similar as those applied for regenerated catalysts. For example, for a process employing several ebullated-bed reactors in a series, rejuvenated catalyst could be cascaded to the most upstream reactor.

CHAPTER 9

New Catalysts from Spent Catalysts

It has been shown in the preceding chapters that spent catalysts in their operating form can be either regenerated or rejuvenated for the next utilization cycle. However, there are numerous options for catalyst preparation when the operating size and form of spent catalysts are altered. Metals which can be isolated from spent catalysts may be returned to the catalyst manufacturing companies and used for catalyst preparation. In this regard, other than hydroprocessing catalysts can also be prepared. Of course, after isolation, there are numerous applications for the pure compounds containing metals such as Mo, W, Ni, Co, and V. For example, the extracts containing Mo, V, and Ni obtained by leaching spent catalysts may be used directly for catalyst preparation by impregnating conventional supports. The same extract can be introduced to heavy feeds and as such act as catalyst during slurry bed hydroprocessing. The processes, which can carry metal reclamation to the "extinction", produce pure Al_2O_3 which can be reused as the support for catalyst preparation. The production of refractories is another potential application for the reclaimed Al_2O_3. There is a potential for the use of the crushed spent hydrodesulfurization (HDS) catalyst for the hydrocracking (HCR) of vacuum residue in a slurry bed reactor. However, its commercial application may not possible because of the limited availability. Nevertheless, either reprocessing spent hydroprocessing catalysts to obtain the desirable particle size and shape or crushing for use in the slurry bed hydrocracking appear to be the most attractive options. Non-petroleum applications have been receiving attention, although it is believed that this opportunity has not yet been fully explored.

9.1 Petroleum Applications

Regeneration and rejuvenation of spent hydroprocessing catalysts have been conducted to recover as much as possible of the original activity with the aim to reuse the catalyst in the original operation. If this cannot be achieved, the regenerated and/or rejuvenated catalysts can be cascaded to the less demanding operation. Otherwise, the spent catalysts may still be subjected to non-conventional treatments, before they can be returned to the operation. In this regard, significant advances have been made in reprocessing spent hydroprocessing catalysts, in addition to several other non-conventional methods.

Table 9.1: Properties of spent catalysts [From ref. 522. Reprinted with permission].

Catalyst	Composition (wt.%)							Surface area (m^2/g)	Pore volume (mL/g)
	Mo		Ni		V		C		
	1	2	1	2	1	2			
A	6.3	8.2	3.4	4.4	3.4	4.4	23.1	90	0.15
B	5.6	7.0	3.9	4.9	5.9	7.4	19.2	73	0.16
C	4.5	5.3	4.0	4.7	9.7	11.4	15.3	18	0.10

1: determined by analysis; 2: assuming most of carbon was removed on decoking.

9.1.1 Reprocessing

Among several options, re-utilization of the metal-fouled non-regenerable catalysts in a reprocessed form in hydroprocessing operations has been explored [9,522–524]. In this case, spent catalysts, either de-oiled or decoked, are crushed to 20–90 μm particle size. According to the observation made by Santhyia and Ting [525], such level of crushing favors the liberation of metals from the support matrix. Then, the active metals such as Ni and Mo as well as V became more freely accessible for the interaction with boehmite used as a reprocessing agent. The reprocessed mixture is then formed into the size and shape of particles approaching those of commercial catalysts. Promising results on hydrodemetallization (HDM) and HDS of an atmospheric residue confirmed that reprocessing is a viable option for re-utilization of spent hydroprocessing catalysts. Apparently, there have been attempt to reprocess a smaller-size spent-regenerated catalysts into alternatives sizes and shapes [525]. For example, hollow cylinder offers an optimal combination of void space and activity for grading applications. Top bed grading materials for bed topping were introduced in four diameters ranging from 10 to 25 mm.

9.1.1.1 Procedure and Analysis

In this procedure, spent hydroprocessing catalysts from the atmospheric residue desulfurization (ARDS) process were used. This process comprises four fixed-bed reactors in the series (Fig. 3.13). Two catalysts used for reprocessing included catalysts A and B from the fourth and third reactor, respectively, whereas the most deactivated catalyst C was from the H-Oil reactor. The properties of these spent catalysts are shown in Table 9.1 [522]. These reactors have been part of the processes operated by the Kuwait National Petroleum Company (KNPC). The pretreatment of spent catalysts included either successive extraction in the Soxhlet apparatus using naphtha and toluene or decoking. Catalyst decoking was performed in a horizontal rotary furnace. The temperature was increased in steps to 370, 450 and 500 °C in the flow of 5% O_2 in N_2. Subsequently, the diluted O_2 was replaced by air, while temperature

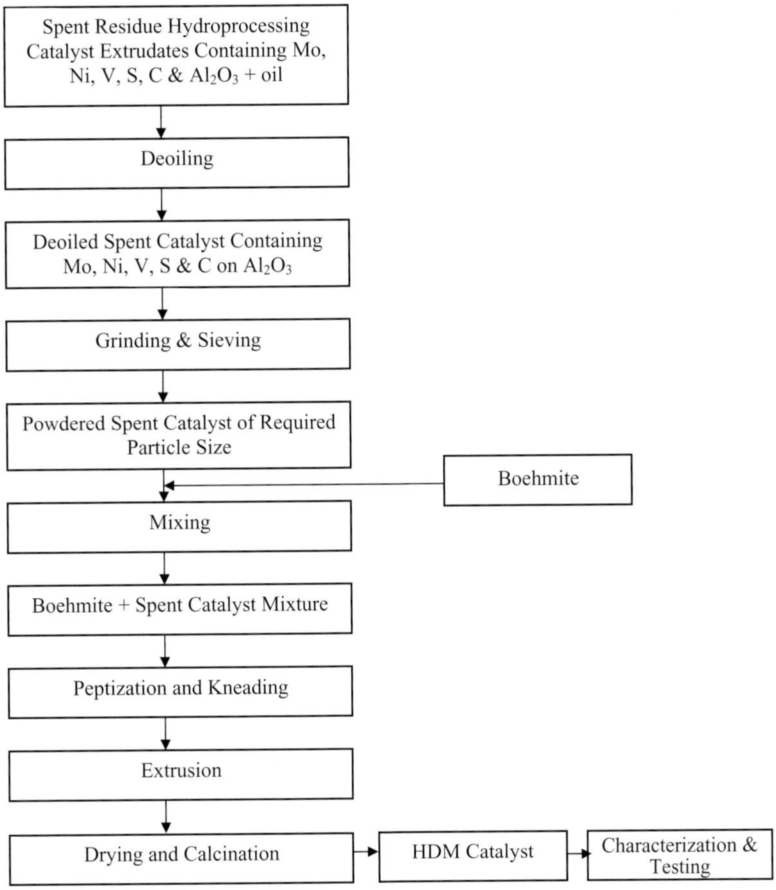

Figure 9.1: Operational steps in the preparation of catalyst extrudates from spent catalyst [From ref. 9. Reprinted with permission].

was maintained at 500 °C. Under these conditions, most of the coke was removed. After de-oiling and/or decoking, catalysts were ground to obtain particle size in the range of 25 to 90 μm, which was similar as that of boehmite. Boehmite used in the study was a commercial product manufactured by Sasol, Germany. It was in a powder form and had particle size in the following range: <25 μm (3%), 25 to 45 μm (9%), 45 to 90 μm (80%) and >90 μm (8%). This solid was used for mixing with the ground spent catalysts to obtain a paste suitable for the extrusion. The sequence of operational steps used for the preparation of extrudates from spent catalyst-boehmite mixtures are shown in Fig. 9.1 [9].

The concentrations of Ni, Mo, and V in spent catalysts A, B, and C (Table 9.1) [522] and laboratory prepared catalysts were determined using the inductively coupled plasma atomic emission spectroscopy (Varian Liberty II ICP-AES). A mercury porosimeter (Quantachrome

Table 9.2: Atomic ratios of catalysts [From ref. 522. Reprinted with permission].

Catalyst	Ni/Mo	V/Mo
A	0.84	0.84
B	0.95	1.67
C	1.08	3.20

Poremaster – 60) was used for pore volume and pore size distribution estimates, while Quantochrome unit was used for the determination of Brunauer-Emmett-Teller (BET) surface area. The content of metals in the decoked catalysts was estimated from the analysis in Table 9.1 assuming that most of the carbon was removed on decoking.

The results in Table 9.1 [522] were used to estimate the V/Mo and Ni/Mo ratios shown in Table 9.2. For catalysts prepared from the same spent catalyst, these ratios remain the same in spite of the different amounts of spent catalysts added to boehmite. These ratios exceed that in the fresh $NiMo/Al_2O_3$ catalysts. For example, for the fresh $NiMo/Al_2O_3$, the Ni/Mo ratio approaches 0.45. At this ratio, the amount of the Ni-Mo-S active phase formed on sulfiding is optimized [53]. Therefore, a direct involvement of the deposited Ni in excess of the optimal amount in hydroprocessing reactions seems to be plausible. It is, however, anticipated that in the presence of a large amount of V, the Ni-Mo-S active phase can be modified.

The preparation of the catalyst extrudates was conducted in the laboratory kneading and mixing machine Type LUK 2.5 AS manufactured by Werrner and Phleiderer Co, Germany. In this case, ~300 g of spent catalyst-boehmite mixture in the desired ratio was added to the mixing chamber. During mixing and kneading, 185 mL of the diluted HNO_3 (2%) was added in drops to the boehmite powder, as peptizing agent. At the end of mixing and kneading, the product was extruded through the die (1.5 mm) to form catalyst extrudates. The extrudates were dried at 110 °C for 24 h, followed by several calcining steps, i.e., 370 °C for 1 h, 450 °C for 2 h, and 500 °C for 2 h. After calcination, the extrudates were cooled in a desiccator.

The radial metal distributions in extrudates were determined using a JEOL scanning electron microprobe X-ray analyzer. The distribution profiles of V for the extrudates prepared from spent catalysts A and B together with the corresponding spent catalysts are shown in Fig. 9.2 [9,524]. It is evident that for both spent catalysts, the V concentration near the outer edge is substantially higher than at the center of particle. The former is the main cause of catalyst deactivation during operation. On the other hand, in reprocessed catalysts, the radial distribution of V is rather even.

The effect of the spent catalyst content in reprocessed catalysts on the pore size and pore volume distribution is shown in Table 9.3 [523]. These results were obtained for the mixture of

Table 9.3: Effect of different amount of spent catalyst on pore size and pore volume distribution in reprocessed catalysts [From ref. 523. Reprinted with permission].

Spent catalyst (wt.%)	<50%	Pore size (A) 50–100		100–200		200–300		300–500		Total pore volume (mL/g)
		mL/g	%	mL/g	%	mL/g	%	mL/g	%	
0	0.1	0.006	0.9	0.519	77.4	0.087	13.0	0.026	3.9	0.671
15	1.3	0.032	5.6	0.418	73.0	0.071	12.4	0.019	3.2	0.573
25	2.1	0.034	6.5	0.262	49.8	0.168	31.9	0.020	3.9	0.525
40	1.0	0.038	7.5	0.100	20.1	0.280	56.2	0.043	8.6	0.500
60	0	0.030	8.3	0.105	28.8	0.189	51.9	0.023	6.1	0.363

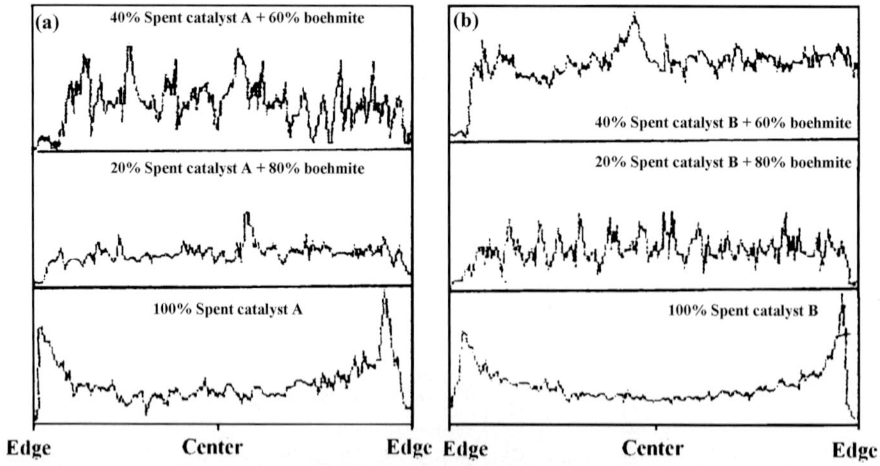

Figure 9.2: Radial distribution of vanadium for catalyst prepared (a) from spent catalyst A; (b) from spent catalyst B [From ref. 9. Reprinted with permission].

boehmite with the most fouled catalysts C having pore volume of 0.10 mL/g (Table 9.1). In spite of this, an adequate pore volume could be achieved up to 40% of the spent catalyst. For most mixtures in Table 9.3, pore volume in 100–300 A pores accounted for more that 80% of the total pore volume. Figure 9.3 [9] shows the trends in surface area and pore volume of the reprocessed catalysts versus the amount of spent catalyst C. It should be noted that the surface area of the latter was 18 m^2/g (Table 9.1), whereas the surface area of boehmite was 165 m^2/g. Again, an adequate surface area was maintained up to about 40 wt.% of the spent catalyst in the reprocessed catalyst.

9.1.1.2 Testing of Reprocessed Catalysts

The reprocessed catalysts were tested for HDS and HDM of the atmospheric residue derived from a Kuwaiti crude. In the feed and product steams, the contents of Ni, Mo, and V were determined without ashing using the same instrumentation as used for catalysts. The analysis of sulfur in liquid products and in the feed was performed using the Oxford Model 3000 XRF analyses. The catalyst activity was determined in the high-pressure fixed-bed microreactor at 370 and 390 °C and ~12 MPa of H_2 using the H_2/feed ratio of 1000. The tests were of 100 h duration. Prior to testing, the catalysts were presulfided in the 1 %CS_2/gas oil mixture. Properties of the Kuwait atmospheric residue used for testing are shown in Table 9.4 [522].

Figure 9.4 [522,523] shows the effect of the amount of spent catalyst A added to boehmite on HDS and hydrodevanadization (HDV) activity at 390 °C. For both functionalities, the activities increased with the amount of catalyst added. Also, the activities were greater for decoked catalyst. These trends were similar for the other catalysts. However, these trends do not reflect accurately the effect of the amount of active metals added to boehmite. For example, based on

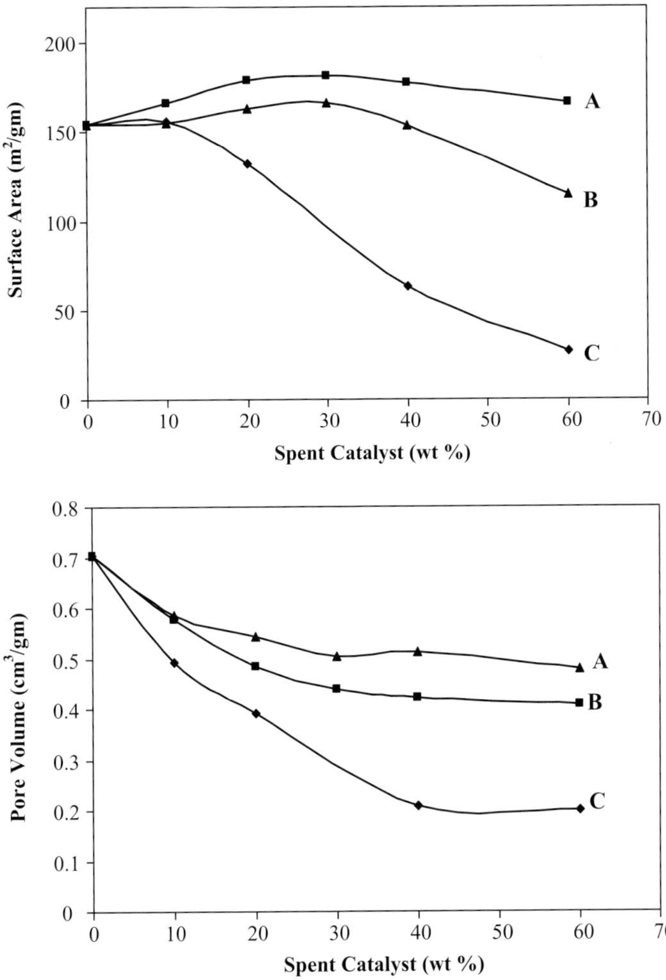

Figure 9.3: Effect of spent catalyst content on surface area and pore volume [From ref. 9. Reprinted with permission].

Table 9.1 [522], adding 20 wt.% of the coked and decoked catalyst to boehmite would result in the addition of about 1.3 and 1.7 wt.% of Mo, respectively. Then, the differences between the coked and decoked catalysts in HDS and HDV activities shown in Fig. 9.4 are expected to be smaller when the conversions are correlated with the content of Mo in reprocessed catalyst. This is indeed shown in Figs 9.5 and 9.6 [522] where the HDS and HDV conversions were correlated with the actual amount of Mo in the catalysts rather than the amount of spent catalysts added to boehmite. In fact, Figs 9.5 and 9.6 suggest that for most part, the differences between the coked and decoked catalysts are much less evident compared with Fig. 9.4 [522,523]. Then, it is the amount of active metals which dominated the activity of reprocessed catalysts.

Table 9.4: Properties of feedstock [From ref. 524. Reprinted with permission].

Density (kg/L)	0.977
Sulfur (wt.%)	4.30
Nitrogen (ppm)	2740
H/C	1.58
CCR (wt.%)	12.20
Asphaltenes (wt.%)	3.78
Metals (ppm)	
Vanadium	69
Nickel	21
Distillation, (°C)	
IBP	391
40%	560
Recovery volume (%)	57.1

The effect of V on HDS activity is rather different than that of Mo, as it is shown in Fig. 9.7 [522]. It is essential that the results in Fig. 9.6 are considered in the context of results in Fig. 9.4. Thus, the change in HDS conversion shown in Fig. 9.7 can be attributed to the combined effects of all metals, i.e., V, Ni, and Mo. Of course, it is not easy to decouple these effects. However, the Ni/Mo and V/Mo ratios in Table 9.2 [522] indicate that for catalyst A, the interference of contaminant metals with active Ni-Mo-S phase should be least evident compared with the other two catalysts. Then, a significant decline in HDS activity from catalyst A to catalyst C suggests that during sulfiding, particularly that of the catalyst C, the active Ni-Mo-S phase was only partially developed. It is believed that for this catalyst, the HDS activity was dominated by V sulfides with a little contribution from the Ni-Mo-S phase. This is in accordance with the high V/Mo ratio (Table 9.2) for catalyst C. This is consistent with the catalytic activity of V sulfides (e.g., V_2S_3, V_3S_4, VS_2) for HDS, although much lower than that of Mo sulfide, observed by Loos et al. [255]. It may be therefore concluded that there is an upper concentration of V (e.g., ~5 wt.%) in spent catalyst above which the HDS activity of the reprocessed catalysts may be only partially recovered.

The effects of Mo content on HDV and hydrodenickelization (HDNi) at 370 and 390 °C are compared in Fig. 9.6 [522]. It is believed that at both temperatures, the activities were similar when accumulated errors are taken into consideration, although at low Mo contents, a scatter of the results was observed. These results also show that the Mo content is determining factor for HDM activity in agreement with the other studies [27]. Thus, the HDM activity tends to level off above 3 wt.% Mo. At Mo content greater than 3 wt.%, the addition of Ni had little

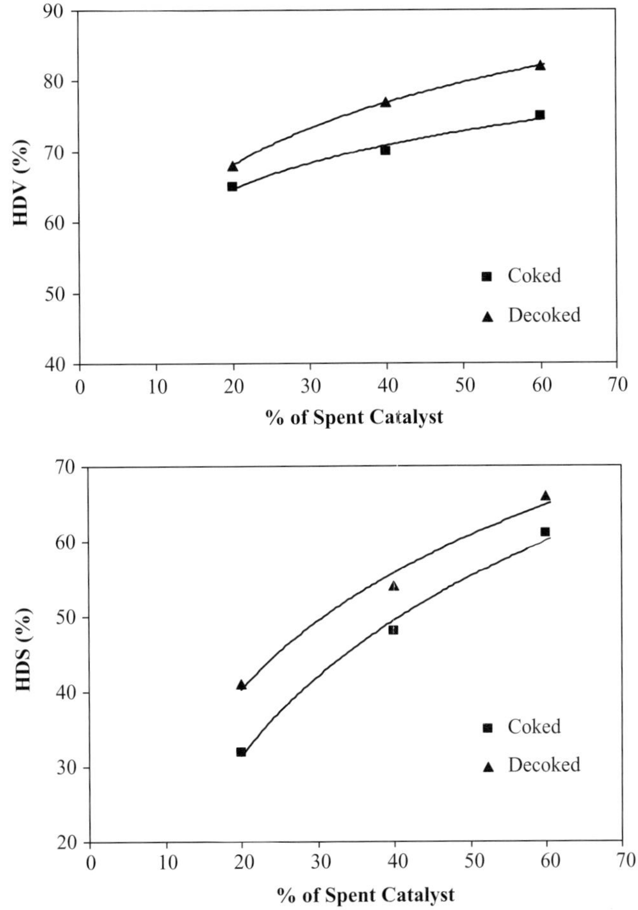

Figure 9.4: Effect of the amount of spent catalyst (catalyst A) added to boehmite on hydrodesulfurization (HDS) and hydrodevanadization (HDV) activities [From ref. 522. Reprinted with permission].

effect on HDM, although it positively influenced HDS and HDN activities. This confirms that the activity of catalysts prepared by reprocessing spent hydroprocessing catalysts exhibit the same general trends, which are observed for the fresh commercial HDM catalysts.

Asaoka et al. [257] showed that V sulfides are active for HDM and hydrodeasphalting (HDAs). For example, HDM was positively influenced by metal deposits until maximum of about 15 wt.% of deposits was formed [160,161]. Even unsupported V_2S_3 exhibited some activity during hydroprocessing reactions [163,164]. It was demonstrated that HDAs is essential for HDM to occur [27]. Thus, in order for porphyrins to undergo HDM, they have to be liberated from large asphaltenic entities to become available for HDM reactions. Because of the high V/Mo ratio of catalyst C, its activity may be governed by V sulfides rather than the Ni-Mo-S

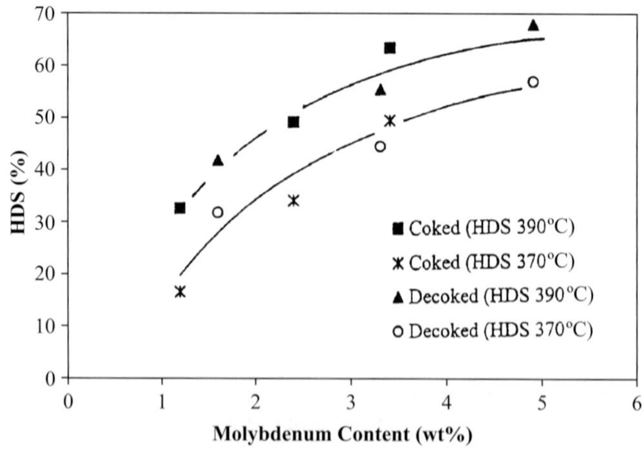

Figure 9.5: Effect of Mo content of the catalyst (catalyst A) on hydrodesulfurization (HDS) activity [From ref. 522. Reprinted with permission].

active phase. Thus, several studies showed that in the presence of V, this phase is gradually converted to V-Mo-S phase which is much less active than the Ni-Mo-S phase [162].

Apparently, the HDS and HDM activities of the reprocessed catalysts prepared from spent catalyst A and B were similar in spite of a greater level of deactivation of the latter. In their activity, these catalysts approached that of commercial $NiMo/Al_2O_3$ catalysts. In fact, reprocessed catalysts prepared from spent catalyst A were more active for both HDS and HDM, as it is shown in Fig. 9.8 [522]. A high content of V in the catalysts prepared from catalyst C was responsible for low HDS activity. However, negative effect of V on HDM activity was less evident. In fact, even the catalysts prepared from the heavily deactivated catalyst C may exhibit adequate activity in less severe applications.

It should be noted that the trends in Figs 9.5 and 9.6 [522] are in a general agreement with those observed in the studies on preparation of hydroprocessing catalysts reported in the literature [27]. Yet, the reprocessed catalysts were prepared from the material which differed significantly from that used for preparation commercial catalysts. Thus, in the former case, the material has undergone transformations of the fresh catalyst during commercial operation lasting almost one year until its replacement was necessary because of the activity decline. After removing from commercial reactor, the spent catalyst could be subjected to either oxidative regeneration or rejuvenation providing that the level of the activity recovery was still acceptable. Otherwise, other options for the spent catalyst, i.e., metal reclamation, disposal in landfills, etc. had to be identified. However, even for a heavily deactivated catalyst, the activity can be restored by reprocessing spent catalysts according to the method shown in Fig. 9.1 [9].

Another attempt to reprocess spent hydroprocessing catalysts was made by Gardener and Dennis [527] who prepared an active catalyst from spent catalyst that contained Co, Mo, and

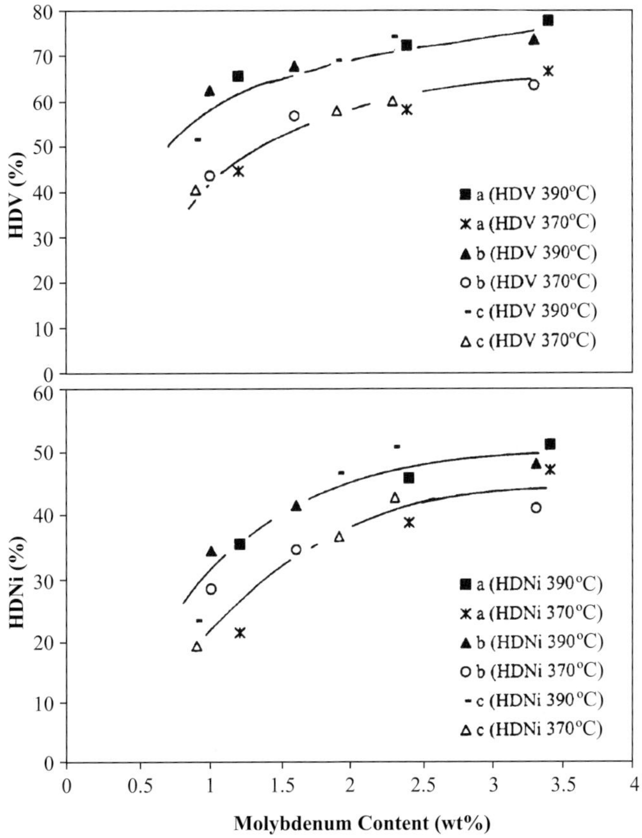

Figure 9.6: Effect of Mo content of catalysts reprocessed from coked spent cata2lysts (a: catalyst A; b: catalyst B; c: catalyst C) on hydrodevanadization (HDV) and hydrodenickelization (HDNi) activities at 370 and 390 °C [From ref. 522. Reprinted with permission].

Al_2O_3 together with carbonaceous deposits. The spent catalyst was mixed with an alumina containing material and extruded to prepare the new catalysts. The prepared catalysts contained 1.7 wt.% Mo, 0.6 wt.% Co and had a higher percentage of pores in the 60–500 Å range. High temperature calcination (at 800 °C) was used for pore enlargement in the catalyst. The catalyst was found to be very effective for HDM residual feeds. In another related study, de Boer [528] used a decoked spent $CoMo/Al_2O_3$ HDS catalyst that contained high levels of MoO_3 (20 wt.%) and CoO (5 wt.%) and small amount of V_2O_5 (0.2 wt.%) and NiO (0.1 wt.%) for preparing a large pore hydroprocessing catalyst. The spent catalyst was mixed with alumina in the form of a fine powder (particle size < 20 μm) and extruded to produce new catalyst extrudates. The prepared catalyst contained 13 wt.% MoO_3 and 3.2 wt.% CoO on alumina support. High temperature calcination treatment was used for pore enlargement. A sample of the catalyst that was calcined at 750 °C had large pores with mean pore diameter of

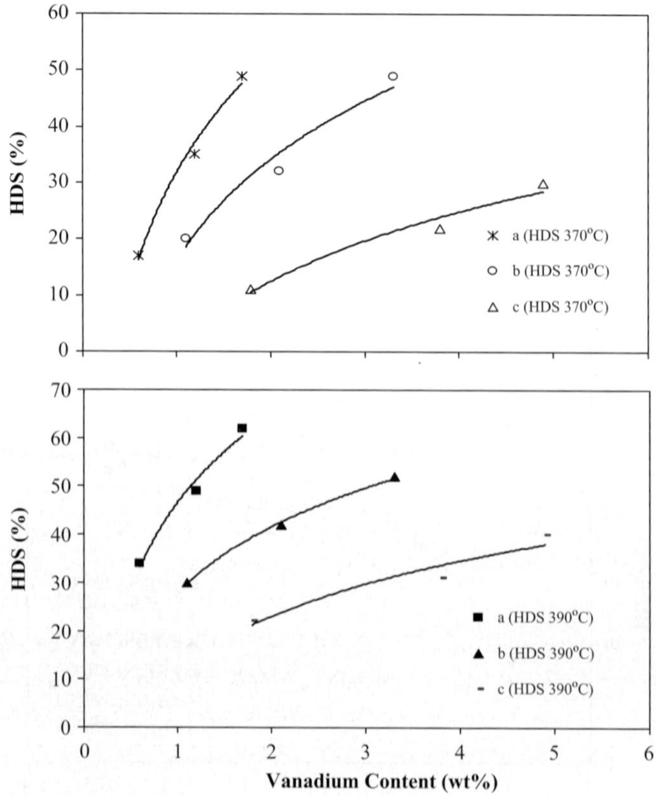

Figure 9.7: Effect of V content of catalysts reprocessed from coked spent catalysts (a: catalyst A; b: catalyst B; c: catalyst C) on hydrodesulfurization (HDS) activity at 370 and 390 °C [From ref. 522. Reprinted with permission].

120 Å. The performance of the catalyst for residual oil hydroprocessing was found to be roughly equal to that of a commercial catalyst.

9.1.1.3 Effect of Hydrothermal Treatment on Reprocessed Catalysts

The activity of the reprocessed catalysts can be significantly enhanced by subjecting spent hydroprocessing catalysts to hydrothermal treatment at 300 °C for 2 h [524]. The pore volume and pore size distribution of the spent catalysts before and after hydrothermal treatment are compared in Fig. 9.9. It is seen that the pore volume after hydrothermal treatment increased from 0.24 to 0.48 mL/g, whereas surface area increased from 37 to 70 m^2/g. The volume of pores in the 100–2000 A diameter range increased from 0.16 to 0.38 mL/g, while the mesopore volume in the 100–500 A increased from 0.12 to 0.24 mL/g.

A hydrothermally treated spent catalyst was used to prepare reprocessed catalysts using boehmite in similar manner as described above (Fig. 9.1), i.e., 20, 40, and 60 wt.% of the

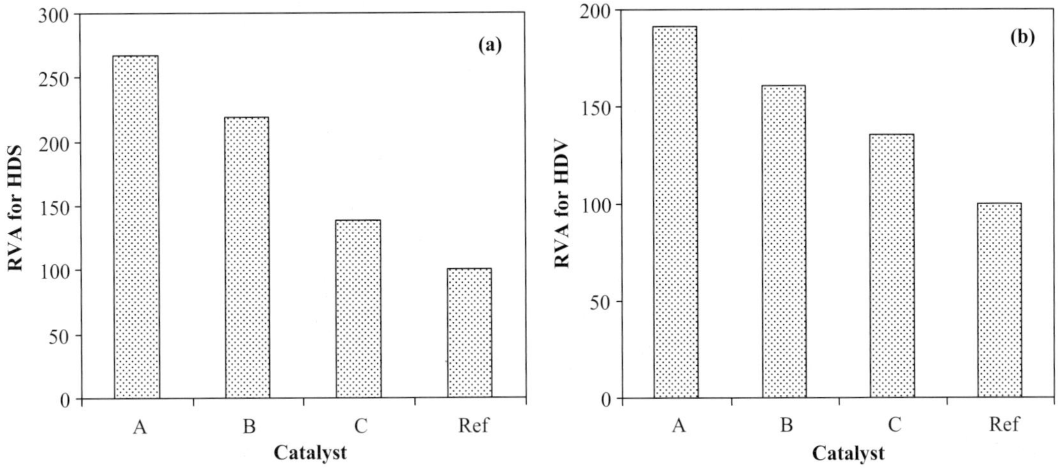

Figure 9.8: Activity comparison between reprocessed and reference (commercial) catalyst: (a) hydrodesulfurization (HDS) activity; (b) hydrodevanadization (HDV) activity [From ref. 522. Reprinted with permission].

sample were mixed with boehmite and extruded. The surface area and pore volume of the reprocessed catalysts shown in Fig. 9.9 [524] indicate a beneficial effect of hydrothermal treatment. For example, for the reprocessed catalyst containing 40% of spent catalyst, surface area and pore volume increased from 70 to 144 m^2/g and from 0.27 to 0.39 mL/g, respectively. The HDS and HDM activities determined under identical conditions as above [9,524,525] shown in Fig. 9.10 [524] confirmed that the additional enhancement in the catalyst activity reprocessed catalysts can be achieved by a hydrothermal treatment of spent catalyst.

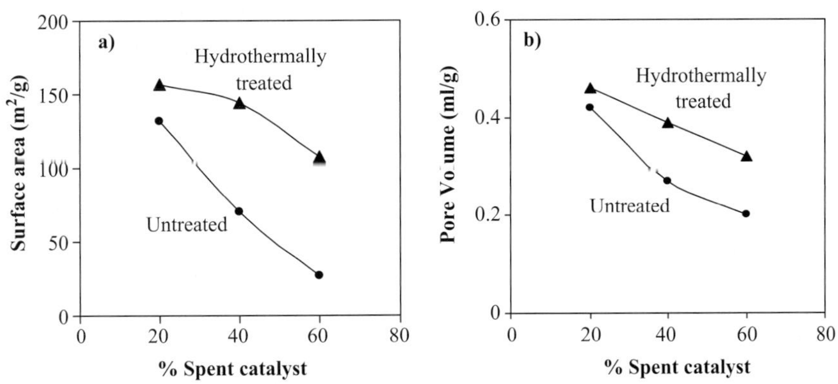

Figure 9.9: Effect of hydrothermal treatment and content of spent catalyst on surface area and pore volume of reprocessed catalyst [From ref. 524. Reprinted with permission].

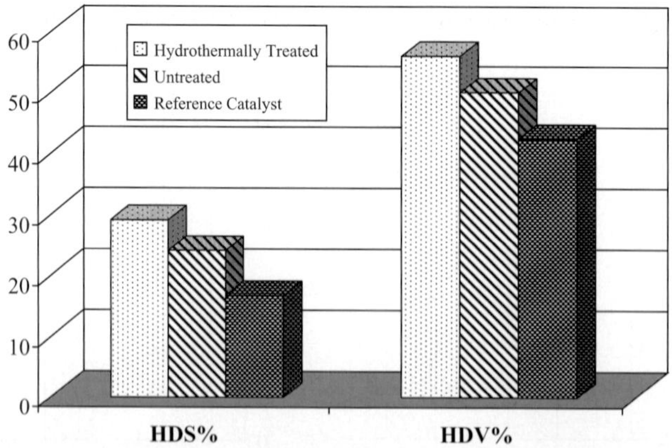

Figure 9.10: Effect of hydrothermal treatment on hydrodesulfurization (HDS) and hydrodevanadization (HDV) activity reprocessed catalyst [From ref. 524. Reprinted with permission].

It was observed that hydrothermal treatment resulted in the conversion of γ-alumina to boehmite according to the following reaction [529]:

$$\underset{\gamma\text{-alumina}}{Al_2O_3} + H_2O \Rightarrow \underset{\text{boehmite}}{2AlO(OH)}$$

Boehmite formed in this way was well crystallized having large size of crystallites. Calcination of boehmite (Fig. 9.1) at 500 °C yielded γ-alumina with large pores. In alumina supports, porosity originates from the space between particles [530]. Therefore, when large size alumina particles are packed to form extrudates, the space between the particles, i.e., pore size and pore volume, will be large. Then, rehydration of γ-alumina in the spent catalyst and its transformation to well crystalized boehmite were responsible for the pore enlargement. Consequently, the activity of the reprocessed catalyst was significantly enhanced.

9.1.2 Other Preparation Methods

Lee et al. [531] found that active reforming catalysts can be prepared using the metals containing extract obtained by selective leaching of spent residue hydrotreating catalysts with either oxalic acid or citric acid. The significant difference between selectivity of these acids is indicated by very low content of Ni and Mo in the extract from oxalic acid leaching, as it is shown in Table 9.5 [531]. Moreover, the selectivity of oxalic acid for V was much greater than that for Ni and Mo. This extract, containing 1.8% V and small amounts of Mo (187 ppm) and Ni (111 ppm), was used for catalyst preparation. High surface area (340 m^2/g) amorphous silica was used as support. The prepared catalyst showed significant dehydrocyclization

Table 9.5: Metal content of acid extraction solutions [From ref. 531. Reprinted with permission].

Metal	Concentration (ppm)	
	Oxalic acid	Citric acid
Fe	20	2650
Mo	167	1530
Ni	111	15000
V	18000	21000

activity to convert paraffins to aromatics. With this catalyst, the octane number of the paraffinic naphtha increased from 55 to almost 87 at 81% yield. The catalyst prepared from the citric acid extraction liquor contained mainly V, Ni and Fe. A high cracking and dehydrogenation activity of this catalyst was attributed to the presence of Ni. This catalyst gave high yields of H_2, coke and light olefins at the expense of aromatics in products.

Cronauer and Bjorklund [532] developed a process for producing fresh molybdenum containing catalysts by depositing molybdenum removed from a spent hydrotreating catalyst that contained Mo, Co or Ni and Al_2O_3 together with some other minor elements such as silicon, iron, and Ti. The spent catalyst was first decoked and sulfided and then treated with anhydrous gaseous hydrogen chloride at temperatures in the range 400 °C to 575 °C. The molybdenum was removed from the spent catalyst as molybdenum oxychloride while the other metals such as Co, Ni, Fe, Ti and the supporting Al_2O_3 or SiO_2 were not affected during chlorination. The vaporized molybdenum oxychlorides was deposited onto a fresh alumina support in the presence of air. The performances of the prepared catalyst for promoting hydrotreating or other reactions were not reported in this study.

A pure MoO_3 suitable for catalyst preparation could be obtained from spent hydroprocessing catalysts using a radiation-thermal treatment [533–536]. The method is based on the observation that active metals in spent hydroprocessing catalysts behaved differently under irradiation by a 1.2 MeV fast electron beam. For example, in the spent $NiMo/Al_2O_3$ catalyst, the Mo-containing phases are selectively destroyed, whereas Ni-containing species are preserved. As a consequence, the sublimation of Mo as MoO_3 is significantly enhanced. Similar observations were also made for the spent $CoMo/Al_2O_3$ catalysts. Potential options for utilization of the Mo free catalysts, i.e., Ni/Al_2O_3 and Co/Al_2O_3 catalysts have not been explored. It is believed that these materials may be suitable for some catalytic applications. However, the method would have to be cost-effective to attract any commercial interests.

9.1.3 Spent Catalysts in Slurry Bed Hydrocracking

The processes employing slurry bed reactors have been developed with the aim to offset the costs of operation associated with a large catalyst inventory required for upgrading problematic feeds via catalytic route. Thus, rather than to use commercial catalysts, these processes employ low-cost solids which contain catalytically active transition metals such as Fe, Ni, V, etc. Such solids are either disposed from industrial operations (e.g., red mud from aluminum production, a high Ni and Cu content solid from nickel and copper production, steel making, etc.) or they are occurring naturally. Therefore, their cost represents only a fraction of that of commercial hydroprocessing catalysts. The best-known processes operating in a slurry mode include Hydrocracking-Deasphalting-Hydrgenation (HDH) process developed by Venezuelan INTEVEP [537] and German Veba Combi Cracking (VCC) process developed by VEBA [538]. They were operated on a demonstration scale (150 barrels per day) and a commercial scale, respectively. Apparently, both processes are suitable for hydrocracking heavy feeds the metal content of which exceeds 1000 ppm of V + Ni (e.g., Boscan). Apparently, there is a number of other slurry bed hydrocracking processes in various stages of development. In this regard, the updated information was published elsewhere [539].

It was disclosed by Beret et al. [540] that fines of the spent catalyst could be used in a process for upgrading heavy petroleum feed. The process comprised a two-stage system. The fines introduced into the first stage had coke-suppressing and demetalizing activity. The effluent from the first stage was directly passed and processed under hydrocracking conditions in the second stage.

In a crushed form, spent hydroprocessing catalysts exhibited a high activity during hydrocracking of vacuum residue in a slurry bed reactor [541]. This is not surprising, because in a powder form of catalyst, diffusion problems encountered during hydrocracking of heavy petroleum feeds are almost completely avoided. During slurry bed hydrocracking, active metals in spent-crushed catalysts (e.g., Ni, Co, Mo, and Fe) are rapidly converted to corresponding sulfides in situ. This results in the enhancement of hydrogen activation on the surface of catalyst particles. As a consequence, active hydrogen converts coke precursors to volatile products before they are converted to coke. Even a heavily deactivated crushed catalyst (without decoking) can be successfully used in the slurry bed reactors. This was confirmed by Matsumura et al. [542] who used a spent catalyst containing 9.6 wt.% V, 6.0 wt.% Ni, and 12.9 wt.% carbon. In spite of the heavy contamination, the catalyst exhibited a good activity during upgrading of vacuum residue. The results published by Sakabe and Yagi [408] support the observations made in these studies [541,542].

It was indicated earlier that the metals loaded extracts from rejuvenation of spent hydroprocessing catalysts may be co-slurried with a heavy feed and as such act as catalysts. Apparently, this option for the utilization of metals from spent catalysts has not yet been explored.

9.2 Catalysts for Non-Petroleum Applications

The utilization of metal-fouled spent catalysts discarded from heavy oil upgrading processes in the preparation of active catalysts for reduction of nitrogen oxides has been reported in some patents [543–545]. The spent catalysts used in these studies contained large amounts of V together with Mo, Ni, and Co on the alumina support. Besides the above components, coke and sulfur were also present in the spent catalyst. According to a process developed by Ku et al. [543], the spent catalyst was first heated in air at a temperature of around 500 °C to remove the coke and sulfur. The resulting coke free materials that contained 11.3% V, 3.4% Ni, 2.7% Mo, 31.7% Al, 0.8% Co, 0.5% Fe was pulverized into a powder with a size of 200 mesh. The spent catalyst powder was mixed with water and acid or alkali to form slurry having a viscosity of 40 to 400 cps and a pH of between 3 and 6. The spent catalyst slurry was deposited in a ceramic comb and then calcined. The catalyst was capable of selectively reducing nitrogen oxides in the exhaust gas with high efficiency. In another related patent, Choi et al. [545] used a tungsten impregnated support (e.g. TiO_2, γ-Al_2O_3) instead of ceramic honey comb for depositing the spent catalyst metals (e.g. V, Mo, Ni). The catalyst prepared by this process was very advantageous in terms of selectivity for reduction of nitrogen oxides as well as better poisoning resistance to sulfur oxides. Further improvements in the performance of such catalysts for selective reduction of nitrogen oxides from exhaust gases was achieved by these authors by using spent catalysts that contained low concentrations of V and high concentrations of Mo [546]. The catalyst preparation procedure comprised the following steps:

(1) Pretreating a spent catalyst discharged from a hydro-desulfurization process of an oil refinery, which comprises 4 wt.% or less of vanadium, 4 wt.% or less of nickel, 5 wt.% or more of molybdenum and 1 wt.% or less of sulfur on an alumina support by thermally treating said spent catalyst followed by washing with water.

(2) Providing a titania impregnated with 3 to 10 wt.% of tungsten on the basis of titania weight.

(3) Pulverizing the pretreated spent catalyst, followed by homogeneously mixing the pulverized spent catalyst with the tungsten-impregnated titania under the addition of water and acid.

(4) Dehydrating the mixture to remove excess moisture and active metal components therein.

(5) Drying the dehydrated mixture at 100 to 200 °C for 9 h or longer, followed by grinding the dried mixture.

(6) Forming a catalyst body by extruding the grinded mixture or coating the grinded mixture to a structure, followed by drying under a constant temperature and humidity condition and then calcining the dried structure.

Catalysts prepared in accordance with the above steps contained 1–3 wt.% of V, 1–3 wt.% of Ni, 2–8 wt.% of Mo and 1–7 wt.% of W and had a specific surface area of 100–150 m^2/g and pore sizes of 120–180 Å. The catalysts were effective for reduction of nitrogen oxides, i.e., 90% or more. In this catalyst, vanadium which facilitates the oxidation of sulfur dioxide to sulfur trioxide was present at a low amount. This was favorable for selective reduction of nitrogen oxides contained in an exhausted gas together with large amounts of sulfur components at high temperature in the presence of ammonia as a reducing agent.

It was reported that the spent-decoked $CoMo/Al_2O_3$ catalyst was active during the reduction of sulfur oxides to elemental sulfur [547]. However, no details on the type of chemical process employed were given in this brief information. However, Zulfuganov et al. [548] reported that in order to achieve an efficient conversion of sulfur oxides to elemental sulfur, the oxidative regeneration of the spent $CoMo/Al_2O_3$ at 500 °C was necessary.

The spent-decoked $CoMo/Al_2O_3$ and $NiMo/Al_2O_3$ catalysts were used for the direct decomposition of H_2S in the mixture with either H_2 or CH_4 [549]. The decomposition to H_2 and elemental sulfur was significantly enhanced in the presence of the spent catalysts compared with little decomposition observed over γ-Al_2O_3. Presulfiding the catalysts did not influence the conversion of H_2S. In every test, the $CoMo/Al_2O_3$ catalyst was more active than $NiMo/Al_2O_3$ catalyst. The H_2S aided reduction of CO_2 was catalyzed by these catalysts as well. However, the spent-decoked catalysts favored the formation of COS, whereas CO was the main product over the corresponding fresh catalysts.

It is believed that the utilization of the spent hydroprocessing catalysts in non-petroleum application has not been fully explored. For example, these spent catalysts can still possess adequate activity for various gas–solid applications. Various types of the Mo- and W-containing catalysts have been used commercially. The presence of V suggests that after an adequate pretreatment the spent catalysts may be suitable for various oxidation reactions, whereas the presence of Ni offers a possibility of utilizing such catalysts in hydrogenation reactions.

9.3 Gas Treatment Sorbents

The presence of various transition metals (e.g., Ni, Co, Mo, W, Fe, V, etc.) indicates the potential of spent hydroprocessing catalysts for preparation of sorbents for removing H_2S from gaseous streams. Such gases are produced during gasification of carbonaceous solids, carbonization of coal, coking of heavy petroleum feeds. In addition, the H_2S containing gases are by-products of refining operations. The removal of H_2S from the gaseous streams is an essential requirement prior to their utilization, i.e., fuel, steam reforming, oligomerization, pyrolysis for olefin production, etc.

Table 9.6: Free energies for reactions of metal oxides with H_2S [From ref. 424. Reprinted with permission].

Reaction	ΔG	
	At 800 K	At 900 K
$MoO_3 + 2H_2S = MoS_2 + 2H_2O$	−21.4	−16.7
$CoO + H_2S = 1/9Co_9S_8 + 1/8S_2 + H_2O$	−14.3	−14.2
$NiO + H_2S = 1/3Ni_3S_2 + 1/6 + S_2 + H_2O$	−11.4	−12.0
$ZnO + H_2S = ZnS + H_2O$	−17.7	−17.7
$Fe_3O_4 + 4H_2S = FeS + FeS_2 + FeO + 4H_2O$	−21.6	−22.9
$FeO + H_2S = FeS + H_2O$	−10.2	−10.3

Traditionally, the gas clean-up has been performed using a combination of Claus and Scott processes which are usually integral part of commercial operations such as petroleum refineries and gasification plants employing coal and petroleum coke as feedstocks. Commercial gas clean-up sorbents contain oxides of Fe, Zn, Ni, Cu, and Ca [550]. The H_2S contaminant is removed from the gases by reacting with metal oxides according to the following general equation in which Me represents metal:

$$MeO + H_2S = MeS + H_2O$$

Therefore, before being used as gas clean-up sorbents, spent hydroprocessing catalysts have to be decoked. The particle size and shape suggest that after decoking spent catalysts can be used directly in the fixed-bed gas cleaning reactors. After saturation with sulfur, the fixed-bed is oxidatively regenerated to be used in next cleaning cycle. The sulfur oxides containing off-gas is contacted with a solution of lime to produce high purity gypsum which is marketable product.

Table 9.6 [424] lists the tentative reactions involving metal oxides in decoked catalysts with H_2S. The large driving forces for the H_2S removal, as indicated by negative ΔG values, confirm the suitability of spent catalysts for gas clean-up. This was experimentally confirmed using three spent-decoked catalysts shown in Table 9.7 [551]. Breakthrough time's correlations which are used to determine cleaning efficiency of the sorbents are shown in Fig. 9.11 [551]. Such correlations are established in a fixed-bed reactor using a simulated coal gasification mixture containing H_2S. The breakthrough times determined in this way are listed in Table 9.8 [551] together with those determined for the best-known commercial sorbent such as zinc ferrite. The first glance comparison indicates superiority of zinc ferrite; however, this difference diminishes significantly when the comparison is normalized to the same amount of active metals. Also, it has to be emphasized that the cost of zinc ferrite is about one order of magnitude greater than that of the spent-decoked catalysts.

Table 9.7: Composition of spent-decoked catalysts [From ref. 551. Reprinted with permission].

Metal (wt.%)	Catalyst A	Catalyst B	Catalyst C
Mo	10.0	7.7	9.7
Ni	< 0.01	3.3	2.9
Co	2.9	< 0.01	< 0.01
Fe	< 0.05	0.06	0.1
V	nd	4.0	nd

9.4 Preparation of Useful Materials from Spent Catalysts

Detailed accounts of the processes reported in patented or published literature on the use of spent hydroprocessing catalysts in the preparation of various types of catalysts and other valuable materials except the recovery of metals and metal compounds were given by Marafi and Stanislaus [22]. Table 9.9 lists useful materials prepared from spent hydroprocessing catalysts and their industrial application. Utilization of spent catalysts as raw materials in the production of a wide range of commercial products is an attractive option for their recycling from both environmental and economic points of view. In this regard, construction materials, e.g., cement, concrete, etc., water treatment applications, abrasive materials and synthetic aggregates have been receiving attention.

9.4.1 Utilization in Cement Industry

Although this outlet is dominated by spent fluid catalytic cracking (FCC) catalysts, after certain pretreatments, spent hydroprocessing catalyst may be suitable as well as long as the

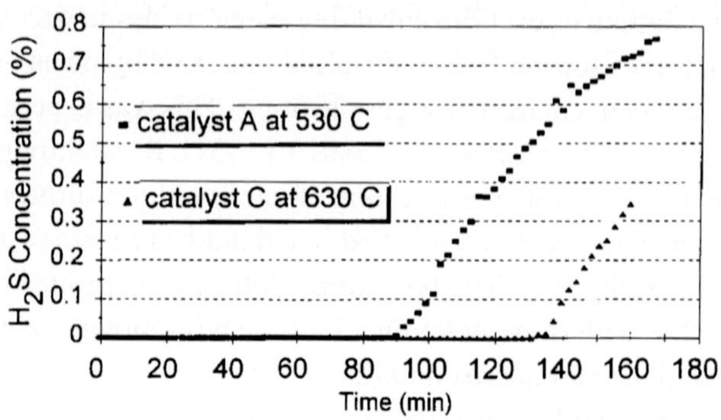

Figure 9.11: Breakthrough curves for catalyst A and catalyst C [From ref. 551. Reprinted with permission].

Table 9.8: Breakthrough times for spent-decoked catalysts and zinc ferrite [From ref. 551. Reprinted with permission].

	Breakthrough time (min)	
	A	A[a]
Catalyst A		
at 530 K	90	101
at 630 K	105	118
Catalyst B		
at 530 K	33	77
at 630 K	24	56
Catalyst C		
at 630 K	47	149
Zinc ferrite		
at 530 K	170	170
at 630 K	180	180

[a] *Normalized to the same amount of active metals.*

content of contaminants is not too high. The use of FCC catalysts in cement and concrete production has been reported in many studies [22,552–554]. These catalysts contain mainly SiO_2 and Al_2O_3 together with some minor impurities. Studies have shown that up to 6% of spent FCC catalysts can be mixed with other raw materials (75% lime stone and 19% clay) for Portland cement production. In the cement Kiln at high temperature SiO_2 and Al_2O_3 react with lime to produce Portland cement.

Extensive studies have been conducted in recent years on utilization of spent FCC catalysts in the production of cement mortars for concrete [22,555,556]. The results in general showed that the spent catalyst could be used as a substitute for sand as well as source of ultra fine material

Table 9.9: Useful materials prepared from spent hydroprocessing catalysts [From ref. 22. Reprinted with permission].

Material produced	Industrial applications
Fused alumina with corrundum crystallite structure	Abrasive material; refractory material
Fused aluminium oxide	Ceramic and refractory industries
Anorthite glass ceramics	Electrical insulating material in electronic industry
Synthetic aggregates	Construction industry; cement concrete
Refractory cement/refractory brick	Refractory material for high temperature furnace

to partially replace cement. Su et al. [552] reported that spent FCC catalyst could substitute up to 15% of cement content in the mortar without sacrificing the quality of concrete. In fact, the substituted concrete was found to show a greater compressive strength than that without substitution [106]. In another study, Su et al. [557] showed that spent catalyst can replace up to 10% of fine aggregate (sand) in cement mortars without decreasing the mortar strength. Pacewska et al. [555] studied the mechanism of interaction of waste FCC catalysts with Portland cement and reported that spent catalyst had pozzolanic properties and its pozzolanic activity depended on its grain size. Addition of the spent catalyst favorably modified the porous structure of the concrete, increased its comprehensive strength, density and reduced water absorption. It was also found that in the presence of the aluminosilicate material of the spent catalyst, the $Ca(OH)_2$ content decreased in the cement pastes due to pozzolanic reaction [558,559]. The surface area of the hydrated paste became higher and the mean pore diameter decreased as compared to reference sample prepared without mixing the spent catalyst.

Conventional spent hydroprocessing catalysts supported on γ-alumina may require some pretreatment to attain suitable cementitious properties before they can be used in cement and concrete industry. However, the advances in hydroprocessing catalysis indicate on the growing consumption of catalysts which are supported on acidic supports. In their properties, these supports (e.g., amorphous silica-alumina, zeolites, etc.) approach FCC catalysts. The increased use of the modified hydroprocessing catalysts is dictated by the steady increase in the supply of heavy crude requiring hydrocracking step to be converted to primary liquids. It is well established that hydrocracking is aided by acidic supports. Currently, such catalysts are used for dewaxing paraffinic feeds to prepare middle distillate fuels or lube base oil from conventional crudes. Similar catalysts have been used for upgrading primary liquids and waxes from Fischer-Tropsch synthesis. Of course, bifunctional catalysts containing noble metals (e.g., Pt and Pd) could only be considered after most of the noble metal was recovered from spent dewaxing catalysts.

9.4.2 Waste Water Treatment Agents

Preparation of a waste water treating agent of superior quality from spent FCC catalysts has been reported in a process developed by Sanga and Nishimura in Japan [560]. The process involves alkali treatment of waste catalyst (zeolite) grains arising from the fluidized bed catalytic cracking units. Cation exchange property similar to that of fresh zeolites imparted to the catalyst by this treatment. Use of spent cracking catalysts (zeolites) containing V, Fe, Ni, copper and/or carbon for biological treatment of waste water from municipal and industrial sources is reported in a US patent [561]. In this process, the waste cracking catalyst is combined with activated sludge and the waste water is contacted with the sludge at conditions at which biological oxidation takes place. The presence of inorganic oxides (from the spent catalyst) in the sludge increased the rate of biological oxidation. Preparation of the active

A and X-type zeolites by hydrothermal synthesis using spent FCC catalysts as raw material has been reported in some recent studies [562,563]. The prepared zeolite material (4A type) exhibited a crystallinity exceeding 90% and had an ion exchange capacity resembling that of commercial products used for water treatment.

For similar applications in water treatment, spent hydroprocessing catalysts would have to be used in a non-decoked form to inhibit leaching of metals. However, in combination with activated sludge unwanted leaching could be minimized. There are also numerous options to minimize the leachability of the metals from spent hydroprocessing catalysts using various methods for the metals immobilization. Therefore, the water treatment applications, as a potential outlet for spent hydroprocessing catalysts, deserve a thorough evaluation.

9.4.3 Other Materials

Although the use of spent hydroprocessing catalyst in cement/concrete production and water treatment appears to be restricted because of its hazardous nature, there are many other useful materials such as fused alumina, synthetic aggregates, anorthite glass-ceramics, refractory cement, and refractory brick, which have been prepared from spent hydroprocessing catalysts. These materials are listed Table 9.9 [22].

9.4.3.1 Abrasives and Alloys

A process for the preparation of abrasive material from spent hydrotreating catalysts (e.g. Co–Mo–Al_2O_3 and Ni–Mo–Al_2O_3) was developed by Zeiringer [564]. The basic steps involved in the process are:

(1) Melting the spent catalyst in presence of a reducing agent such as coke or aluminium metal (powder).

(2) Cooling the melt at a suitable rate to obtain the abrasive material with desired crystallite size.

(3) Mechanically separating the catalyst metal components (Co–Mo, or Co–Mo–Ni–V) accumulating as an alloy residue before or after solidification of the abrasive material containing mainly alumina as corundum crystallites.

The solid product containing alumina as corundum crystallite is an excellent abrasive. Depending on the type and amount of additives as well as the crystallite size, these abrasives may be used for various purposes. The catalyst component metals accumulating as an alloy at the bottom of the melting vessel are separated from the alumina melt before solidification or permitted to solidify therewith in a block. Depending on the type of the catalyst, the alloy is comprised primarily of Mo–Co, W, Ni, Mo–Co–Ni–V and various impurities such as sulphur, carbon, silicon, iron, or titanium. After the alloy has been mechanically separated from the

Table 9.10: Composition of abrasive and alloy from $CoMo/Al_2O_3$ [From ref. 564. Reprinted with permission].

Abrasive (wt.%)		Alloy (wt.%)	
Fe_2O_3	0.10	Co	26.9
CoO	0.15	Mo	62.0
MoO_3	0.06	S	0.04
S	0.04	Fe	5.43
C	0.05	Si	2.15
SiO_2	0.58	C	3.15
MgO	0.05		
Al_2O_3	Balance		

abrasive component, it may be used directly in the steel or alloying industry. If the impurities are present in undesirable amounts, the alloy may be refined in any conventional manner.

Table 9.10 [564] shows the composition of the abrasive material and the alloy prepared from roasted spent Co–Mo–Al_2O_3 catalyst (containing 3.5% Co, 7.7% Mo, 0.05% S, 0.62% Si and the remainder Al_2O_3 carrier) using coke as reducing agent (catalyst to coke ratio = 20:1).

In another preparation, the spent catalyst from residue hydroprocessing operation was used. The roasted catalyst contained, by weight, 3.3% Co, 6.7% Mo, 11.3% V, 3.5% Ni, 0.03% S, 0.54% Si and the remainder alumina. It was mixed with aluminium metal granules and the mixture (spent catalyst to aluminium metal ratio = 6:1) was melted in a three-phase electric arc furnace and cooled. The resultant abrasive and alloy components had the following compositions shown in Table 9.11 [564]. In another patented process developed by Cichy [565] for the preparation of abrasive alumina from spent catalyst, two furnaces were used for smelting spent catalyst. The alumina containing slag furnace is transported to a second furnace for finishing to satisfactory temperature, composition and quality for casting as an abrasive material or cast refractory.

9.4.3.2 Ceramic Materials

Alumina present in spent hydroprocessing catalysts is converted to fused alumina by a smelting process at Gulf Chemical and Metallurgical Corporation (GCMC). The smelting process is actually integrated with metal recovery operations at GCMC for complete recovery of all useful products (i.e. metals and Al_2O_3) from spent catalyst wastes [566,567]. The alumina residue, containing the unleached metals left over after removing the metals (Mo and V) from the spent catalyst by roasting with Na_2CO_3 and leaching processes, is mixed with a reductant material and melted in an electric arc furnace to produce fused alumina for ceramic

Table 9.11: Composition of abrasive and alloy prepared from spent catalyst [From ref. 564. Reprinted with permission].

Abrasive (wt.%)		Alloy (wt.%)	
Fe_2O_3	0.06	Co	12.0
CoO	0.07	Ni	12.9
NiO	0.05	Mo	25.0
MoO_3	0.06	V	41.5
V_2O_5	0.31	C	1.4
S	0.04	Al	1.6
C	0.04	Fe	2.8
MgO	0.05	Si	1.2
Si	0.04		
Al_2O_3	Balance		

Table 9.12: Composition of Gulf Chemical and Metallurgical Corporation (GCMC) smelting products [From ref. 567. Reprinted with permission].

Weight %	Fused alumina	Mixed alloy
Al	47–51	0.1–0.3
Ni	0.1–0.3	37–43
Co	<0.02	12–17
Mo	0.1–0.3	3–17
V	0.1–0.3	4–14
Si	0.1–0.5	1.11
Fe	0.5–1.2	6–13
Na	0.2–1.5	-

and refractory applications. The unleached metals present in the alumina residue form a mixed alloy containing Ni, Co, Mo, and V during smelting process. Composition of the fused alumina and mixed alloy materials produced by the smelting process are presented in Table 9.12 [567].

Recently, Su et al. [557] prepared anorthite glass-ceramic material using the alumina residue left after recovering the metals from spent hydrotreating catalysts. Small quantities of unrecovered heavy metals (Co, Mo, V) present in the alumina base were effectively stabilized in the anorthite glass-ceramic product, which can be used as an electrically insulating material in the electronic industry.

The use of alumina containing industrial wastes (e.g. spent catalysts) in preparing stressing cements with low self-stressing energy is reported in a Russian paper [568]. In the recent

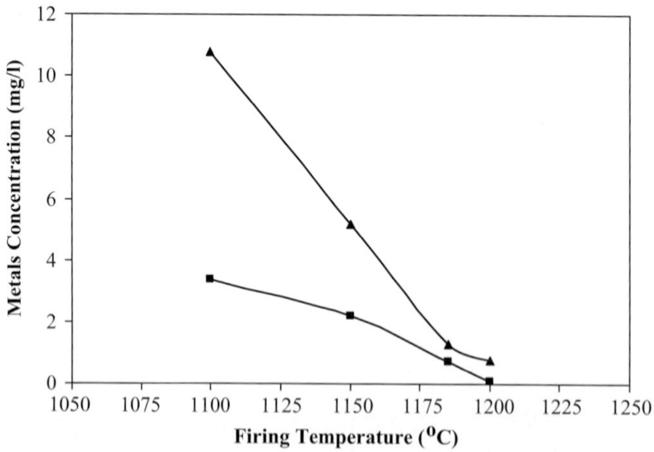

Figure 9.12: Effect of firing temperature on metal concentration in leachate [From ref. 427. Reprinted with permission].

study, Sokolov et al. [569] showed that NiO-containing spent catalysts with alumina carrier was suitable for manufacturing alumina bearing refractory cement.

9.4.3.3 Synthetic Aggregates

Al-Fulaij et al. [570] investigated the possibility of producing non-leachable materials of high compressive strength such as synthetic aggregates from spent hydroprocessing catalysts. The potential for the use of such synthetic aggregates in cement concrete production was also evaluated as part of the study. The process involved mixing about 4 wt.% of the spent catalyst in the form of a fine powder (particle size less than 180 μm) with clay (20 wt.%), gatch (68 wt.%), sand (8 wt.%) and water, shaping the wet mix into small balls of about 20 mm diameter, drying the balls at 110 °C for 12 h and then heating them at high temperature in the range 1150–1300 °C. The effects of temperature on leachability and compression strength of the synthetic aggregate are shown in Figs 9.12 and 9.13 [427], respectively. The results revealed that synthetic aggregate materials with minimum leaching (<1 ppm) and maximum compressive strength were produced when the firing temperature was around 1175 °C [360,478]. The cement concrete prepared from the spent catalyst-based synthetic aggregate had a compressive strength of 461 kg/cm^2, while that of natural aggregate was 485 kg/cm^2. These results clearly indicated that the spent catalyst-based synthetic aggregates were stable and non-leachable and they could be used in the construction industry for concrete production by incorporating in a cement matrix.

9.4.3.4 Bricks Production

The study conducted by Acchar et al. [571] indicates on a potential of utilization of spent hydroprocessing catalysts in manufacturing of red-clay based materials such as bricks.

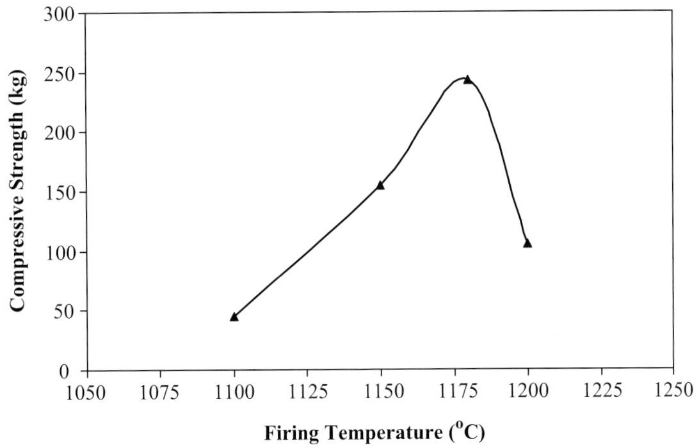

Figure 9.13: Effect of firing temperature on compressing strength of synthetic aggregate [From ref. 427. Reprinted with permission].

Samples containing as much as 20 wt.% spent catalysts were extruded and fired between 700 and 1150 °C in air. The properties of the resultant material, such as firing shrinkage, water absorption and mechanical strength, confirmed that bricks production may be a suitable outlet for spent hydroprocessing catalysts.

CHAPTER 10

Spent Catalysts from Dewaxing Operations

Apparently, n-paraffins are desirable components of diesel fuels because their cetane number is the highest among all hydrocarbon groups. Similarly, n-paraffins possess the highest viscosity index and, as such, they appear to be suitable components of lubricating oils. Unfortunately, n-paraffins have to be removed from both fuels and lubricants because of their undesirable cold flow behavior, i.e., high cloud point, pour point, and freezing point. By converting n-paraffins to i-paraffins, cetane number of diesel fuel, and viscosity index of lube base oil are affected, but they are still much higher than those of aromatic and naphthenic compounds. At the same time, melting point of n-paraffins is significantly decreased, i.e., from +35 °C to −15 °C for C_{20} n-paraffin for the C_{20} iso-paraffin containing 5-methyl substituent, respectively. To achieve desirable cold flow properties, the n-paraffins must be removed from the corresponding petroleum fractions.

Conventional dewaxing involved the removal of aromatics by the extraction using solvents such as furfural, phenols, etc. prior to the removal of n-paraffins. In a concentrated form, aromatics were separated from the solvent by distillation. The extraction of aromatics usually followed by the removal of long chain paraffins. Removal of the former was necessary to improve viscosity behavior of lube base oil. The paraffins were removed by mixing the feed with a solvent, e.g., acetone-benzene, barisol (ethylenedichloride-benzene), trichloroethylene, etc., and cooling the mixture until most of the paraffins solidified in a crystalline form. Subsequently, the solid paraffins were filtered off in rotary drum filters and solvent distilled off for reuse.

During catalytic dewaxing, n-paraffins are converted to i-paraffins. The main reactions occurring during dewaxing include hydrocracking (HCR) and hydroisomerization (HIS). The relative contribution of these reactions to the overall dewaxing depends on the catalyst structure, temperature, and the origin of the feed. For example, HCR is more important reactions when production of diesel fuel and gasoline from vacuum gas oil (VGO) and deasphalted oil (DAO) are considered. A high activity of the catalyst for HIS is desirable to maximize the conversion of VGO and DAO to the lube base oil by converting n-paraffins to iso-paraffins. It is advantageous to use a catalyst possessing a high HIS activity when naphtha fraction is used for gasoline production. In fact, in some cases, the coproduction of transportation fuels (gasoline, jet fuels, and diesel) with lubricants may be part of the refinery strategy [572]. The catalyst possessing a good HCR activity besides the adequate HIS activity would be more suitable for such applications. To meet specifications for aromatic contents in fuels and viscosity index of lube base oil, a hydrofinishing step may be necessary, whereas a dehazing step is sometimes the last step during lubricants production. Therefore, although the most important, dewaxing is not the sole step when lube base oil and fuels are the products of interest.

On a commercial scale, catalytic dewaxing became part of the petroleum refining schemes only since middle 1980s [573]. Catalytic dewaxing is conducted under conditions which only slightly differ from those employed during conventional hydroprocessing (e.g., 260–430 °C and 2–5 MPa of H_2, 250–450 m^3 of H_2 per meter cube of the feed). Typically, catalytic dewaxing is carried out in the trickled-bed reactor over either a bi-functional, zeolite catalyst or in the fixed-bed comprising several reactors and/or layers/sections of different catalysts (Fig. 3.11). For VGO and DAO feeds, the structure of dewaxing catalysts used in such operations differs from that of the conventional hydroprocessing catalysts. Different types of catalysts may be necessary for upgrading the products from Fischer-Tropsch synthesis in which n-paraffins and n-olefins account for most of the hydrocarbon groups present.

Apparently, hydroprocessing step is necessary prior to catalytic dewaxing to remove sulfur, nitrogen, resins, as well as the traces of metals and asphaltenes. This step needs to be conducted using conventional catalysts with the aim to protect dewaxing catalyst in the subsequent step, particularly if the latter catalyst contains noble metals (Pt, Pd, etc.) and acidic supports (e.g., zeolite, amorphous silica-alumina, silica-alumina phosphates, etc.). Thus, such a catalyst may be sensitive to NH_3 and H_2S. In the case of lubricants, the primary product of dewaxing may require the additional hydrofinishing and/or dehazing to ensure that the properties of base oil are in line with specifications of commercial products. Because of high hydrogenation (HYD) activity, the noble metals containing catalysts are the most suitable for the final refining step. Therefore, besides conventional hydroprocessing catalysts, other types of catalysts are being as part of the integrated dewaxing operations.

10.1 Conventional Catalysts

Hydroprocessing step is an integral part of the overall dewaxing process. The objective of this step is the removal of contaminants, which would affect catalyst life in the downstream reactors. For this purpose, conventional hydroprocessing catalysts used for upgrading distillate fractions have been used. The catalysts are selected to match properties of the feed and those of the anticipated products. Because of only trace quantities of metals and asphaltenes present in the feeds, the regenerability of spent catalysts from the hydroprocessing step of dewaxing units is usually high. Thus, several utilization–regeneration cycles should be achieved without any difficulties.

Coke deposition and poisoning by N-bases are the main causes of catalyst deactivation. Under optimal operating conditions (e.g., temperature, H_2 pressure, H_2S/H_2 ratio, etc.), a continuous operation lasting more than two years should be achieved. The properties of spent hydroprocessing catalysts from dewaxing operations are identical as those of spent catalysts used for upgrading distillate fractions. Detailed accounts of these properties were given in preceding chapters. Environmental and safety aspects are identical as well.

10.2 Dewaxing Catalysts

The HIS/IS (isomerization) are among the most important reactions for converting n-paraffins and n-olefins to corresponding isomers. For the latter, double bond IS may be part of the overall mechanism. To various degrees, cracking reactions occur either in parallel with HIS or as consecutive reactions because of the acidic nature of catalyst employed [574,575]. Such situation is desirable in the case of paraffinic feeds with boiling range exceeding that of the fuels of interest. The catalysts used for HIS and HCR reactions must exhibit a desirable acidity [576]. For HCR, the acidity must be regulated to prevent excessive formation of the unwanted gaseous by-products and coke. At the same time, the acid strength needed for HIS depends on the length of chain. Thus, the HIS of light paraffins (e.g., butane and pentane) requires very strong acidic sites, i.e., such as those encountered in the oxo-anion promoted zirconia. For long chains, the acidic strength required for branching is lower. This suggests that a bi-functional catalyst, exhibiting a good activity for simultaneous HIS and HCR, requires the optimization of the acidic sites. For example, a catalyst with the dominance of medium and weak acid sites may exhibit a high activity for HIS, but its activity for HCR may be rather low [576]. To various degrees, other hydroprocessing reactions occur in parallel with HIS and HCR reactions. This suggests that origin of the feed and properties of anticipated products have to be taken into consideration while designing catalysts for HIS/HCR.

10.2.1 Composition of Dewaxing Catalysts

A wide range of catalysts have been developed and tested for IS and HIS of n-paraffins and n-olefins. These catalysts were dominated by different combinations of active metals (e.g., Pt and Pd) with zeolites, although during the early stages of research amorphous silica-alumina and active clays were also receiving attention. Silica-alumina phosphates (SAPO) molecular sieves and sulfated-zirconia-based catalysts have been attracting interests as well. In addition, active metals supported on different supports (SiO_2, Al_2O_3, and various carbon supports) were used in several studies. The activity determination involved both model compounds and real feeds. The studies in which different types of catalysts were tested under identical conditions are of a particular importance for comparison of catalyst performance [577,578]. The H_2 pressure varied between atmospheric up to about 6 MPa, whereas temperatures was between

100 and 400 °C. It has been noted that some catalyst testing was done in the IS mode, e.g., in the presence of N_2 as carrier gas.

Among zeolites, a high selectivity was achieved using the medium pore size ZSM-5 zeolite. Bendoraitis et al. [579] observed that the ZSM-5 zeolite and Pt/mordenite exhibited the shape selective properties. The latter had larger pores than ZSM-5 zeolite. The pour points of lube base oil were similar over both catalysts, whereas viscosity index over ZSM-5 catalyst was higher than that over Pt-mordenite [579]. A modification of the ZSM-5 zeolite was tested by Chen and Garwood [580] for dewaxing a middle distillate fuel. In this case, the zeolite was exchanged with Zn/CH_4Cl solution to obtain 0.9 wt.% of Zn. The activity of this catalyst was in the range of that of the Ni/kieselguhr catalyst [581]. In view of the similarity of the process, the observations made during dewaxing of the middle distillate feeds may also be relevant for dewaxing of VGO and DAO.

A series of non-zeolitic catalysts was patented by Gillespie et al. [582–584]. The catalyst development was based on the observation that catalysts containing lanthanide series and platinum-group components provide a superior performance and stability during HIS of the full boiling range of n-paraffins to i-paraffins. In this case, the catalyst comprised the tungstated support zirconium oxide, a lanthanide element, and/or yttrium as the first component, and a second component being a metal from platinum-group. The first component consisted of a single lanthanide-series element or yttrium, while the second component of a single platinum-group metal. Preferably, the first component was ytterbium, holmium, yttrium, cerium, europium, or their mixtures, whereas the second component was Pt. The catalysts contained the inorganic-oxide binder, such as alumina. Another solid-acid HIS catalyst patented by Gillespie [583] consisted of the sulfated ZrO_2 as well as a group III A component, and Pt. The catalyst was active for the conversion of a paraffinic feed to an iso-paraffin rich product with the significantly enhanced cold flow properties. The catalyst comprising of the support of the tungstated zirconia, at least one lanthanide element, preferably ytterbium or holmium and Pt, was active for the selective upgrading a paraffinic feed to iso-paraffins rich product [584]. Other suitable supports included HfO_2, TiO_2, and SnO_2. Another catalyst consisted comprised a combination of the sulfated ZrO_2 support with one lanthanide element or yttrium component, and the Pt deposited Al_2O_3. This was very active catalyst for HIS as well. In this case, the Al_2O_3 was used as the binder [585].

Vigorous research activities in the development of novel HIS catalysts by several research groups should be noted. This research has been carried out in line with a growing interest in synthetic fuels from Fischer-Tropsch synthesis. Detailed accounts of these studies will be given by de Klerk and Furimsky [586]. A cursory account is only given in this book to indicate fundamental differences between the structure of conventional hydroprocessing catalysts and catalysts comprising novel phases.

10.2.2 Deactivation

It is believed that the life of catalysts used for dewaxing, as well as those for HYD, hydrofinishing, and dehazing, which may be integral parts of the overall dewaxing process, should approach at least two years providing that no unexpected events occurred during the operation. In most cases, the feed for dewaxing is free of contaminates because it was already pretreated by hydroprocessing. Therefore, the coke deposition and poisoning by N-compounds should be the main causes of catalyst deactivation. However, N-bases would have to be removed down to few parts per million (ppm) of nitrogen in the case that the dewaxing catalysts contain noble metals. Such catalysts can also be poisoned by H_2S. Therefore, the deep HDS and HDN of the feed may be necessary before dewaxing. To a certain extent, recrystallization of the catalytically active phases to less active phases during the prolonged exposure to hydroprocessing conditions could occur as well. This could be the cause of permanent catalyst deactivation.

In spite of rather clean systems being used, catalyst deactivation during HIS/IS could not be avoided. This was evidenced by declining conversions and selectivity with time on stream during the HIS of a paraffinic feed obtained from Fischer-Tropsch synthesis free of sulfur and nitrogen [587]. In this case, oxygenates present in the feed were the major contributor to the activity decline. In the absence of oxygenates, the coke deposition was the main factor contributing to deactivation, although recrystallization of active phase affecting catalyst activity could not be ruled out. It is expected that the structure of coke formed during the dedicated HIS operations will differ markedly from that observed on the spent catalysts used in hydroprocessing of petroleum feeds. For example, Cowley [587] reported that the coke formed during the HIS of pentenes over an acidic non-zeolitic molecular sieve catalyst was not aromatic, and did not comprise hydrogen deficient polynuclear aromatics, i.e., hard coke, but rather paraffinic, olefinic, or polyolefinic structures with the H/C ratio exceeding 1.0. However, the structure of coke depends on the type of catalyst, and operating conditions. Therefore, it is possible that the formation of an aromatic coke during the HIS of hydrocarbons occurs as well, though to a much lesser extent.

The influence of H_2 pressure (from atmospheric to 0.5 MPa) on the HIS of n-octane over Pt/SAPO-5 and Pt/SAPO-11 was investigated by Campelo et al. [588] at 648 K. The deactivation with time on stream was more pronounced for the Pt/SAPO-5 catalyst. However, gradual increase in H_2 pressure (to 0.3 and 0.5 MPa) resulted in an enhancement in catalyst activity with time on stream, although during the subsequent time on stream, the activity continued to decline. The activity increase coincided with the decreasing amount of coke deposited on the catalyst with increasing H_2 pressure. For SAPO-11, the same change in H_2 pressure resulted in a significant increase in the activity. Moreover, no catalyst deactivation during the entire run (almost 16 h on stream) was observed. It is therefore believed that the coke formed under conditions of HIS is rather "soft" and can be reversibly removed by an increase in H_2 pressure and temperature. Because of its potential benefit to operation, this should be the focus of additional investigations.

10.2.3 Environment and Safety

The environmental and safety aspects of spent hydroprocessing catalysts used in the upstream and downstream units of dewaxing units were discussed in details in Chapter 5. After being unloaded from reactor, spent dewaxing catalysts have to be handled according to the procedures prescribed by environmental and health authorities, similarly as spent hydroprocessing catalysts. The properties of spent catalysts from dewaxing may differ from those of typical hydroprocessing catalysts. First of all, milder operating conditions and only trace quantities of contaminants (sulfur, nitrogen, asphaltenes, and metals) in the feeds, combined with a high HYD activity of dewaxing catalysts, ensure a low aromaticity of coke on catalyst surface [587]. Most likely, such coke possesses a high concentration of sites, which are very active for oxygen adsorption. Consequently, a pyrophoric nature of coke is much greater that that of coke on spent conventional hydroprocessing catalysts. In addition, if present, noble metals are generally in a reduced form and, as such, are active sites for oxygen chemisorption as well. Some noble metals compounds form complexes with hydrogen [55]. Such complexes may be present in the coke, unless the spent catalyst was subjected to a special pretreatment. Then, if present, these complexes would react rapidly on the exposure to air. Therefore, much more precautions have to be taken during handling, storage, transportation, and disposal, particularly during the unloading of such catalysts from reactor, compared with spent hydroprocessing catalysts. A total lack of information on the pyrophoric nature of spent catalysts from dewaxing operations should be noted. In this regard, a database of properties of spent dewaxing catalysts is desirable.

Similarly as for pyrophoric properties, there is little information on the behavior of spent dewaxing catalysts during the contact with water alone and/or in combination with air and water. The leachability of spent noble metals containing catalysts and their toxicity levels while in a solution have not yet been determined. This may be attributed to a high value of noble metals. Thus, rather than to dispose of, such spent catalysts are of a much greater interest to both refining companies and metal reclaiming companies compared with spent hydroprocessing catalysts.

10.2.4 Regeneration

First of all, the amount of coke on spent dewaxing catalysts may approach ~5 wt.%, whereas for some spent hydroprocessing catalysts, this amount may exceed 20 wt.%. As it was already indicated, the coke on spent dewaxing catalysts is much more reactive (less refractory) than that on the latter catalysts. These differences suggest that a desirable level of catalyst activity recovery may be achieved in situ by an increase in H_2 pressure and temperature, as it was reported by Campelo et al. [588]. These observations indicate that coke deposition on dewaxing catalysts is a reversible process. However, this can be concluded using only a rather limited database. In view of the potential for extending life of the catalyst, this method of catalyst reactivation deserves additional attention. Hydrofinishing and dehazing steps, which may be part of the overall catalytic dewaxing schemes, are usually conducted under milder conditions over noble metals containing catalysts suggesting that the deposited coke would be even more reactive than that deposited on dewaxing catalysts. Such spent catalysts would be even more suitable for reductive regeneration and/or reactivation.

In view of the above facts, regenerability of the catalysts from all stages of catalytic dewaxing (e.g., HYD/HCR, HIS, hydrofinishing, and dehazing) for reuse should be rather high in the case that oxidative regeneration method is used for decoking. Because no contaminant metals were present in the feed, the catalyst regeneration would only require decoking, using the oxidative burn-off. This would involve the same procedures, which were discussed in the Chapter 6 for the conventional hydroprocessing catalysts. However, a special attention must be paid to the temperature control because of a higher reactivity of coke. Otherwise, temperature excursions, which could damage catalysts, could not be avoided. According to Jacobsen [589–591], the indirectly fired kiln used for the oxidative regeneration of spent noble metal containing catalysts must comprise extremely accurate, multiple temperature zones programmed for a specific type of spent catalyst. As it is required for regeneration of spent hydroprocessing catalysts, the kiln used for regeneration of spent catalysts containing noble metals must be integrated with downstream air pollution equipment, i.e., scrubbers, baghouses, and incinerators, to ensure the compliance with environmental regulations.

The presence of noble metals (e.g., Pt and Pd) in some dewaxing catalysts suggests that the conditions of their regeneration may approach those applied during regeneration of the reforming catalysts [592,593]. For the latter more than 100, and in one case almost 230 regeneration–utilization cycles could be performed before the catalyst replacement was necessary almost after five years [594]. The large number of the utilization–regeneration cycles attained for the reforming catalyst may be partly attributed to rather clean feed (e.g., hydroprocessed naphtha) used for reforming compared with the hydroprocessed VGO and DAO. Therefore, it is believed that for dewaxing catalysts, the number of the regeneration–utilization cycles will be less than that for reforming catalysts, but much greater than that for typical spent hydroprocessing catalysts. The regenerability of the Pt/Pd containing catalysts used for hydrofinishing, and dehazing as part of the overall middle distillates, and lubricants production, is expected to be much greater than that of the typical dewaxing catalysts. In fact, the duration of these catalysts in the operation may approach that of the reforming catalysts. Thus, the temperatures employed during the hydrofinishing and dehazing operations are usually about 100 °C lower than that used during dewaxing. With respect to the HYD equilibrium of aromatics, such conditions are favorable for maintaining a high rate of hydrodearomatization.

The regeneration of noble metals containing spent catalysts can rely on the experience gained during regeneration of spent reforming catalysts, which usually contain Pt on γ-Al_2O_3 [15]. In some cases, regeneration is an integral part of the overall reforming operation. For regeneration, spent catalyst is withdrawn from reactor and transferred to the regenerator where coke is burnt off and catalyst reactivated. It has been noted that a small portion of catalyst, so called non-flowing catalyst, remains on the walls and at the bottom of reactor. The coke on this catalyst may continuously build up to approach 50 wt.%. For such catalyst, metal recovery may be the only option. However, it was reported by Blashka et al. [595], that a light fraction could still be recovered from the heavily deactivated catalyst by density separation. The separated light fraction can then be reactivated and returned to the operation.

10.2.5 Metal Reclamation

Compared with the conventional spent hydroprocessing catalysts, entirely different approach is used for the non-regenerable catalysts containing noble metals. Thus, because of the high value of the noble metals, refineries must pay special attention to the inventory and handling of such catalysts. This involves an accurate chemical analysis determining the content of noble metal, which has to be performed before shipping spent catalysts for metal reclamation. This was confirmed by Rosso et al. [596,597]. For example, in the case of Pt, a tolerance of $\pm 0.5\%$ may represent a significant monetary value. The accuracy of the analysis can be significantly improved by the removing contaminants (e.g., coke, sulfur, moisture, etc.) using preburning [589–591]. Both on-site and off-site preburning has been practiced, although with respect to the final settlement with the metal reclaiming company, the former appears to be more favorable for the refinery.

There are a number of processes operating on the commercial scale for the recovery of precious metals from various types of spent catalysts [15]. These methods can be applied towards spent dewaxing catalysts after minor modifications. One of the methods is based on the dissolution of the γ-Al_2O_3 support, while leaving precious metals in a solid form. The dissolution can be achieved using either NaOH or H_2SO_4 [591,597,598]. For NaOH, decoked catalyst is treated at 200 °C and under pressure of 10 bar [598]. During this treatment, the support is converted to soluble sodium aluminate, which is separated from Pt by filtration. Then, the Pt can be dissolved using the treatment with HCl and Cl_2. Once in the solution, Pt higher than 99.9% purity can be obtained. In similar manners, Ir can be isolate from spent catalysts [598].

On the other hand, the dissolution of metals, while leaving the support undissolved, can be achieved by the mixture of HCl and an oxidant, i.e., HNO_3, H_2O_2, $NaClO_3$, NaOCl, etc. [599,600]. The process can operate either in a batch mode or a continuous mode. In the study of Mastny et al. [601], the decoked catalyst containing Pt and Re was treated either with 5 M HCl/HNO_3 acids or ammonia. The temperature had a pronounced effect on metals dissolution. For example, using HCl at 90 °C, almost complete dissolution of Pt and Re was achieved after 2 and 4 h, respectively. When aqua regia was used, the rate of dissolution was higher. When diluted ammonia was used at 60 °C, after 1 and 2 h, 80 to 85% and 85 to 91% of Re, respectively, was dissolved, while almost all Pt remained in the catalyst. An alkaline cyanide solution was found to be suitable for the recovery of Pd from spent catalysts [602]. Once in the solution, precious metals can be isolated in the pure form by precipitation [603,604], solvent extraction [605], and membrane separation [606].

Other metal reclamation process is based on the gas phase volatilization of precious metals using the Cl-containing agents such as $AlCl_3$, CCl_4, mixture of $CO + Cl_2$, and phosgene [599]. In this method, Pt and Pd are selectively chlorinated to volatile products, which on cooling condense. Apparently, main constraints of this method are the handling of toxic agents, and products.

The pyrometallurgical process patented by Japanese inventors is based on the mixing spent Pt containing catalyst with the mixture of metallic copper and copper oxide, as well as a flux and reductant components [605]. The melt of this mixture consisted of two layers, i.e., one containing Pt and copper metal, and the other containing slag. These layers could be readily separated. The Pt enriched mixture was heated in air to partially oxidize copper metal to copper oxide. This resulted in the formation of two layers, one containing oxide and the other consisting of copper metal significantly enriched with Pt. These two layers could be readily separated for further processing.

For noble metal catalysts supported on carbon, the total oxidation of carbon and carry-overs may be the method of choice for noble metals recovery from spent catalysts [607]. For this purpose, a two-stage method may be suitable. In the first stage, reactive carry-overs are removed by controlled oxidation. This follows by a prolonged oxidation of carbon support, which is much less reactive than carry-overs. After removal of all carry-overs and carbon support, the remaining material was very homogeneous. Also, its volume was substantially decreased.

This chapter was incorporated in anticipation of more advanced hydroprocessing catalysts containing precious metals and acidic supports entering the market in a near future. As it has been indicated, the information on development and testing of such catalysts is rather extensive, whereas the properties of the corresponding spent catalysts (e.g., regenerability, toxicity, flammability, leachability, etc.) are unknown. Additional research is necessary to fill this gap.

CHAPTER 11

Metal Reclamation from Spent Hydroprocessing Catalysts

Considering stringent environmental and safety regulations (e.g., unlimited liability), metal recovery from spent catalysts is more attractive an option than landfilling. Previous chapters indicated that spent catalysts, discarded as solid wastes from the hydroprocessing units of petroleum refining industries, contain alumina and metals, such as Mo, Ni, Co, and V, in appreciable concentrations [9,13,27,121]. These metals are highly valuable and are used extensively in the steel industry and in the manufacture of special alloys. These metals are usually manufactured from the ores and minerals containing them. Spent hydroprocessing catalysts could be used as a cheap source for these valuable metals. This will result in recycling and reutilization of the waste catalysts. Consequently, environmental problems associated with handling of spent catalysts can be eliminated. In view of the environmental and economic benefits, increasing attention has been paid to the development of processes for recovering metals and other valuable materials from spent non-regenerable hydroprocessing catalysts. This is supported by a growing number of articles appearing in scientific and technical literature.

Several methods, such as chlorination, acid leaching, alkali leaching, bioleaching, roasting with soda salts, etc., have been studied. The detailed accounts of these studies have been given in both open and patent literature. Focus has been on recovery of Mo, Ni, V, and Co from the spent hydroprocessing catalysts. Several companies have also been established for large-scale reclamation of metals and metal compounds from spent hydroprocessing catalysts. Table 11.1 [608] lists pure compounds which can be isolated from spent catalysts using various reclamation methods. In the following sections, the information available in the literature both on the laboratory studies and industrial scale processes for recovery of metals from spent hydroprocessing catalysts are reviewed.

11.1 Laboratory Studies on Metal Recovery from Spent Hydroprocessing Catalysts

Most of the studies on recovery of metals from spent hydroprocessing catalysts involve leaching with the solutions of both inorganic and organic agents. Leaching with the aid of microorganism, i.e., bioleaching has been attracting attention as well. The dissolution of

Table 11.1: Typical compounds recovered from spent hydroprocessing catalysts [From ref. 608. Reprinted with permission].

Alumina trihydrate	Alumina-silica aggregate
Calcium tungstate	Ammonium molybdate
Chromium oxide	Ammonium vanadate
Molybdenum oxide	Nickel-cobalt concentrate
Molybdenum trisulfide	Sodium vanadate
Vanadium hydroxide wet cake	Sodium molybdate
Vanadium pentoxide	Ferro molybdenum
Ferro vanadium	

metals in water may also be enhanced by roasting spent catalysts with compounds containing alkali metals, such as sodium and potassium. Two-stage processes may employ both leaching and roasting. The volatilization and/or dissolution of metals of interest can be enhanced by chlorination. Attempts have been made to develop novel methods which could be competitive with conventional methods for metal reclamation.

Once in the solution, the metals can be isolated in a pure form using established methods based on selective precipitation. There have been decades of experience with metals separation from various solutions using extracting agents of organic origin. In this case, a high selectivity of extraction can be achieved by merely adjusting the pH of the solution containing an extracting agent. The structures of several extracting agents which have been used commercially are shown in Fig. 11.1 [622].

11.1.1 Leaching Studies

Aqueous solutions of ammonia and ammonium salts, various concentrations of inorganic acids (e.g., HCl, H_2SO_4, and HNO_3) and alkalis have been used as leaching solutions. Among organic agents, water-soluble organic acids have been attracting most of the attention. Compared with inorganic acids and ammonia salts solutions, aqueous solutions of organic acids ensure an environment requiring much less safety precautions. A wide range of microorganisms and fungi have been tested for their suitability and selectivity for bioleaching.

11.1.1.1 Leaching with Ammonia and Ammonium Salt Solutions

Ammonia and ammonium salt solutions have been used by many researchers to extract metals, such as Mo, V, Co, and Ni, from spent catalysts. In the patented process, Gutinikov [609] used ammonium carbonate solution for leaching of metals from spent catalysts and recovered more than 90% of Mo and V, and 60–70% of Ni. A similar process was reported by Millsap and Reisler [610] for extracting metal values from the spent Ni-Mo/Al_2O_3 catalyst. The catalyst

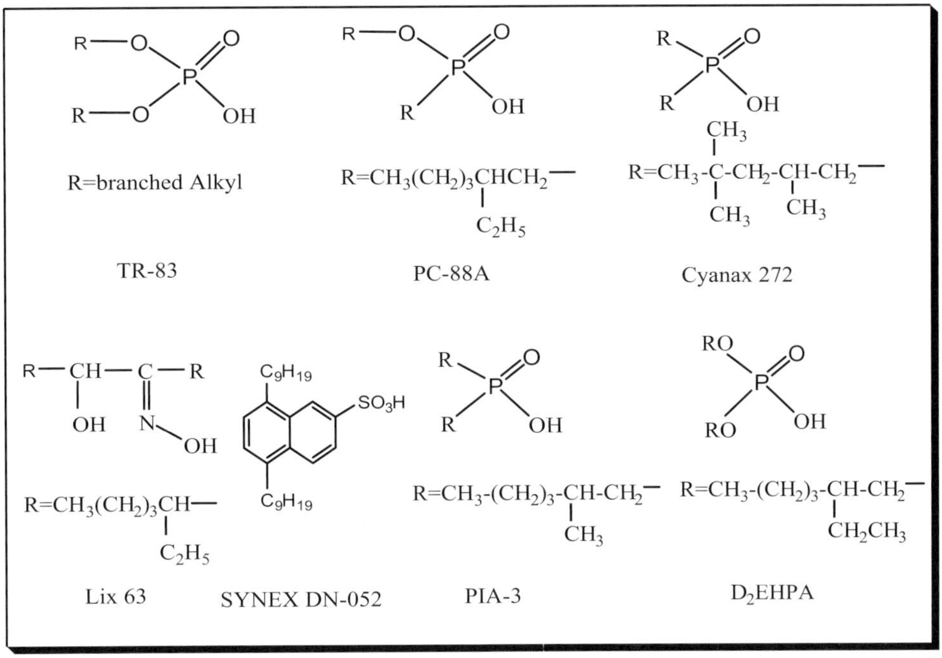

Figure 11.1: Some commercial solvent extraction reagents [From ref. 622. Reprinted with permission].

was first roasted and then leached with the solution of ammonium hydroxide and ammonium carbonate to extract Mo and Ni leaving Al_2O_3 as a residue. Ni was precipitated as carbonate after stripping the excess ammonia from the leach solution. Molybdenum was recovered by solvent extractions and precipitated as calcium molybdate.

The use of ammonium salt solutions for metal extraction from spent catalyst is also described in many patents assigned to Chevron Research Company [610–612]. For example, Hubred and Van Leirsburg [612] were able to extract at least 85% of Mo, 75–85% of V, 75–80% Ni, and 45% of Co from decoked spent hydroprocessing catalysts by leaching with an aqueous solution of ammonia containing an ammonium salt, such as ammonium carbonate or ammonium sulfate. The pH of the ammonia solution was in the range of 9.5–11 and the extraction temperature was in the range of 85–95 °C. Decoking temperature and leaching time were found to have a significant effect on the extracting of Ni and Co. Nickel extraction was affected as the roasting temperature during the decoking process was increased above 600 °C. Leaching time had a pronounced effect on the Co recovery, which reached the maximum at about 5 to 10 min, and subsequently decreased. At the same time, the alumina extraction was negligible (<0.1%). Separation of the individual metals from the leach-solution containing the mixed metals was achieved by solvent extraction. For example, Mo and V could be separated from Ni and Co by extraction with the quaternary amine at a pH of 10.4. The LIX 64N and

LIX 51 agents were used to extract Ni and Co from the solution after Mo and V extraction [612].

In another related study, Marcantonio [613] used an aqueous solution of ammonia, ammonium salt, and hydrogen peroxide to extract metals (V, Mo, Ni, and Co) from spent hydroprocessing catalysts. The ammonium salts used in the study included $(NH_4)_2CO_3$ and $(NH_4)_2SO_4$. The initial pH and the concentrations of ammonia, ammonium salt, and H_2O_2 in the mixed reagent strongly influenced the extent of extraction of metals. The V and Co extractions were remarkably enhanced by H_2O_2 in addition to the ammonia + ammonium salt reagent, while little effect of H_2O_2 on the Mo and Ni extractions was noted. About 93% of Mo, 88% of V, 80% of Ni, and 78% of Co present in the spent catalyst were extracted by leaching with a reagent having an initial pH of 10.4, and ammonia, ammonium salt and H_2O_2 concentrations, 2 M, 0.5 M and 0.14 M, respectively.

The selective recovery of Mo from the unsupported spent MoS_2 catalyst contaminated with Ni and V, by reducing the oxidized catalyst with a reducing agent, such as hydrazine, before extraction with ammonia, is described in another patent assigned to Chevron [614,615]. The spent catalyst was first oxidized to convert the metal sulfides to oxides and then treated with hydrazine to reduce the high valence state of vanadium to a lower valence state. It was then leached with aqueous ammonia. The use of a reducing agent before the ammonia extraction step resulted in superior separation of Mo from V and Ni. The recovered Mo can be sulfided to produce an active MoS_2 catalyst for recycle.

In another study patented recently, Marcantonio [616] used ammonia pressure leaching technique in the presence of oxygen to produce ammonium molybdate, ammonium metavanadate (AMV), and nickel ammonium sulfate from an unsupported spent hydroprocessing catalyst that contained sulfides of Mo, V, and Ni. The reactions that occur in the process are shown below:

$$MoS_2 + 4.5O_2 + 6NH_3 + 3H_2O \Rightarrow (NH_4)_2MoO_4 + 2(NH_4)_2SO_4$$

$$V_2S_3 + 7O_2 + 8NH_3 + 4H_2O \Rightarrow 2NH_4VO_3 + 3(NH_4)_2SO_4$$

$$NiS + 2O_2 + 6NH_3 \Rightarrow Ni(NH_3)_6SO_4$$

The ammonium molybdate and the nickel ammonium sulfate remained in solution while the AMV precipitated out as a solid in the leach slurry which was separated by filtration and converted to V_2O_5 by calcination. The solution containing Mo and Ni was subjected to further solvent extraction steps for separation of these metals. Vanadium pentoxide, ammonium molybdate, and nickel sulfate were the final marketable products in this process.

Villarreal et al. [617] were able to recover selectively 98% of vanadium as AMV (NH_4VO_3) from the spent hydrodemetallization (HDM) catalyst, that contained 27.3 wt.% V_2O_5,

1.92 wt.% MoO_3, 3.2 wt.% NiO, and 39.6 wt.% Al_2O_3, by leaching with the aqueous NH_3 (15–17 M) solution at room temperature. Mo and Ni were not leached from the spent catalyst under these conditions used in the study [617]. In a recent study, Yoo et al. [618] used ammonium sulfate solution for leaching Ni from the spent hydrodesulfurization (HDS) catalyst. About 94% of Ni was leached from the catalyst with 2.6 mol/L $(NH_4)_2SO_4$ solution at the temperature 368 K. The extent of leaching of other metals from the spent catalysts was not reported in this study.

11.1.1.2 Leaching with Acids

Extraction of the metals present in spent hydroprocessing catalysts by leaching with acids has been studied extensively by many researchers. It became evident that various types and concentrations of inorganic acids, such as HCl, H_2SO_4, and HNO_3, have been used as leaching solutions. Among organic agents, water-soluble organic acids have been attracting most of the attention. Compared with inorganic acids, aqueous solutions of organic acids ensure an environment requiring much less safety precautions. A generalized scheme for leaching with acid is shown in Fig. 11.2.

11.1.1.2.1 Inorganic acids

In the study of Valverde et al. [619], more than 95% of Co, Ni, and Mo were leached out using 9 M H_2SO_4 at 90 °C within 70–90 min at stirring rate of 200 rpm. The acid/catalyst ratio, estimated from the stoichiometric amount of acid required for the conversion of all metals to

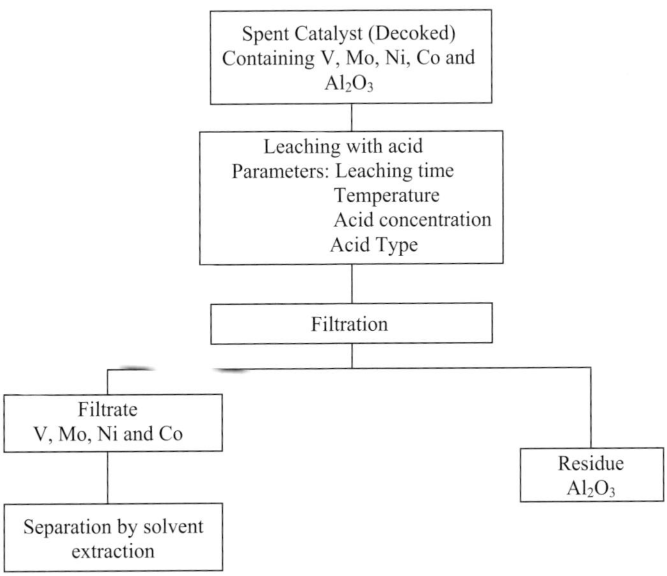

Figure 11.2: Metal extraction using acid.

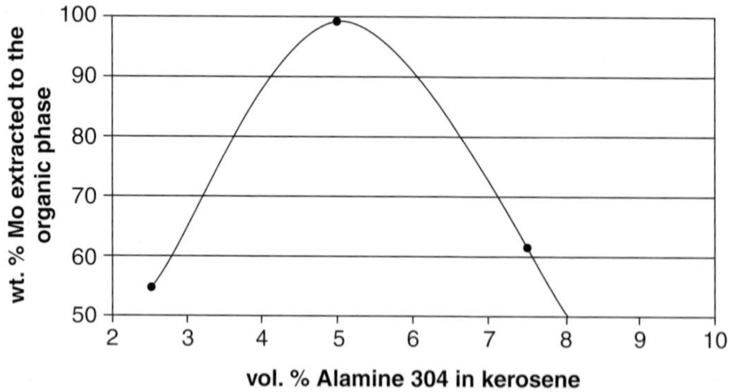

Figure 11.3: Effect of alamine 304 concentration on Mo extraction (pH 1.5, 25 °C) [From ref. 619. Reprinted with permission].

corresponding sulfates, was 2.1:1. Preoxidation of spent catalyst influenced the leaching efficiency. A higher efficiency was achieved when preoxidation was carefully controlled to avoid sintering of metals. Otherwise, diffusion and/or reaction of metals with the support during preoxidation affected subsequent leaching. After leaching, the residual solid was filtered off and filtrate used for metal separation using combination of solvent extraction, selective precipitation, and ion exchange column. Figure 11.3 [619] shows the effect of concentration of extracting agent, such as alamine 304 in kerosene on the extraction of Mo. Besides alamine 304 concentration, pH of the solution had a pronounced effect on leaching. As it is shown in Fig. 11.4 [619], the best extraction efficiency was attained when pH of 2 was approached. The raffinate was used for recovery of the Co and/or Ni. This was achieved by adding 0.3 mol/L solution of $(NH_4)_2C_2O_4$ until the solution became colorless. Solid oxalates of Co or Ni were filtered off and washed with 0.1 mol/L $(NH_4)_2C_2O_4$ and dried.

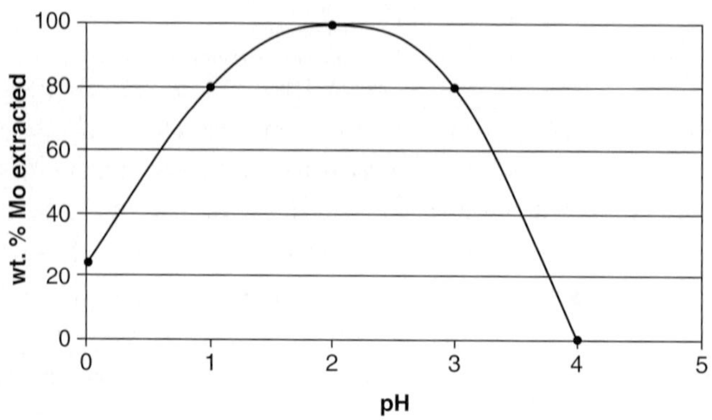

Figure 11.4: Effect of pH on Mo extraction with 5 vol.% alamine 304 at 25 °C [From ref. 619. Reprinted with permission].

In the patented process developed by Hyatt [620], the spent $CoMo/Al_2O_3$ catalyst was treated with sulfuric acid in the presence of H_2S under pressure (7.5–15 atm) in an autoclave at 100–200 °C. The presence of H_2S resulted in the precipitation of Mo and Co as sulfides while the Al_2O_3 was converted to soluble $Al_2(SO_4)_3$. The metal sulfides were separated from the $Al_2(SO_4)_3$ solution and subjected to oxidation under pressure in an autoclave to convert MoS_2 to solid molybdic acid and the CoS to $CoSO_4$. The molybdic acid was separated by filtration from the $CoSO_4$ solution. The cobalt was recovered by ion exchange. A schematic flow diagram of the process is shown in Fig. 11.5 [620]. The extraction and recovery of valuable metals, such as Mo and Co together with the alumina as $Al_2(SO_4)_3$, could be achieved in this process. Recovery of V is not reported in this acid leaching study.

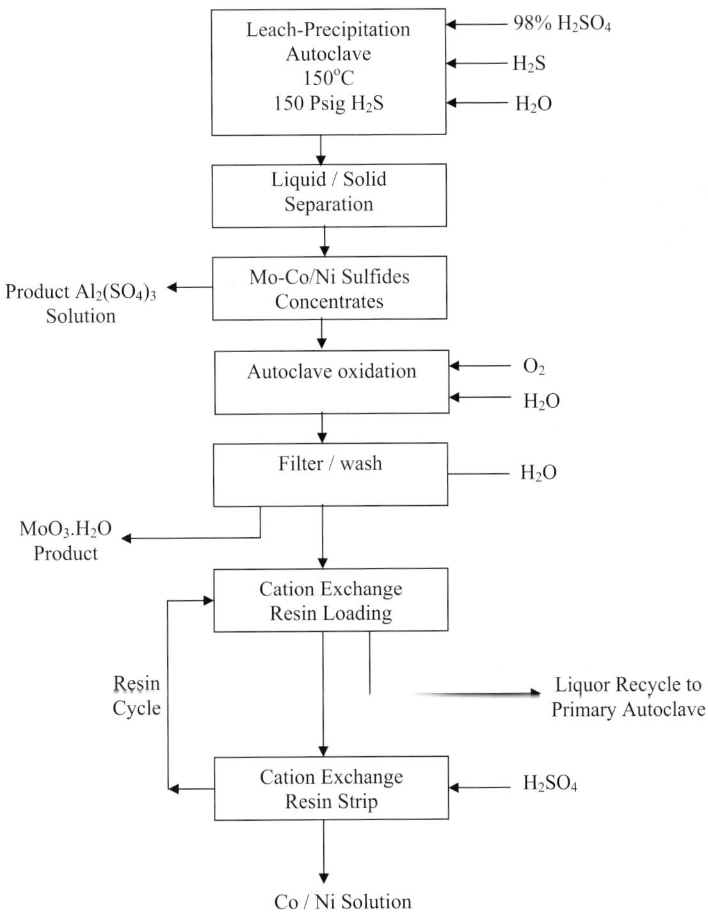

Figure 11.5: Primary steps for the separation of metals from spent hydrodesulfurization (HDS) catalyst by H_2SO_4 in the presence of H_2S [From ref. 620. Reprinted with permission].

Toyabe et al. [621] developed an improved acid (H_2SO_4) leaching process for the complete extraction and recovery of all components (including Al) of the spent hydroprocessing catalyst that contained Mo, V, Ni, Co in an alumina carrier. The process included the following steps:

(1) Roasting the waste catalyst at a temperature range of 400 °C to1000 °C to obtain a roasted product.

(2) Preparing a reduction dissolution by dissolving the roasted product with sulfuric acid in the presence of a metal (Al) as a dissolution catalyst.

(3) Separating a large part of the aluminum from the reduction dissolution solution and recovering aluminum as ammonium aluminum sulfate from the solution, optionally after subjecting the reduction dissolution solution to a treatment of removing iron.

(4) Extracting molybdenum as a molybdate by solvent extraction from the solution after separating and recovering aluminum from the solution.

(5) Extracting vanadium as a vanadate by solvent extraction from the solution obtained as a residue after extracting molybdenum.

(6) Recovering nickel and cobalt each as a hydroxide from the extraction residue after recovering vanadium.

The use of H_2SO_4 for extraction of metals and Al_2O_3 present in spent hydroprocessing catalysts was studied in detail by Inoue et al. [622,623]. The spent catalyst that contained coke, Co, Mo, V, Ni, and Al_2O_3 was first decoked and then treated with H_2SO_4 (63%) and evaporated to dryness. The evaporated mass containing the metals (Co, Mo, V, Ni, and Al) as sulfates was dissolved in water. Individual metals, including Al, were then separated from the solution by solvent extraction using different pH levels [624]. A series of extraction agents, such as TR-83, PC-88A, Cyanex 272, Lix-63, and SYNEX DN-052, were used. The details of these studies, as they apply to the separation of metals from solution by solvent extractions, are given in a later section of this chapter.

In a process developed by the Hall Chemical Co (Ohio), hydrochloric acid was used for leaching of metals from spent hydrotreating catalysts [625]. The spent catalyst was dissolved completely in HCl and then molybdenum was precipitated first, as a sulfide, by adding H_2S. Next, vanadium was taken out by solvent extraction, reclaimed by adding caustic soda, and precipitated as vanadium pentoxide. Nickel was removed by solvent extraction, and then recovered as a sulfate by adding sulfuric acid. Cobalt was salvaged in a similar manner as nickel.

High recoveries of Mo, Ni, and Al, i.e., 93, 98, and 96%, respectively using H_2SO_4 were reported by Siemens et al. [626]. Similarly, high recovery of Mo and Co, such as 92 and 97%, respectively, were achieved by Rastas et al. [627] using the mixture of $H_2SO_4 + Al_2(SO_4)_3$ and $H_2SO_4 + (NH_4)_2SO_4$. Rabah et al. [628] used a mixture of the concentrated sulfuric and nitric acids (3:1) to extract Mo and Co from the spent HDS catalysts. After leaching, the metals were separated as hydroxides which were roasted in oxygen atmosphere to obtain the respective oxides. Reduction of Mo and Co oxides in hydrogen at 1100 °C and 900 °C, respectively, produced metal powders with high purity. The maximum recovery efficiency was claimed to be 96%.

The two mixtures of strong acids, i.e., one comprising 50 mL of 70% HNO_3 + 50 mL of 96% H_2SO_4 (mixture A) and the other 50 mL of 70% HNO_3 + 25 mL of 96% H_2SO_4 + 25 mL 37% HCl (mixture B), were used in the process described by Lai et al. [629]. The spent-decoked $NiMo/Al_2O_3 \cdot SiO_2$ catalyst was treated at 70 °C. The leachates were filtered to remove solid residue and then diluted with 0.5 M HNO_3. The isolated solid was leached again in the second leaching stage and then in the third leaching stage. To recover metals from the filtrate, electrolysis experiments were conducted using a moving bed electrolytic cell installed with non-conductive glass beads. Fluidization of the bed was maintained by recirculating the medium between the storage tank and the electrolytic cell. Before the electrolytic experiments, the pH of filtrate was adjusted to 2–4 by 10 N NaOH. A parallel set of experiments was carried out using a static bed. The schematics of both static bed and fluidized bed systems used to carry out electrorefining are shown in Fig. 11.6 [629]. The electrolytic recoveries of Mo, Ni, and V achieved in this study as well as potentials of the cell versus time correlations are shown in Fig. 11.7.

The use of aqueous solutions of SO_2 for leaching metals from the spent $CoMo/Al_2O_3$ HDS catalyst was studied by Raisoni and Dixit [630]. Dissolution of the metals was strongly influenced by the pH of the solution as shown in Fig. 11.8. Both Co and Mo were readily dissolved from the calcined spent catalyst by aqueous SO_2 having pH of 1 and 100% extraction of both metals was obtained in 8–10 min at ambient temperature. The spent catalyst used in this study contained Co_3O_4 6.12%, MoO_3 10.8%, Al_2O_3 65.2%, SiO_2 17.7%, S 0.10%, and C 0.35%. The following reactions of Co and Mo with SO_2 solution were reported to occur in the dissolution process.

$$2Co + 3SO_2 + 4H_2O \Rightarrow CoSO_4 + CoS_2O_6 + 8H^+$$

$$4Mo + 12SO_2 + 16H_2O \Rightarrow 2Mo(SO_4)_2 + 2Mo(S_2O_6)_2 + 16H^+$$

The extent of dissolution of Al_2O_3 and the separation of the metals from the mixed metal solution were not reported in this study.

11.1.1.2.2 Organic acids

The leaching of metals present in spent hydroprocessing catalysts with organic acids has been studied by many researchers. Beuther and Flinn [479] compared the efficiencies of many

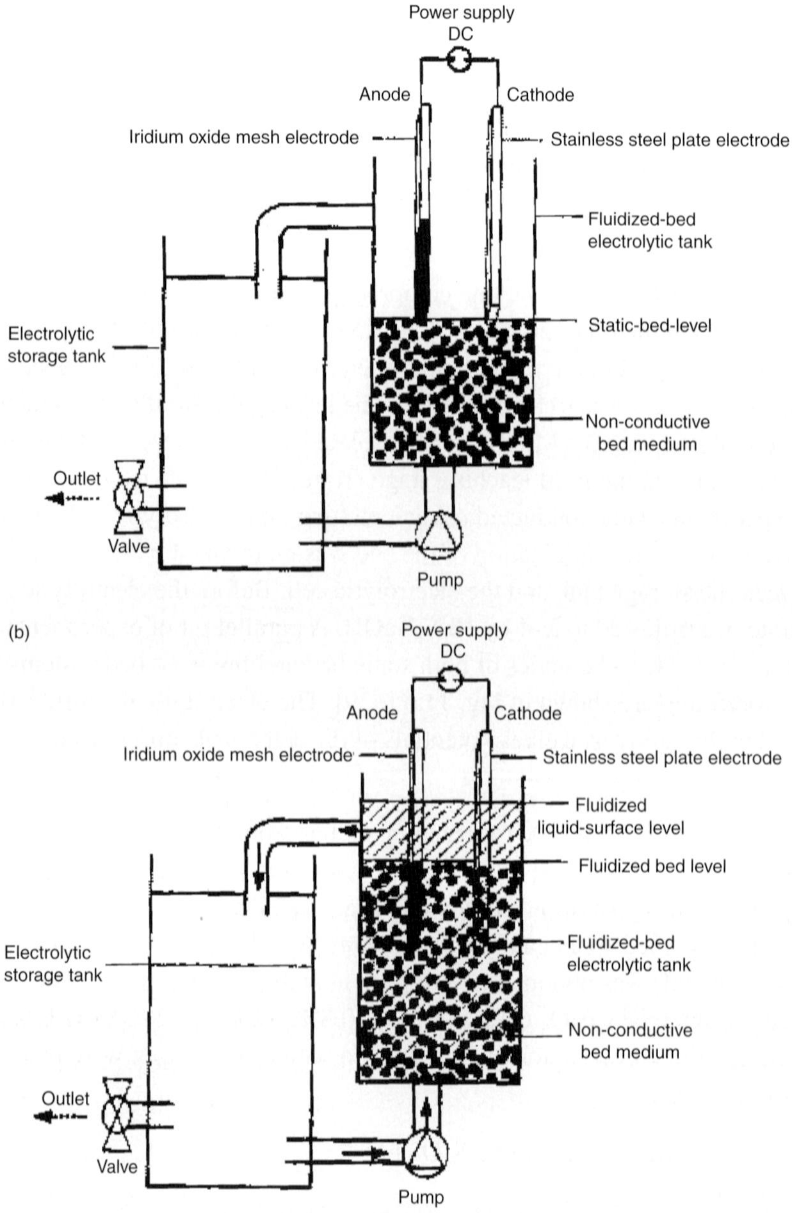

Figure 11.6: Experimental systems: (a) static bed and (b) fluidized used for electrolytic experimental [From ref. 629. Reprinted with permission].

organic acids, such as oxalic acid, lactic acid, citric acid, glycolic acid, phthalic acid, malonic acid, succinic acid, salicylic acid, and tartaric acid, for leaching metals (V, Mo, Ni, and Co) from a spent hydrotreating catalyst. The 1% aqueous solutions of the acids were used at ambient temperature and pressure in this study. Oxalic acid was found to show the highest

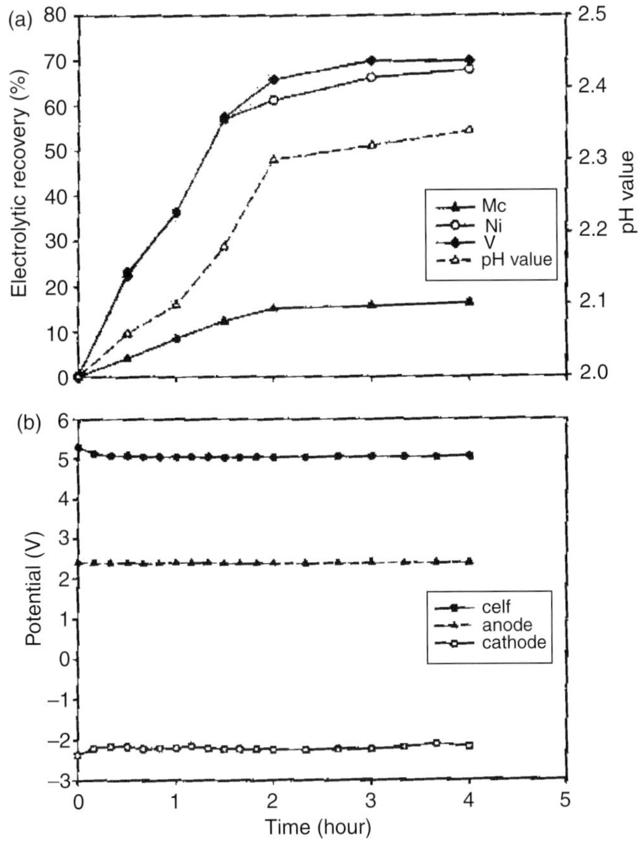

Figure 11.7: Electrolytic recoveries of Mo, Ni and V (a) and potential of cell anode and cathode versus time (b) [From ref. 629. Reprinted with permission].

efficiency for leaching all four metals, e.g., 47% V, 59% Mo, 60% Co, and 56% Ni were leached by 1 wt.% oxalic acid solution. The effect of reagent concentration and other conditions of leaching were not investigated and optimized in this study.

Marafi et al. [480,488,490] used organic acids, such as oxalic acid, citric acid, and tartaric acid, for selective leaching of metal contaminants, such as V and Ni, present in spent residue hydroprocessing catalysts. Although the objective of these studies was rejuvenation of catalyst for reuse, they are briefly summarized here, because of their relevance to metal reclamation. Decoked (oxidic) and coked (sulfidic) form of the catalysts were used in leaching experiments. The selectivity for removal of the major metal foulant (V) was higher for the catalyst leached prior to decoking. However, the addition of an oxidizing agent, such as $Fe(NO_3)_3$, $Al(NO_3)$, H_2O_2, was found necessary to improve the leaching efficiency of organic acids when the spent catalyst with coke (i.e. prior to decoking) was used [475,477,488]. Vanadium, which is the

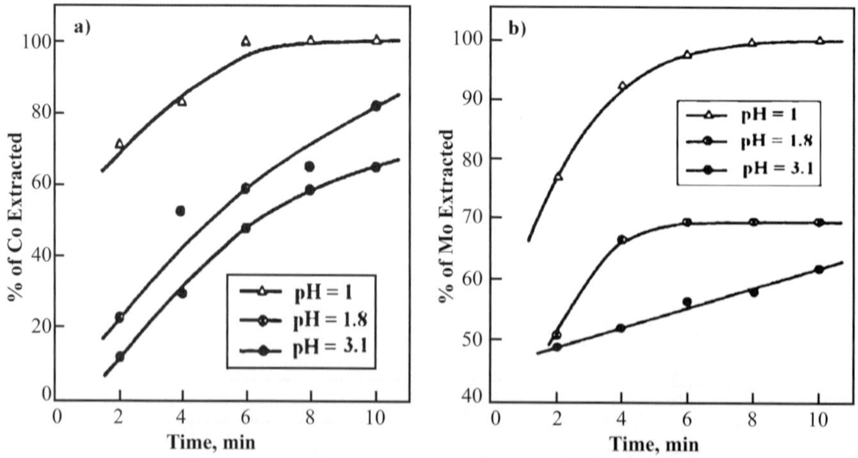

Figure 11.8: Effect of pH on metal extraction: (a) Co extraction and (b) Mo extraction [From ref. 630. Reprinted with permission].

major metal foulant deactivating the catalyst by blocking the pores, was selectively removed in large amounts from the catalyst pores increasing the surface area and porosity that led to better activity recovery [476]. Both metal recovery and catalyst recovery for reuse are possible in this process.

In another related study, Marafi and Furimsky [631] compared the efficiency of four organic acids, water and an aldehyde for extracting metals from a spent HDM catalyst. In this study, the objective was to maximize leaching of metals (Mo, Ni, V, and Fe) using the same organic agents which were used for catalyst rejuvenation. Prior to leaching experiment, the coke deposited on the catalyst was removed by combustion using 4% O_2 in N_2 at temperatures in the range 350–450 °C. The decoked catalyst contained 4.4 wt.% Mo, 11.6 wt.% V, 5.9 wt.% Ni, and 2.5 wt.% Fe, and the balance Al_2O_3. Leaching experiments were conducted at 50 °C by ultrasonic agitation in an ultrasonic bath. The following order in leaching efficiency was found for the six reagents:

$$\text{Tartaric acid} = \text{citric acid} > \text{glyoxylic acid} \sim \text{lactic acid} > \text{glycolic acid} > \text{water} > \text{glyoxal}.$$

Almost 95% of Mo and V as well as more than 85% of Ni were dissolved during the four hours agitation of decoked spent catalyst in an ultrasonic bath with a solution containing 5 wt.% of either tartaric acid or citric acid. The leaching efficiency could be further improved by optimization of leaching conditions, such as temperature, catalyst/solution ratio, contact time, etc.

A recent study published by Mulak et al. [632] focused on optimizing the parameters for the extraction of Mo, Ni, V, and Al from a spent HDS catalyst using aqueous oxalic acid solution

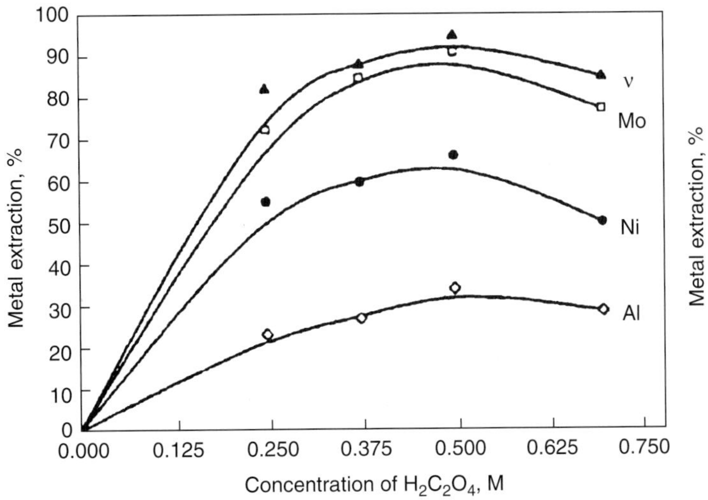

Figure 11.9: Effect of $H_2C_2O_4$ concentration on metal leaching efficiency at 50 °C in 3.0 M H_2O_2 solution after 4 h leaching [From ref. 632. Reprinted with permission].

mixed with H_2O_2. The effects of oxalic acid and hydrogen peroxide concentrations and the stirring speed on the rate of metal leaching were studied. The results revealed that addition of hydrogen peroxide to oxalic acid up to 3.0 M H_2O_2 concentration enhanced leaching of metals remarkably, and thereafter remained relatively constant. Change in the concentration of oxalic acid solution within the range 0.25–0.7 M caused a gradual increase in the extraction of all metals up to 0.5 M and then a slight decrease with further increase to 0.7 M (Fig. 11.9). The highest extraction of metals from the spent catalyst (at 50 °C with a solution of 0.5 M $H_2C_2O_4$ with 3.0 M H_2O_2) was found to be 90% Mo, 94% V, 65% Ni, and 33% Al in four hours leaching.

Table 11.2 summarizes the studies on recovery of metals from spent hydroprocessing catalysts using either acids or mixtures of acids with different reagents. In most studies, Mo was the metal of primary interest. For this metal, more than 90% recovery was achieved regardless of the type of reagents. Similarly, a high recovery of V and Co should be noted.

11.1.1.3 Alkali Leaching

Leaching of spent catalyst with aqueous alkaline solution (NaOH or Na_2CO_3) has been reported in a few studies. For example, Rokukawa [633] described a process in which an aqueous alkaline solution of sodium carbonate mixed with hydrogen peroxide as an oxidizing agent was used to extract Mo and V selectively from unroasted spent hydroprocessing catalysts. More than 99% of Mo and about 85% of V were recovered in the process while the extraction of Ni, Co, and Al_2O_3 were minimum. This work was recently advanced by Park et al. [634,635] for selective extraction of high purity MoO_3 from spent HDS catalyst using a

Table 11.2: Metal recovery from spent catalysts by acid leaching.

Metals in spent catalysts	Reagents	Recovery of metals (wt.%)					Reference
		Mo	V	Ni	Co	Al	
Mo, V, Ni, Co, Al	H_2SO_4	99.7	92.5	–	–	–	[622,623]
Mo, Ni, Al_2O_3	H_2SO_4	93	–	98	–	96	[626]
Co, Mo	Aqueous SO_2 solution	100	–	–	100	–	[630]
Co, Mo	$H_2SO_4 + HNO_3$ (3:1)	96	–	–	96	–	[630]
Co, Mo	$H_2SO_4 + Al_2(SO_4)_3$ or $(NH_4)_2SO_4$	92	–	–	97	–	[627]
Mo, Co, Ni	$H_2SO_4 + H_2S$	–	–	–	–	–	[620]
Mo, V, Co, Ni	H_2SO_4	> 90	–	–	> 90	–	[311]
Mo, V, Ni, Al	Oxalic acid + H_2O_2	90	94	65	–	33	[632]
Mo, V, Ni	Tartaric acid	93	94	83	–	–	[631]
Mo, V, Ni	Citric acid	94	94	85	–	–	[631]
Mo, V, Co, Ni, Al	H_2SO_4	–	–	–	–	–	[621]

mixture of Na_2CO_3 and H_2O_2. The results indicated that the recovery of molybdenum was largely dependent on the concentrations of Na_2CO_3 and H_2O_2 in the reaction medium, which controls the acidity (pH) of the leach liquor and carry over of impurities, such as Al, Ni, P/Si, and V. In this case, the leaching process was exothermic and leaching efficiency of molybdenum decreased with increasing solid to liquid ratio. For large scale leaching of spent catalyst, under optimum conditions (20% pulp density, 85 g/L Na_2CO_3, 10 vol.% H_2O_2, and 1 h reaction), it was possible to recover 84% Mo. The obtained leach liquor contained (g/L): Mo: 22.0, Ni: 0.015 and Al: 0.82, P: 1.1, Si: 0.094 and minor quantities of V: 8 mg/L, As and Co $<$ 1 mg/L. Purification of the leach liquor by carbon adsorption at pH around 0.75 and desorption of the adsorbed molybdate with 15% NH_4OH produced high purity ammonium molybdate from which MoO_3 with 99.4% purity was obtained by calcination. The process details are schematically shown in Fig. 11.10 [635].

Villareal et al. [617] compared two solutions of NaOH, i.e., one containing 10% and the other 40% of NaOH. For the former, the amount of Mo and V leached out approached 92 and 89%, respectively, whereas for the concentrated solution, 79 and 72%, respectively. For both solutions, very small amounts of Ni and Al could be leached out.

11.1.1.4 Two-Stage Leaching

The process tested by Jong et al. [636] comprised two stages. The first stage involved leaching spent $NiMo/Al_2O_3$ catalyst with NaOH solution to extract Mo and some Al, while the rest of Al and Ni remained in the residue. The Mo was isolated from the NaOH leach by precipitation with $CaCl_2$. In the second stage, the residue from the first step was leached with sulfuric acid to extract Ni. The leachate from this step was treated with NH_4OH and $(NH_4)_2SO_4$ to precipitate Al and other metal impurities. At about pH 10, Ni was recovered by solvent extraction. The NaOH leaching step extracted from 81 to 91 wt.% of the Mo. The H_2SO_4 leaching step extracted from 92 to 98 wt.% of the Ni. In the final recovery step, over 84 wt.% of the Mo and 99% of Ni were recovered.

Two-stage leaching processes involving leaching with an alkali solution in the first stage and an acid or ammonia in the second stage or the other way around have also been reported in many other studies. Angelids et al. [637] used NaOH in the first stage and H_2SO_4 in the second stage to leach Mo, Co, and Ni from spent HDS catalysts. The process is based on the difference of solubility of MoO_3 and CoO or NiO in acidic and basic media. Both $CoMo/Al_2O_3$ and $NiMo/Al_2O_3$ type spent catalysts were first roasted in air using controlled conditions at temperatures around 450 °C for 2 h to remove the coke and sulfur present in the catalyst. The calcined $CoMo/Al_2O_3$ catalyst contained 10.3% Mo, 3.5% Co, and 39.7% Al. The spent $NiMo/Al_2O_3$ catalyst contained 10% Mo, 2.4% Ni, and 31.7% Al. The effects of reagent concentration, leaching time, and temperature were investigated for both first stage leaching with NaOH and second stage leaching with H_2SO_4. Leaching temperature significantly influenced both Mo as well as Co and Ni extractions in the first and second

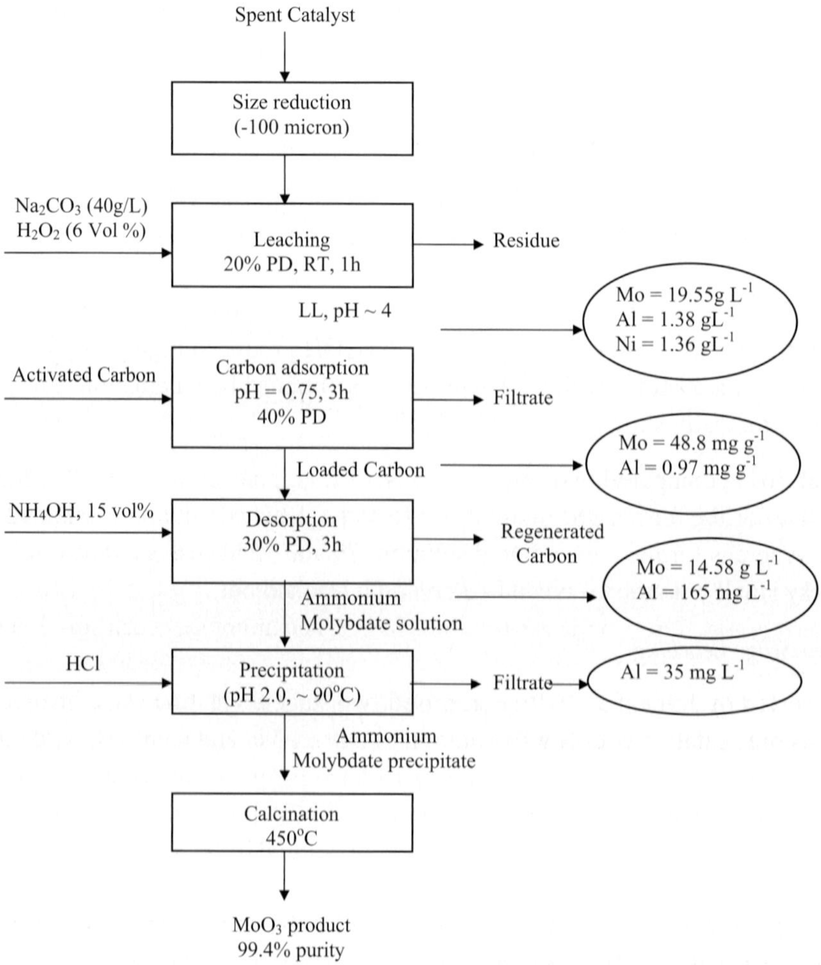

Figure 11.10: Flow sheet of the process with material balance for the recovery of molybdenum from spent catalyst by leaching with $Na_2CO_3 + H_2O_2$ [From ref. 635. Reprinted with permission].

stages, respectively. Maximum recoveries of 97% Mo in the first stage and 90–93% Co and Ni in the second stage were achieved in this process within two hours of leaching at 100 °C. The Co or Ni dissolution was negligibly small (<0.1%) in the first stage alkali leaching. Similarly, Mo leaching was negligible (<0.3%) in the second stage acid leaching. However, a substantial amount of Al was leached in both stages (12–14% in the first stage and 55–65% in the second stage). Sun et al. [361] used concentrated ammonia solution in the first stage and 10% H_2SO_4 in the second stage to leach Mo and Co from the spent $CoMo/Al_2O_3$ catalyst. About 83% Mo and 77% Co together with 4% Mo were recovered, respectively, in the first and second stages. In another related study, these authors were able to recover 83% Mo and 44% Ni in the first and second stages respectively from a spent $NiMo/Al_2O_3$ catalyst using ammonia in the first stage and H_2SO_4 in the second stage [638].

In the patented process, Veal et al. [639] used caustic leach to extract Mo and V in the first stage and aqueous ammonia/ammonium carbonate in the second stage to extract Ni from a de-oiled but not decoked spent HDS catalyst that contained the metals (Mo, V, and Ni) in a sulfided form. The caustic leach was conducted in two steps: first step at atmospheric pressure in the presence of caustic and air at a temperature less than 60 °C and a pH range of 10–13. About 50–70% of V and Mo sulfides were converted to soluble thiosulfate and sulfate species while Ni and Co were not reactive under these conditions. After filtration, the solids from the atmospheric caustic leach were subjected to pressure leach with caustic at a higher oxygen pressure and at a higher temperature to solubilize the remaining V and Mo sulfides. About 97% of the Mo, 92% of the V, and 98% of the sulfur were solubilized in the atmospheric and pressure leaching steps of caustic leaching. The soluble metals (V and Mo) were then separated by solid/liquid separation. The final recovery of the soluble V and Mo from the liquid stream was accomplished by solvent extraction. The residual solid that contained nickel was leached with an aqueous ammonia/ammonium carbonate solution at a pH range of about 10.5–12 and at temperatures in the range of 40–80 °C. The ammonia leach solution containing the solubilized nickel amine complex was then stripped and heated to remove ammonia. As the ammonia was removed, the pH of the solution decreased to about 10 and a basic nickel carbonate was precipitated. Nickel was finally recovered as nickel carbonate with high purity.

11.1.1.5 Bioleaching

Bioleaching offers a novel approach for metal recovery from various solids. It is based on the ability of some microorganisms to transform solid compounds to extractable entities [640]. In this case, the microorganisms can secret either organic or inorganic acids which are participating during the metal dissolutions. Bioleaching was used industrially for recovery of various metals from low-grade mineral resources [641]. In addition, the article published by Santhiya and Ting [525] lists the waste materials, such as fly ash, sewage sludge, spent batteries, and electronic scrap materials, as well as the fluid catalytic cracking (FCC) and hydroprocessing catalysts as potential solids which can be processed by bioleaching. The most common microorganisms, which are capable of the metal solubilization, include bacteria, such as *Thiobacillus ferrooxidans* and *Thiobacillus thiooxidans*, as well as the fungi, such as *Aspergillus* and *Penicillium genera* [642].

Earlier studies on potential application of bioleaching for metal recovery from spent refinery catalysts were reviewed by Furimsky [27]. It was evident that the first attempt to apply the bioleaching method to spent refinery catalysts was made in early 1990s. For example, Blaustein et al. [643,644] used *T. ferrooxidans* and *L. ferrooxidans* for leaching Mo from a coal liquefaction catalyst. Bioleaching experiments were conducted in an autoclave at about 30 °C with shaking at 170 to 200 rpm for about six weeks. The amount of the solubilized Mo increased with decreasing particle size of the catalyst. A series of studies was conducted on another coal liquefaction catalyst ($NiMo/Al_2O_3$) under auspices of US DOE using *T. ferrooxidans*, sulfolobus and thermophilic cultures. In these studies, Joffe and Sperll [645] observed that *T. ferrooxidans* was efficient for solubilizing Ni but the presence of Mo had adverse effect on leaching. Consequently, the leaching process had to be extended to achieve desirable level of leachability. Similarly, leaching of Ni from spent catalysts using *T. ferrooxidans*, denitrifiers and sulfolobus required special attention because of the presence of Mo and W. It was concluded that only microorganisms which are tolerant to Mo and W can be used. However, an efficient leaching of both Ni and Mo was achieved by using heterotrophic denitrifying bacteria. In a similar study, thermophilic cultures, such as *Bacillus stearothermophilus* and *Metallosphera sedula*, were grown and used by Sanback and Joffe [646] and Sanback [647]. *T. thiooxidands* was used by Briand et al. [648] to extract vanadium from spent catalyst.

More recently, Santhiya and Ting [525] and Aung and Ting [642] described one-step and two-step procedures applied for bioleaching of metals from spent $NiMoP/Al_2O_3$ and FCC catalyst. For the latter, both methods were tested. In the one-step method, the spent catalyst and the fungus were incubated together with the medium (sucrose) at pH of 5.5. In this case, the system was agitated in an incubator at 30 °C and 120 rpm. At regular time intervals, the samples were withdrawn, filtered, and analyzed to determine the extraction efficiency. The two-step method involved bioleaching the spent catalyst in the second step after preculturing the fungus for two days, in the first step. For a spent FCC catalyst, the leaching efficiency during the two-step leaching was consistently higher than that from the one-step leaching. Also, bioleaching was more efficient compared with the leaching with organic acids (oxalic, citric, and gluconic) which are secreted during the growth of *A. niger* fungus. The methodology for the growth of the fungus *A. niger* used by Santhiya and Ting [525] was described by Aung and Ting [642] and Strasser et al. [649].

Little difference in the bioleaching efficiency between one-step and two-step methods was observed in the case of the spent $NiMo/Al_2O_3$ catalyst. Thus, for the spent catalyst as received, the extractions of Al, Ni, and Mo were similar, as it is shown in Fig. 11.11 [650]. However, for smaller particles (e.g., 100–150 μm and <37 μm), the dissolution of Ni and Mo was higher than that of Al. The lowest extraction of metals for the 2.97 μm average particles diameter was due to the lower fungal biomass generation and secretion of organic acids. Based on bioleaching lasting 70 days for <37 μm particles, the following order of the rate of leaching can be obtained:

$$Mo >> Al > Ni.$$

The concentration profiles in Fig. 11.12 [650] coincided with the change in the concentration of citrate, oxalate and gluconate ions, which are being secreted during the growth of the *A. niger* fungus. Therefore, the rate of secretion may be the factor which determines the rate of

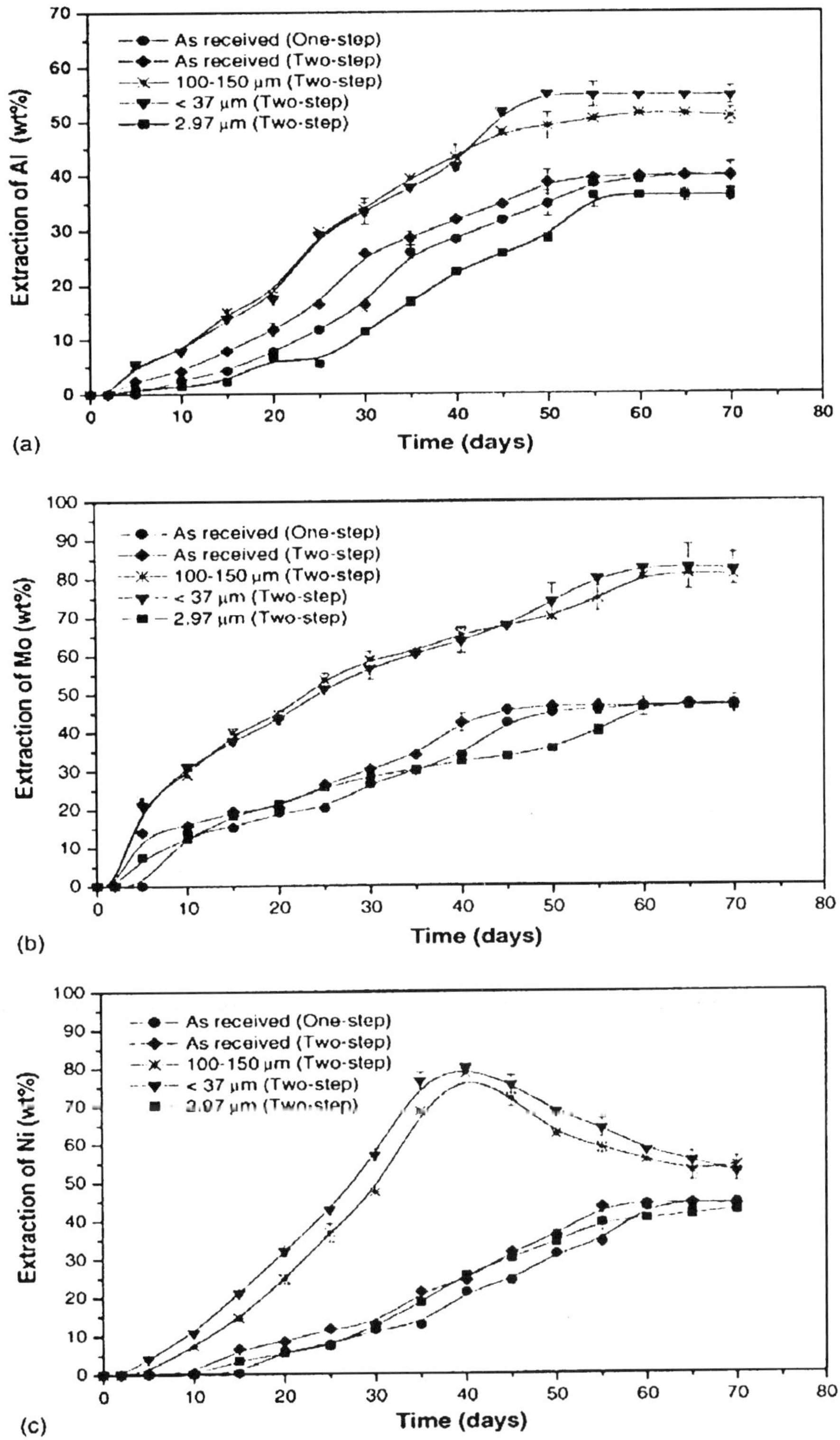

Figure 11.11: Kinetics of bioleaching: (a) aluminium; (b) molybdenum and (c) nickel from 1% spent catalyst of various particle sizes using *A. niger* [From ref. 650. Reprinted with permission].

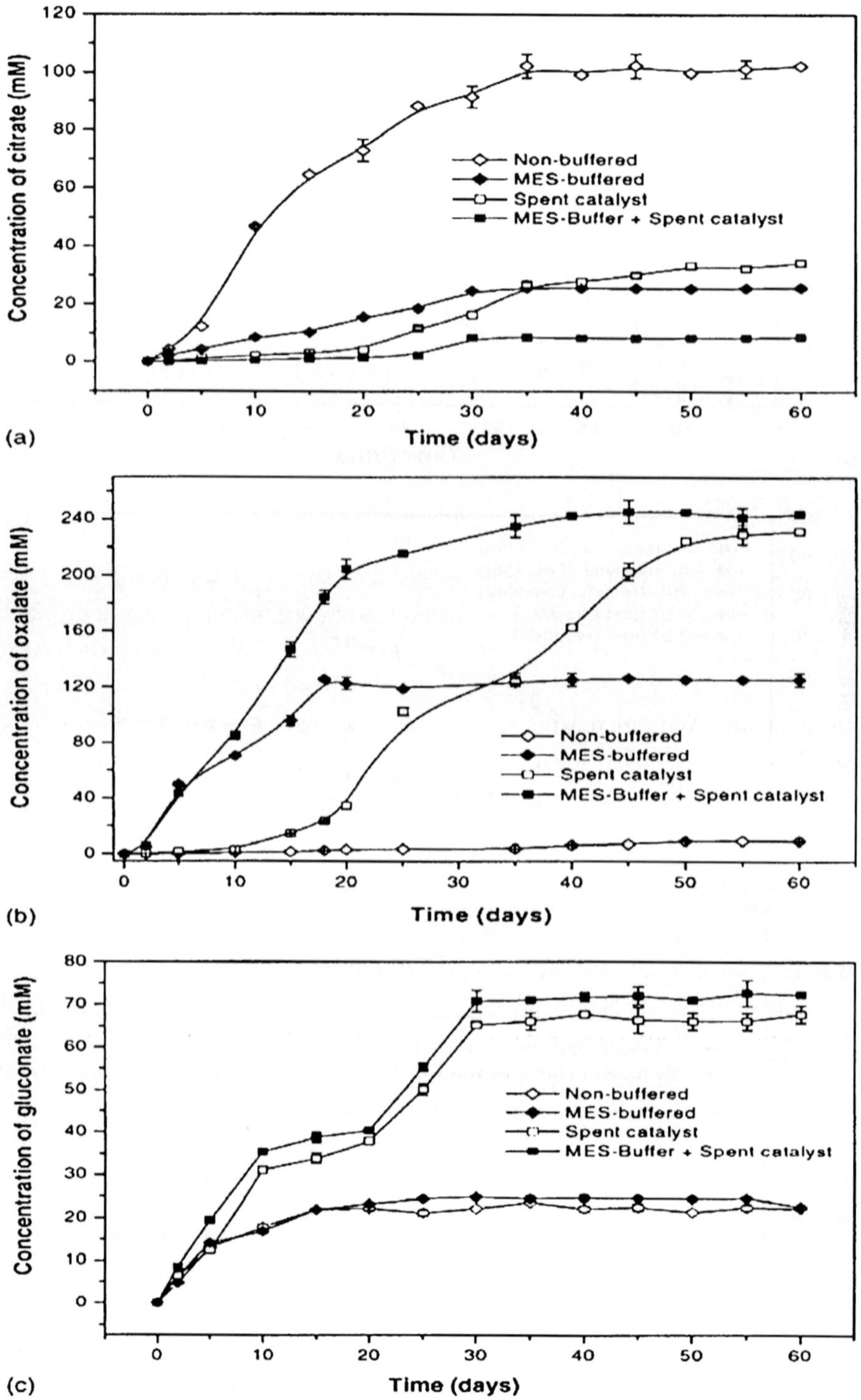

Figure 11.12: Concentration of (a) citrate, (b) oxalate and (c) gluconate secreted during the growth of *A. niger* in non-buffered and buffered sucrose medium in the absence and presence of 1% spent catalyst of particle size <37 μm [From ref. 650. Reprinted with permission].

leaching. In contrast with spent FCC catalyst, little difference in leaching efficiency was observed between bioleaching and leaching with the individual organic acids produced during the growth of *A. niger* fungus.

In another study published by Santhiya and Ting [650], bioleaching of spent hydroprocessing catalysts was conducted using an adapted *A. niger* fungus. Adaptation experiments with single metals revealed that the fungus could tolerate 1000, 1200, and 2000 mg/L of Ni, Mo, and Al, respectively. For the mixtures of these metals, the tolerance was decreased to 100, 200, and 600 mg/L of Ni, Mo, and Al, respectively. In contrast to the adapted strain, in the presence of catalyst, the unadapted strain exhibited no growth. For the adapted strain, leaching efficiency for Ni, Mo, and Al approached 79, 82, and 65%, respectively over 30 days.

The mixed acidophilic culture in the presence of pyrite was used to study recovery of metals from spent hydroprocessing catalysts [651]. This culture as initially grown in the 9K-medium (absence of 9 g/L Fe(II)) where $FeSO_4$ was replaced by pyrite (FeS_2) and then applied for bioleaching experiments. Bacterial action on pyrite catalyzed formation of the Fe^{+3}, H^+, and SO_4^{-2} ions in the solution. This facilitated leaching of metals (Ni, Mo, and V) from the spent catalyst. The study was conducted under a wide range of experimental conditions, i.e., reaction time, amount of spent catalyst and pyrite, and temperature. After seven days with 30 g/L of spent catalyst and 50 g/L of pyrite, the leaching of Ni, V, and Mo into the solution was 85, 92, and 26%, respectively. With increasing spent catalyst loading, the extent of metal dissolution was decreased, probably due to the enhanced removal of Fe^{+3} ion with the residue. Under all conditions tested, Mo showed recovery due to its precipitation with leach residues as MoO_3. This was confirmed by applying spectroscopic techniques, such as EDAX and X-ray diffraction spectroscopy (XRD).

Mishra et al. [652,653] used acidophilic bacteria, which in the presence of sulfur produced H_2SO_4 in the first stage. In the second stage, the acidic medium was used for leaching Ni, Mo, and V from spent catalyst. At 50 g/L of spent catalyst and 20 g/L of elemental sulfur, the amount of Ni, Mo, and V after seven days approached 88, 46, and 95%, respectively. In a one-stage method, bacteria were grown in the presence of spent catalyst. In this case, the maximum of extraction for Ni, Mo, and V approached 88, 58, and 32%, respectively.

11.1.2 Roasting with Alkali Compounds

The solubility of some metals (e.g., Mo, W, and V) in water can be significantly enhanced by roasting spent catalysts with alkali metals containing compounds. After roasting, most of the Ni and Co remain in the solid residue left after water extraction. A wide range of methods has been used for the separation of metals from aqueous solutions. Both sodium and potassium containing compounds (e.g., hydroxides and carbonates) have been used as roasting agents.

11.1.2.1 Roasting with Sodium Salts

Roasting of the spent catalysts with sodium salts (e.g. Na_2CO_3, NaCl, NaOH) followed by leaching and separation of the metals has been reported in a number of studies. A detailed description of the soda roasting process was presented by Sebenik and Ference [654] at the ACS conference in 1982. The spent catalyst was first ground to approximately 100 mesh and roasted in air at 600 °C in a rotary kiln or fluid bed calciner to remove carbon and sulfur deposits. Then, the calcined material was mixed with soda ash (Na_2CO_3) and the mix was roasted again in air at 750 °C to convert Mo and V oxides to Na_2MoO_4 and $NaVO_3$, respectively. The sodium carbonate roaster product was leached with water at 100 °C in a continuous stirred tank to solubilize Mo and V. Insoluble Co-Ni-Al_2O_3 residue was readily separated from the solution containing Na_2MoO_4 and $NaVO_3$ by filtration. The $NaMoO_4$–$NaVO_3$ solution was treated with NH_4C1 at 80 °C to precipitate NH_4VO_3, which was then filtered, dried and calcined at 500 °C to V_2O_5. The filtrate from the vanadium precipitation tank was treated with calcium chloride or lime at ambient temperature to recover the molybdenum as calcium molybdate. Mo and V recoveries ranging between 90 and 95% were achieved in the process. The Co-Ni-Al_2O_3 residue was digested with caustic soda in a stirred pressure autoclave at approximately 250 °C and 500 psi. The sodium aluminate solution was filtered and then treated with CO_2 to precipitate sodium dawsonite ($NaAl(OH)_2CO_3$) which could be used as a Bayer alumina feed. The insoluble NiO and CoO present in the residue were subjected to further processing to recover Ni and Co.

Wang et al. [655] studied recovery of Mo, Co, and alumina from spent CoMo/Al_2O_3 hydrotreating catalyst using caustic roasting followed by ammonia and acid leaching treatments. The spent catalyst was first decoked and then roasted with caustic soda. The roasted material was leached by concentrated ammonia to obtain the solution containing mainly molybdenum and a solid containing essentially alumina and cobalt. Then, the solution was treated with nitric acid and concentrated ammonia to obtain highly pure ammonium molybdate. The residue was dissolved by sulphuric acid, and aluminum sulfate in the solution was crystallized with ammonium sulfate to form ammonium alum. Cobalt was converted into a complex ion by excess ammonia and then recovered as highly pure metal cobalt powder by replacing with zinc powder. The recovery rate for molybdenum, aluminum, and cobalt were higher than 90, 85, and 90%, respectively.

Chen et al. [656] used alkali (NaOH) roasting for recovering metals from the spent hydroprocessing catalyst that contained V, Mo, Ni, Co, and Al_2O_3. A schematic flow diagram of the process is shown in Fig. 11.13. Roasting of spent catalyst with NaOH (mole ratio of Na_2O:Al_2O_3 = 1:2) was carried out at 750 °C for 30 min. The roasted product was leached with water. Vanadium and molybdenum were dissolved in sodium aluminate solution in the leaching process leaving Ni and Co as residue. Barium hydroxide and barium aluminate were added to the solution to recover V and Mo, respectively. After the recovery of V and Mo by

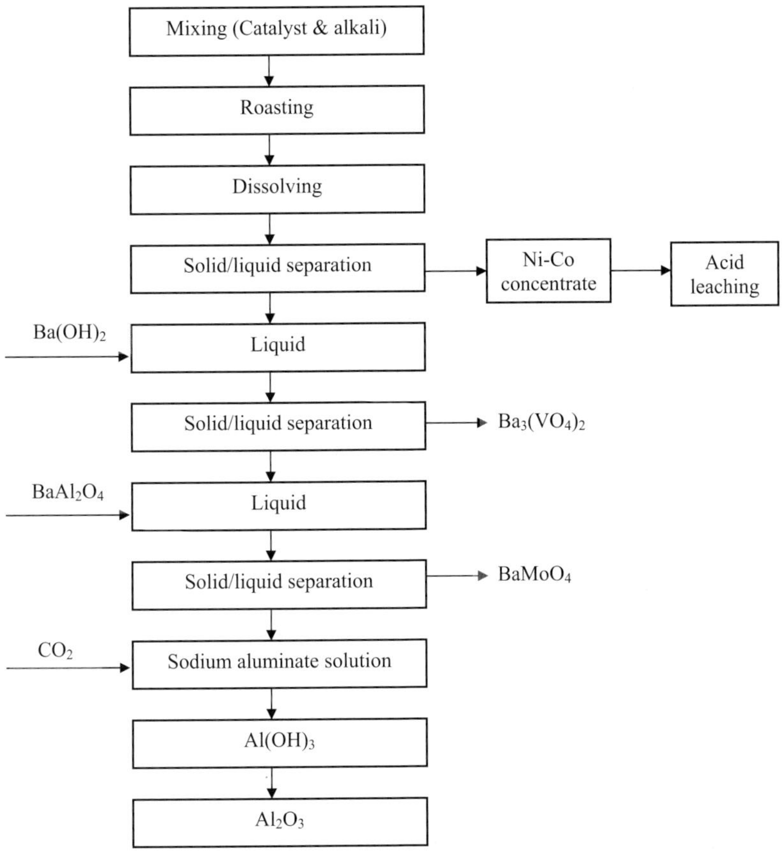

Figure 11.13: Recovering flow sheet of spent catalyst [From ref. 655. Reprinted with permission].

precipitation, the sodium aluminate solution was purified and the carbonation decomposition process was applied to produce $Al(OH)_3$. The $Al(OH)_3$ was then calcinated at 1200 °C to produce α-alumina. The Ni and Co metals were recovered from the residue by leaching with 30% H_2SO_4. Recoveries of V, Mo, Ni, and CO were 95.8, 98.9, 98.2, and 98.5%, respectively. Alumina recovery was 90.6% and the purity of alumina was 99.9%.

Karr et al. [657] conducted a series of experiments to optimize the operating conditions for maximum recovery of Mo from spent hydrotreating catalysts by soda ash roasting process. The spent catalyst used in this study contained 21 wt.% MoO_3, 1.52 wt.% Co_3O_4, 0.85 wt.% NiO, 60.4 wt.% Al_2O_3, and 4.67% SiO_2. The effects of temperature, soda ash (Na_2CO_3) concentration and roasting time on Mo recovery percentage, are shown in Figs 11.14a–c, respectively.

Under selected experimental conditions, i.e., roasting of the spent catalyst with 12 wt.% soda ash at 600 °C for 30 min followed by leaching of the roasted mass with water, 92% Mo was

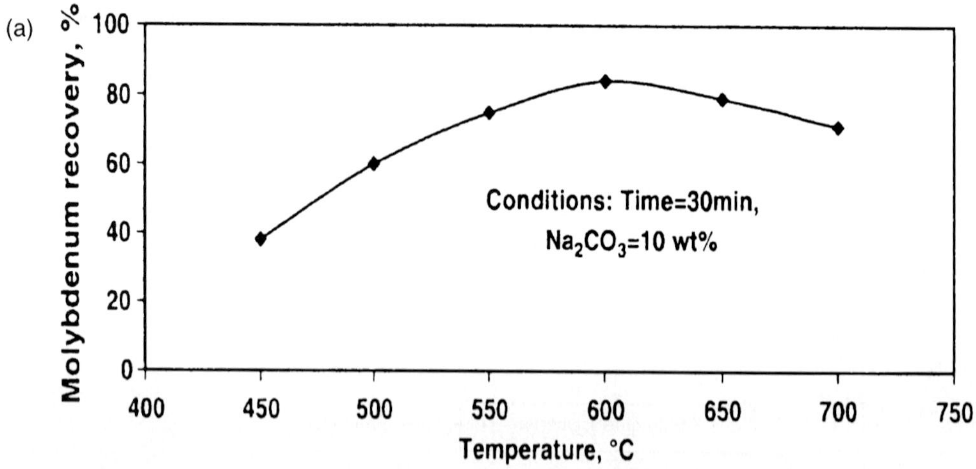

Figure 11.14a: Effect of temperature on molybdenum recovery.

extracted as sodium molybdate. The sodium molybdate solution was treated with HCl to adjust the pH to 2 and then purified by adsorption on activated charcoal with ammonia and the solution was heated at 90 °C to produce ammonium polymolybdate which on calcination at 450 °C produced highly pure (99.9%) MoO_3. A schematic diagram of the overall process is shown in Fig. 11.15 [657].

In another related study published recently by Karr et al. [658], roasting of spent catalyst with NaCl was also found to be effective for Mo recovery from a spent hydrorefining catalyst that contained 21.8 wt.% MoO_3, 4.97 wt.% NiO, 1.54 wt.% Co_3O_4, 60.4 wt.% Al_2O_3, and

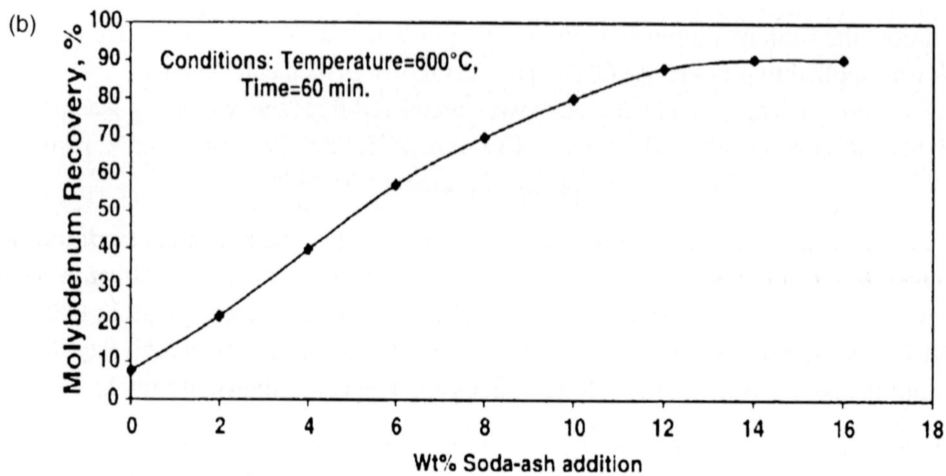

Figure 11.14b: Effect of soda ash addition on recovery of molybdenum.

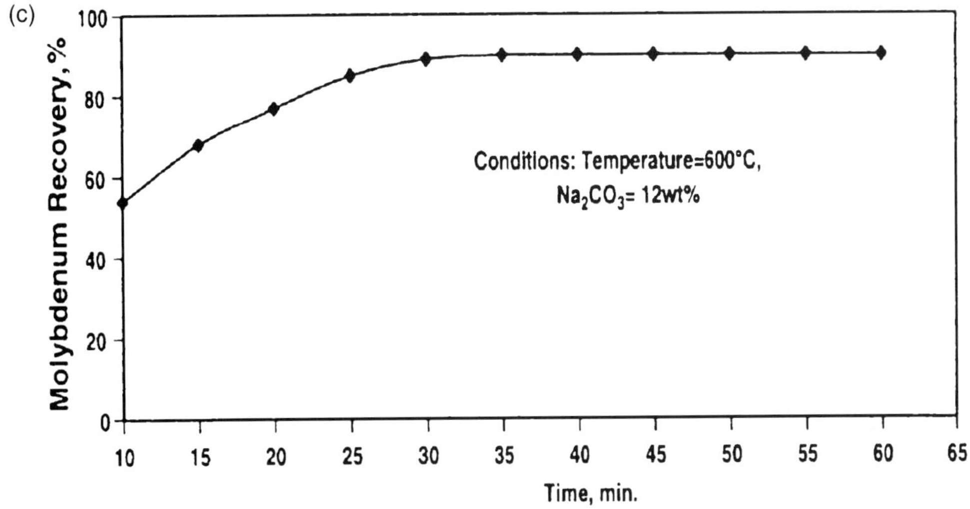

Figure 11.14c: Effect of roasting time on molybdenum recovery [From ref. 657. Reprinted with permission].

5.06 wt.% SiO_2. A flow diagram of the process is shown in Fig. 11.16 [658]. Under optimum experimental conditions (i.e., roasting temperature at 900 °C, 20 wt.% addition of NaCl to the feed, and a roasting period of 60 min), it was possible to extract up to 90% of molybdenum from the spent catalyst. Recovery of the other metals was not attempted in this study.

Recently, Chen et al. [659] investigated extraction of molybdenum and vanadium from a spent catalyst (main chemical composition: 2.05% Mo, 0.42% V, 62% Al_2O_3, and 10.3% SiO_2) by roasting the residue with soda carbonate, followed by hydrometallurgical treatment of the roasted products. In the roasting process, more than 91% of molybdenum and 90% of vanadium could be extracted when a charge containing a sodium carbonate to spent catalyst ratio of 0.15 was roasted at 75 °C for 45 min and the roasted mass was leached with water (liquid to solid ratio of 2) at 80–90 °C for 15 min. After the purification of leach liquor, an extraction solvent consisting of 20 vol.% trialkylamine and 10 vol.% secondary octyl alcohol dissolved in sulfonated kerosene was used to extract molybdenum and vanadium in leach liquor, 10 wt.% ammonia water was used as stripping agent. Adding 30 g/L NH_4NO_3 to the stripping solution and adjusting the pH to 7–8.5, over 99% of vanadium can be crystallized as AMV. Over 98% of molybdenum can be crystallized as ammonium polymolybdate when pH is between 1.5 and 2.5 (pH is adjusted by HNO_3). AMV and ammonium polymolybdate were calcinated at 500–550 °C, the purity of MoO_3 and V_2O_5 was 99.08 and 98.06%, respectively. The overall recoveries of Mo and V were 88.2 and 87.1%, respectively.

An integrated process for the recovery of all components of spent hydroprocessing catalysts including the alumina as commercially marketable products without leaving any residue material as solid waste has been reported by Llanos and Derering [567,660,661]. In this process

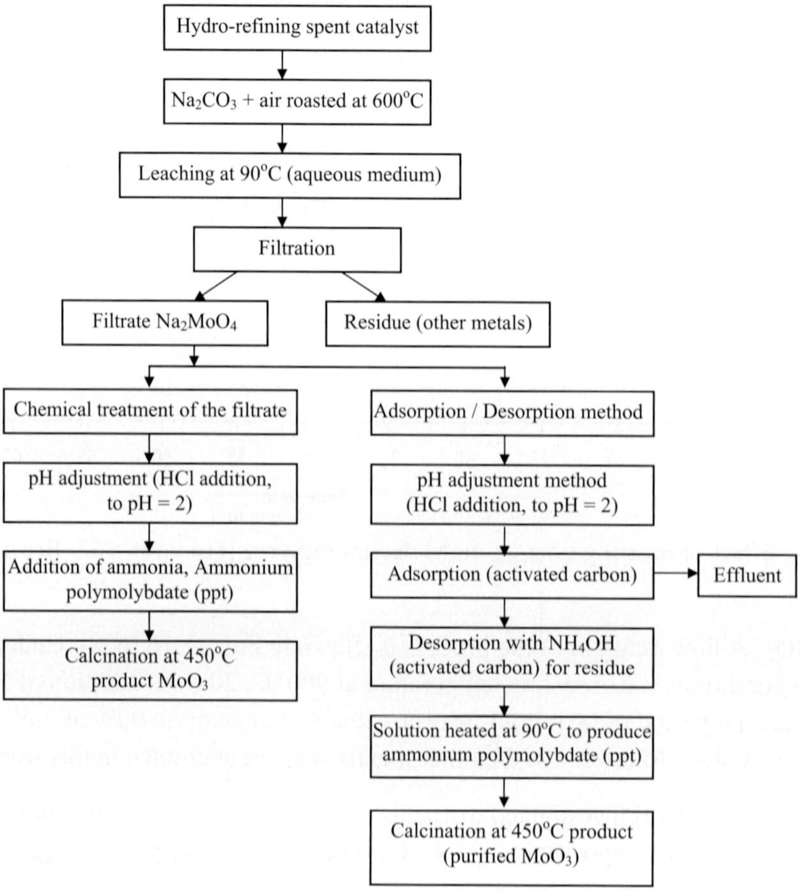

Figure 11.15: Schematic diagram for the production of high purity molybdenum oxide from hydrorefining spent catalyst [From ref. 657. Reprinted with permission].

(Fig. 11.17) [567], spent catalysts are mixed with alkaline materials, such as sodium carbonate and roasted at temperatures between 1400 and 1800 °F (650–900 °C), to convert molybdenum, vanadium, and part of the sulfur into their respective soluble sodium salts. The roasted material is leached with water to obtain a solution laden with molybdenum and vanadium, and a residue containing a portion of the molybdenum and vanadium, all the alumina, nickel, cobalt, silica, and some sodium in the form of sodium aluminum silicate. The solution is treated with ammonium salts to separate vanadium metavanadate, which may be converted into vanadium pentoxide. The remaining solution is further acidified to precipitate molybdic acid, which may be converted into molybdic trioxide. The leach residue is dried and reduced in the presence of carbon to produce a high grade alloy of nickel and cobalt, fused alumina, substantially free of sodium, and to vaporize most of the sodium. The sodium is recovered in the form of sodium hydroxide and is recycled to the roasting and leaching operations.

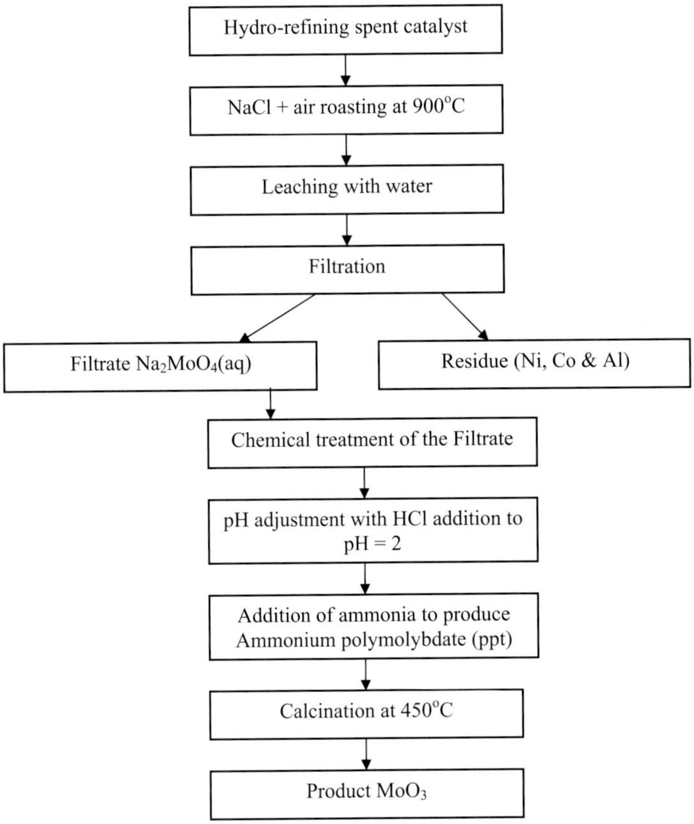

Figure 11.16: Flow sheet for the extraction of Mo from spent catalyst by salt-roasting [ref. 658].

11.1.2.2 Roasting with Potassium Salts

Fusion with $KHSO_4$ followed by water extraction has been used by Busnardo et al. [662] for recovering Mo, Ni, Co, and Al from $CoMo/Al_2O_3$ and $NiMo/Al_2O_3$ type spent catalysts. In this process, Mo, Ni, Co, and Al are converted into water-soluble salts of the metals during high temperature treatment (500 °C) with $KHSO_4$ according to the following reactions:

$$NiO + 2KHSO_4 \rightarrow NiSO_4 + K_2SO_4 + H_2O$$

$$CoO + 2KHSO_4 \Rightarrow CoSO_4 + K_2SO_4 + H_2O$$

$$Al_2O_3 + 6KHSO_4 \Rightarrow Al_2(SO_4)_3 + K_2SO_4 + H_2O$$

$$MoO_3 + 2KHSO_4 \Rightarrow MoO_2SO_4 + K_2SO_4 + H_2O$$

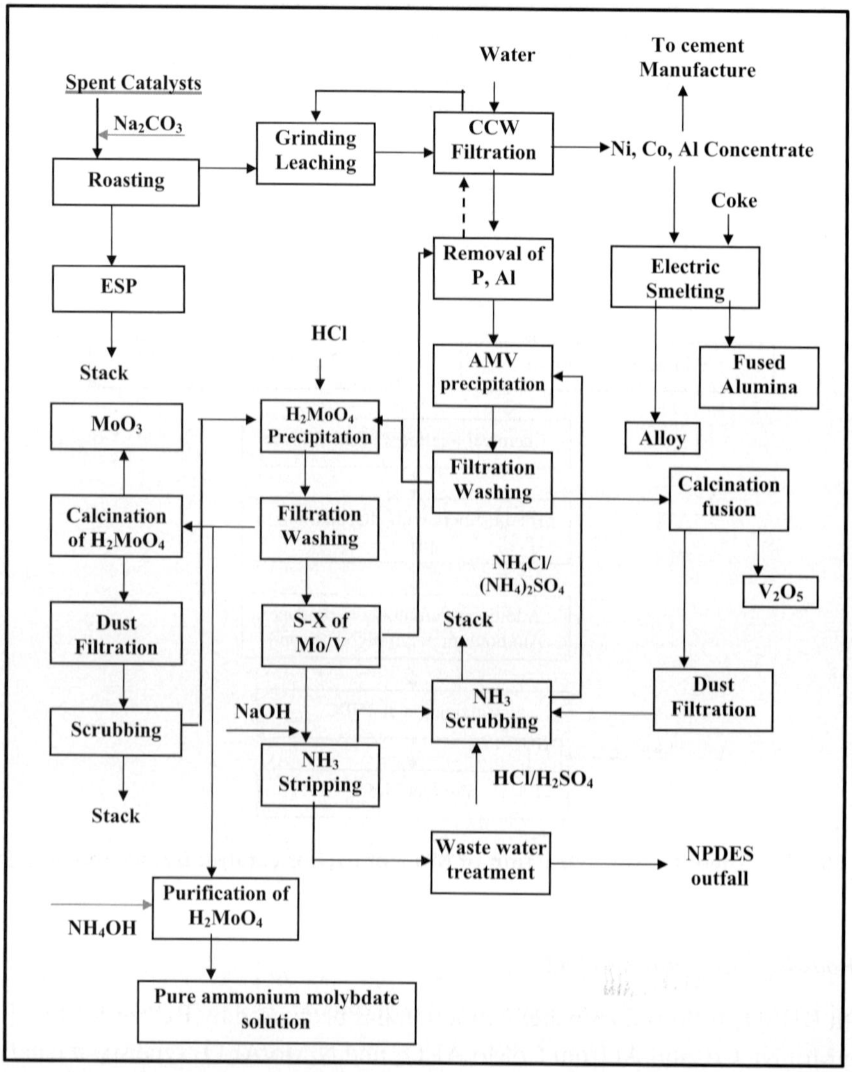

Figure 11.17: Gulf chemical and metallurgical process [From ref. 567. Reprinted with permission].

The solubilized metals are then recovered by conventional precipitation or solvent-extraction techniques. It is reported that >96% of the Mo in the spent catalyst was recovered as crystalline ammonium hepta molybdate $[(NH_4)_7Mo_6O_{24}{\cdot}4H_2O]$. The recoveries of other metals (Al, Ni, and Co) are >90 wt.%. Figure 11.18 [662] shows a general scheme for the recovery of Al, Ni (or Co), and Mo after fusion with $KHSO_4$ and dissolution of the fused mass in water. The yields of Ni, Co, Mo, and Al recovered in this process are presented in Table 11.3 [662]. It is claimed that the final wastes generated in this process have low toxicity and can be sent to industrial dumps or co-processed as mineralizing agent.

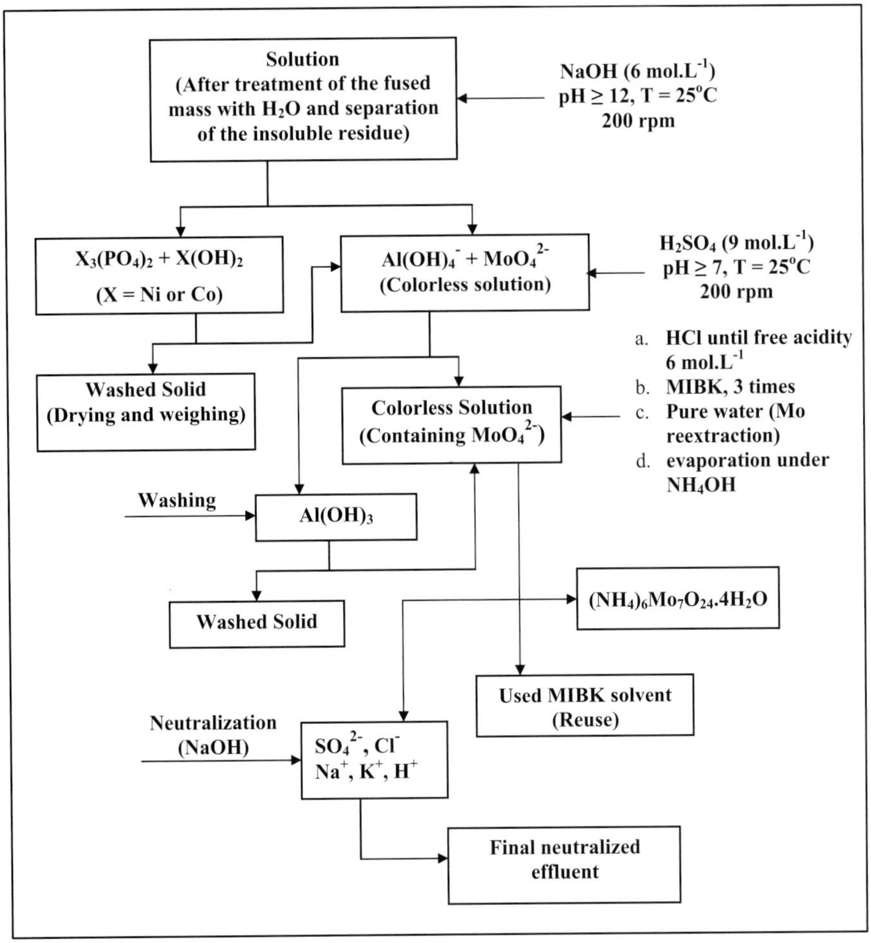

Figure 11.18: General scheme for recovery of Al, Ni (or Co) and Mo after catalyst fusion with $KHSO_4$ and dissolution of the fused mass in water [From ref. 662. Reprinted with permission].

11.1.3 Chlorination

Recovery of metals from waste catalyst materials by chlorination has been studied extensively and reported in many patents and scientific papers. A patented process developed by Gravey et al. [663] comprises transforming the metals into volatile and non-volatile chlorides by carbochlorination and then separating the chlorides in a selective manner. The residual oil present in the spent catalyst was first removed by extraction with a solvent. After the hydrocarbon extraction, the waste catalyst containing metal sulfides, Al_2O_3 and coke was treated with gaseous chlorine at a temperature between 500 and 600 °C. V, Mo, and Al were converted to volatile chlorides under these conditions. Separation of $AlCl_3$ from other chlorides was accomplished by passage through granules of anhydrous NaCl. $MoCl_5$ was

Table 11.3: Recovery of Ni, Co, Mo and Al present in NiMo and CoMo spent catalysts treated with fusion with $KHSO_4$ at 350 and 500 °C [From ref. 662. Reprinted with permission].

Catalyst	Product	Metals recovered (wt.%)			Temperature (°C)
		Ni or Co	Mo	Al	
CoMo	$Co_3(PO_4)_2 + Co(OH)_2$ (blue precipitate)	87.6 ± 0.2	Not detected	Not detected	350
		90.9 ± 0.5	Not detected	Not detected	500
NiMo	$Ni_3(PO_4)_2 + Ni(OH)_2$ (green precipitate)	91.6 ± 0.4	< 0.05	Not detected	350
		94.2 ± 0.8	Not detected	Not detected	500
CoMO	$Al(OH)_3$ (white solid)	0.05 ± 0.01	Not detected	94.5 ± 1.1	350
		Not detected	Not detected	95.6 ± 0.7	500
NiMo		0.05 ± 0.01	Not detected	90.2 ± 0.9	350
		Not detected	Not detected	92.8 ± 0.7	500
CoMo	$(NH_4)_7Mo_6O_{24} \cdot 4H_2O$ (white solid)	Not detected	96.2 ± 0.5	Not detected	350
		Not detected	99.2 ± 0.2	0.05	500
CoMo		Not detected	95.9 ± 0.6	< 0.05	350
		Not detected	97.4 ± 0.5	Not detected	500

separated by passage through granules of crystallized KCl. Finally, VCl_4 was recovered from the residual gases by condensation at 60 °C. The process is particularly applicable for the recovery of Mo, V, and Al present in spent hydroprocessing catalysts. The recovery of Mo, V, and Al exceeded 90 wt.% in the process. The nickel and cobalt chlorides, which are also formed, are generally not volatile at the process conditions and they remain in the solid residue within the column. Their recovery can be effected, for example, by forming an aqueous solution with subsequent precipitation of the corresponding hydrates or the carbonates.

Yoshida et al. [664] reported that metals, such as V and Ni, present in the spent residue hydrotreating catalysts can be recovered by chlorination without subjecting the catalyst to prior oxidation to remove coke and sulfur. The metals (V and Ni) present in the form of sulfides are converted to chlorides on chlorination with a chlorinating agent (e.g. Cl_2, HCl, CCl_4, S_2Cl_2) at a temperature below 600 °C. Vanadium forms VCl_4 which is volatile while nickel forms the non-volatile $NiCl_2$. Vanadium tetrachloride can be recovered from the gas stream by condensation and nickel chloride can be recovered from the solid residue by solvent extraction. Welsh et al. [665] used the mixture of gaseous chlorine, hydrogen chloride, and water vapor for chlorination of the metals present in spent catalysts at temperatures in the range 200–400 °C. The Mo and V were volatilized as oxychlorides and aluminium is volatilized as chloride. The metal chloride vapors were absorbed in an aqueous hydrochloric acid solution at temperatures between about 20 and 70 °C. Ni and Co in the spent catalyst were converted to water-soluble chlorides.

The application of anhydrous chlorination technique for recovery of valuable metals from spent hydroprocessing catalysts was investigated in more detail by Jong et al. [636] for US Bureau of Mines. The chlorination studies were carried out in a batch fluidized bed reactor (diameter, 1.25″ and length 48″) made of Vycor heat resistant glass. The chlorinator was connected to a primary condenser and receiver, and then to the exhaust outlet (Fig. 11.19). The exhaust outlet was connected to a bubbler containing H_2SO_4 to prevent backflow of air into the reactor. Three types of spent catalysts ($NiMo/Al_2O_3$, NiW/Al_2O_3, and $CoMo/Al_2O_3$) were used in their experiments. Vanadium was not present in these catalysts. The catalysts were heated in a flow of N_2 at 400 °C for 30 min to remove moisture from the catalyst before starting chlorination. Since chlorination of metal oxide requires a reductant, such as carbon or CO, waste catalysts containing more than 6.7 wt.% carbon, were chlorinated with only $Cl_2 + N_2$ or $Cl_2 + N_2$ and air; other waste catalysts were chlorinated with $Cl_2 + N_2$ and CO or $Cl_2 + N_2$, CO and CO_2. Chlorination was carried out at 450 °C for 30 min to separate volatile MoO_2Cl_2 or WO_2Cl_2 from less volatile $NiCl_2$ or $CoCl_2$, and $AlCl_3$. The MoO_2Cl_2 or WO_2Cl_2 was hydrolyzed to recover MoO_3 or WO_3 as a final product. Hydrolysis recovered 65 to 84 wt.% of Mo and 90 wt.% of W. The chlorinated spent charge was leached with water. The leach liquor was purified with an $NH_4OH–NH_4Cl$ solution to precipitate Al and other metal impurities. The Ni and Co were recovered by solvent extraction. The Ni or Co was precipitated

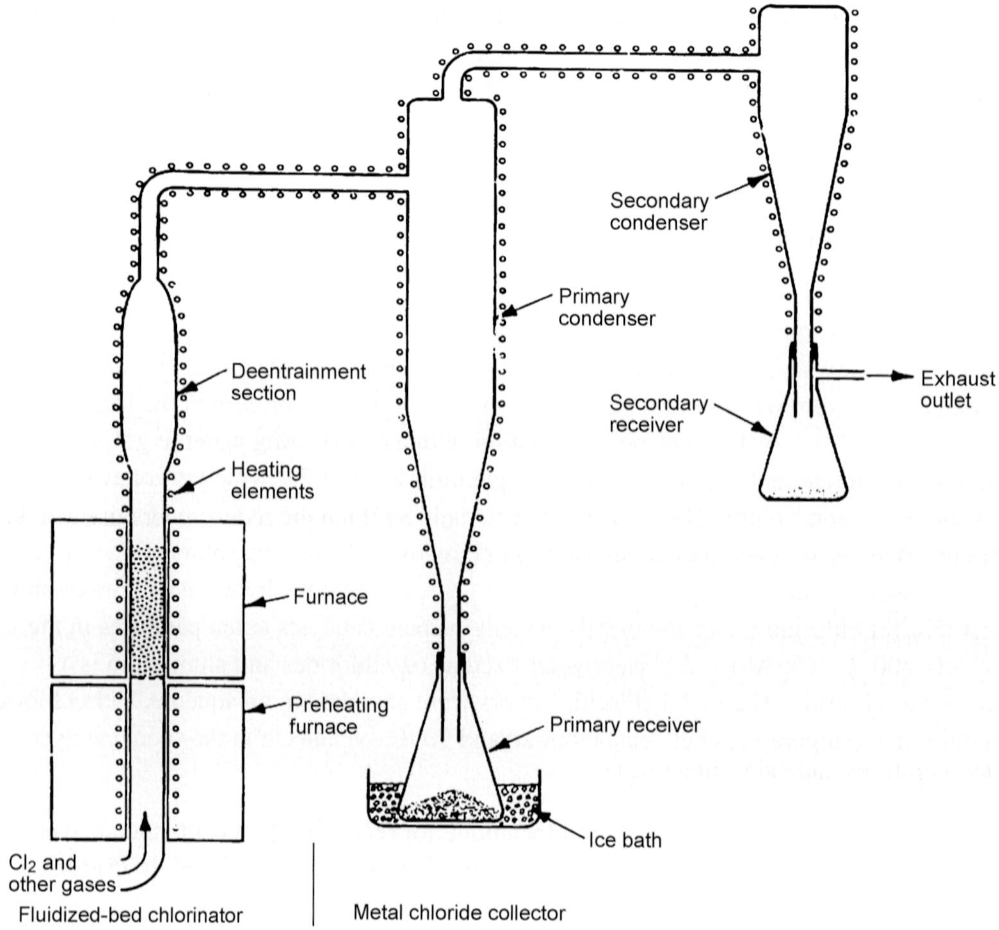

Figure 11.19: Apparatus for chlorination of waste catalysts [From ref. 636. Reprinted with permission].

from the purified solution at about pH 5 with Na_2S. The chlorination step extracted 73 to 99 wt.% of the Ni, 61 to 99 wt.% of the Co, 73 to 95 wt.% of Mo, and 82 to 98 wt.% of the W. In the final recovery step, solvent extraction recovered over 99 wt.% of the Ni from the purified solution. Over 98 wt.% of the Ni or Co was recovered by sulfide precipitation. The maximum overall recoveries for Ni, Co, Mo, and W in weight percent were 98, 97, 80, and 88, respectively.

Further investigations on the recovery of metals (V, Mo, Co, and Ni) from spent hydroprocessing catalysts without Al by selective chlorination were carried out by Gaballah et al. [666,667] in the mid-nineties. Both roasted (decoked) and unroasted (coke containing) spent catalysts were used in the chlorination study and the efficiency of metal recovery from both types of spent catalysts were compared. Chlorination of the unroasted sample at 500 °C for a reaction time of 30 min, with a gas mixture containing Cl_2/N_2 or Cl_2/O_2 equal to one lead to the extraction of more than 98% of molybdenum and 80% of vanadium, respectively. The volatilization of the cobalt and nickel chlorides was negligible. In addition, the chlorination of the alumina was limited to 3 and 9%. The carbochlorination of the roasted sample, that was free from carbon and sulfur, in the same conditions with a gas mixture having a Cl_2/CO ratio equal to one, lead to the extraction of 97, 82 and $\leq$3 of the Mo, V, and Al, respectively. Less than 5% of the cobalt and nickel compounds were transported by the vapor phase. The Mo and V chlorides were recovered separately from the condensates, while chlorides of cobalt and nickel were recovered from the chlorination residues by leaching with acidified water. The final residue composed essentially of alumina and is environmentally safe. Clearly, relatively good selectivity is obtained with a reasonable recovery rate of the valuable elements.

Ojeda et al. [668] studied the effect of different chlorination procedures on the recovery of metals from a spent $CoMo/Al_2O_3$ catalyst. The following three procedures were used in their study:

(1) Mo extraction by chlorination at low temperatures and subsequent chlorination at elevated temperatures to recover Co, employing a Cl_2–CO mixture.

(2) Extraction of the three metals by chlorination with Cl_2–CO gaseous mixture at elevated temperatures and subsequent separation by precipitation of Co at controlled pH.

(3) Mo extraction by chlorination of the catalyst with CCl_4 vapors in air current and subsequent purification of the recovered Mo solutions by precipitation of impurities (Fe and Al) at controlled pH.

In every case, almost 100% Mo was recovered, while Co extraction was 90% in procedure I and 95% in II. In all the assayed methods, hydrated aluminum oxide was obtained as by-product.

Recently, Ojeda et al. [669–671] investigated the chlorination of molybdenum trioxide in the presence of carbon with the aim of determining the experimental conditions at which chlorination occurs and to propose a mechanism for this reaction. The effects of several variables, such as carbon content, mixing time, flow rate, temperature, chlorine mole fraction, and reaction time on the chlorination of MoO_3 in the presence of carbon, were studied. The results showed that carbochlorination starts at 543 K, significantly increasing with temperature and reaction time, and slightly affected by chlorine molar fraction and flow rate, while direct chlorination starts at 803 K. The following global reaction was obtained by identification of the reaction products:

$$2MoO_3(s) + 2Cl_2(g) + C(s) \Rightarrow 2MoO_2Cl_2(g) + CO_2(g).$$

On the basis of the analysis of experimental data, thermodynamic results and observations by other authors, three basic stages were distinguished:

(1) Formation of the chlorinating agent.

(2) Chlorination of molybdenum trioxide.

(3) Carbon oxidation.

Based on the kinetic treatment of the experimental data, a "nucleation and growth" type model was proposed for the carbochlorination reaction of MoO_3.

11.1.4 Metal Recovery by Carbothermic Reduction

Karr et al. [672] investigated the lime aided carbothermic reduction of spent catalyst. For this purpose, they used stoichiometric amounts of carbon and different mole ratio of $CaCO_3$. The pelletized mixture was reduced in a carbolyte furnace at temperature range of 1000–1200 °C under reduced pressure. After cooling, the reduced mass was separated from slag (CaS) by elutriation. The metal powder was again compacted and sintered for subsequent purification step using a fused salt electrolysis. The efficiency of electro deposition was enhanced by optimizing experimental parameters, such as temperature and cathode current density. The schematic of electrolytic cell used for electrorefining is shown in Fig. 11.20 [672].

11.1.5 Metal Recovery Using Electrolytic Cells

The Batelle process represents the first attempt to recover metals from spent hydroprocessing catalysts using electrochemical dissolution [673,674]. In this process, the spent catalyst which contained sulfides of Mo, V, and Ni with carbonaceous material and sulfur is added to the solution containing a strongly oxidizing ion, such as Ce^{4+}. The solution is added at the anode side of the electrochemical cell where aqueous Ce^{3+} is oxidized to Ce^{4+}, which in turn oxidizes coke to CO_2 and H_2O, and metals to soluble oxides. Thus, the oxidation/solubilization of the metal sulfides was achieved by this method. However, separation of the dissolved metals by electrolysis was not studied in this work.

Recently, Lai et al. [675] used a combined acid leaching and electrolysis process to recover valuable metals from spent HDS catalyst. The electrolytic cell was equipped with a glass bead medium, an iridium oxide mesh anode, and a stainless steel plate cathode. An acid solution consisting of concentrated $HNO_3/H_2SO_4/HCl$ with a volume ratio of 2:1:1 was found to be better than the other tested solution ($HNO_3/H_2SO_4 = 1:1$) to leach the metals. For the three-acid mixture, the best solid/liquid ratio and leaching time were 40 g/L and 1 h, respectively, at 70 °C. Under these conditions, the leaching yields of target metals, such as Mo,

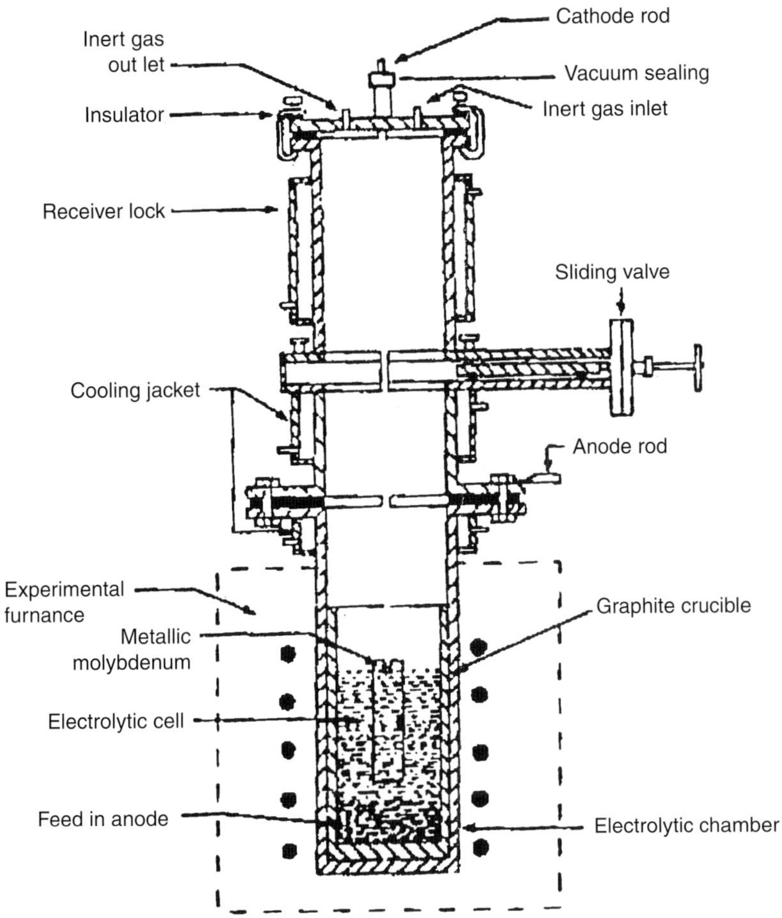

Figure 11.20: Electrolytic cell for carrying out electrorefining [From ref. 672. Reprinted with permission].

Ni, and V, reached 90, 99, and 99%, respectively. When this acid leachate was electrolyzed for 2 h at 2 A constant current (current density = ~35.7 mA/cm^2), the electrolytic recoveries of Mo, Ni, and V were 15, 61, and 66%, respectively. Extending the electrolysis time from 2 to 4 h did not increase the recoveries. For this operation, the total recoveries (leaching yield x electrolytic recovery) of Mo, Ni, and V were around 14, 60, and 65%, respectively.

11.1.6 Metal Recovery by Applying Thermal Plasma

Recently, a process for recovery and reduction of metals (Co and Mo) from spent hydroprocessing catalysts by thermal plasma was developed by Wong et al. [676]. In this process, the spent catalyst was first sintered at >1500 K under plasma conditions and Co-Mo oxide was reduced to Co-Mo. This was confirmed by spectroscopic techniques. The coke remaining on the surface of spent catalyst was decomposed and converted to CO, CO_2, and H_2. Apparently, CO and H_2 produced in the process acted as reducing agents.

11.1.7 Summary of Laboratory Studies

Apparently, pretreatment prior to metal recovery, i.e., deoiling, decoking, and crushing, is common for all methods used for metal recovery from spent hydroprocessing catalysts. The efficiency of recovery of metals is only one of the parameters used to compare different leaching methods. Table 11.4 shows that for leaching methods, recovery is either similar or the reported recovery ranges are overlapping. Therefore, other parameters have to be taken into consideration for the overall comparison. The information on cost of raw materials used for recovery, number of steps involved, associated emissions, severity and required safety precautions, material requirements, etc. may play an important role for identifying the most suitable method.

Among leaching methods, the solution of ammonia or ammonium containing compounds used for leaching can be readily prepared and leachates obtained from these methods can be processed using established procedures. The increased requirements on materials and level of safety have to be considered for the leaching solutions containing strong acids. In this case, special procedures have to be applied to neutralize some by-product liquids before their safe disposal. These problems are almost completely eliminated by using aqueous solutions of organic acids.

A direct comparison of the leaching spent $NiMo/Al_2O_3$ catalyst using the solutions of ammonia with that of the NaOH (10%) solution was conducted by Villareal et al. [617]. The former solution was suitable for the selective leaching of V, while leaving most of the Mo behind. This observation differs from other studies [613,616] in which both V and Mo were leached out. It should be, however, noted that the catalyst used by Villareal et al. [617] was heavily deposited by V, i.e., it had the atomic V/Mo ratio of about 12. Moreover, leaching was conducted at a nearly room temperature compared with higher temperatures used by the other authors. The leachate was heated and cooled to recrystallise and separate NH_4VO_3 before being heated to 450 °C to form pure V_2O_5. The NaOH (10%) solution leached out both V and Mo.

Jong et al. [636] compared the two stages NaOH–H_2SO_4 leaching process with the chlorination process in which solvent extraction was used to recover Ni. The overall operating cost of the leaching process was about twice that of the chlorination process. The cost of raw materials (e.g., NaOH, H_2SO_4, $CaCl_2$, NH_4OH, etc.) used for the leaching process was the main contributor to the different operating costs. During this comparison, the environmental issues were not addressed in details.

Bioleaching of a FCC catalyst was compared with the chemical leaching using organic acids which are secreted during the growth of the *A. niger* fungus [642]. Generally, bioleaching gave

Table 11.4: Summary of metal recovery by different leaching and roasting methods.

Metal present in spent catalyst	Reagent	Metal recovered (wt.%)					Reference
		Mo	V	Ni	Co	Al	
V, Mo, Ni, Co, Al	$Na_2CO_3 + H_2O_2$	99	85	–	–	–	[633]
Mo, Ni, Al	$Na_2CO_3 + H_2O_2$	85	–	65	–		[635]
V, Ni, Mo, Al	NaOH (10%), pH 8.8	92	89	20	–	35	[617]
V, Ni, Mo, Al	NaOH (10%), pH 8.4	99	93	10	–	20	[688]
V, Ni, Mo	NaOH (atmospheric pressure leaching in two steps)	97	92	–	–	–	[639]
V, Ni, Mo, Al	Aqueous NH_3 solution (17 M)	–	99	–	–	–	[617]
Ni, Mo, Al	NaOH (1st stage) and H_2SO_4 (2nd stage)	84	–	98	–	–	[636]
Ni, Co, Mo, Al	NaOH (1st stage) and H_2SO_4 (2nd stage)	97	–	92	92	65–75	[637]
V, Mo, Co, Ni, Al	Na_2CO_3 (roasting + leaching with H_2O)	90–95	90–95	–	–	–	[654]
Co, Mo, Al	NaOH roasting (1st stage) H_2SO_4 leaching (2nd stage)	90	–	–	90	85	[655]
V, Mo, Ni Co, Al	NaOH roasting (1st stage) and H_2SO_4 leaching (2nd stage)	99	96	98	99	91	[659]
Mo, Ni, Al	NaCl roasting + water extraction	90	–	–	–	–	[658]
Mo, Ni, Co, Al	KHSO fusion + water extraction	96	–	90	90	90	[662]
V, Mo, Ni, Co	Aqueous $NH_3 + NH_4CO_3 + H_2O_2$	93	88	80	78	–	[613]
Mo, V, Ni	Citric acid	94	94	85	–	–	[631]
V, Mo, Ni, Al	Oxalic acid + H_2O_2	90	94	65	–	33	[632]

3–20% higher metal extraction efficiency than the chemical leaching. Apparently, bioleaching may not be competitive with the other methods used for leaching. A significantly longer leaching time required during bioleaching to achieve extraction efficiencies, which are comparable to those observed using other leaching methods may be the main drawback of bioleaching. Environmental issues may play a decisive role in evaluating and/or comparing commercial processes for metal reclamation. For example, the roasting and chlorination processes generate gaseous effluent which requires treatment. No waste gases are generated in the case that leaching step is conducted as the first step instead of roasting and chlorination. This may be an important advantage of the leaching methods compared with other two methods. The separation of metals from leachates generates liquid streams, which require purification before disposal. It should be noted that decoking of spent catalysts generates gaseous emissions, such as SO_X, NO_X, NH_3, etc., that have to be removed from the waste gas. This step is common for all processes used for metal reclamation. Apparently, the information on environmental aspects of the commercial processes used for metal reclamation from spent catalysts is the scarce. Therefore, a speculative discussion of this issue can only be afforded.

In recent years, attempts have been made to use rather novel techniques for metal recovery from spent catalysts. This included carbothermic reduction of metal oxides to metals [672]. Electrolytic cells have been employed as well [673,674]. Energy intensive method, such as thermal plasma received attention as well [676]. There is little information indicating commercial use of these methods for metal recovery from spent hydroprocessing catalysts.

11.2 Separation of Metals from Solution

In the preceding sub-chapters, frequent reference has been made to the isolation of metals and/or metal compounds in a pure form from leaching solutions, without providing details. In this regard, extensive information on the separation of metals from solutions can be found in the literature dealing with various aspects of hydrometallurgy. This experience can be applied to spent hydroprocessing catalysts. In this case, more than one metal are usually present in the solution after leaching and/or dissolution of spent catalysts. Additional treatments are required to isolate metal compounds in their pure form. For these purposes, numerous extraction agents of organic origin, with a high selectivity for a metal of interest have been available and used in commercial applications. The structure of some commercial agents used for the extraction of metals from various solutions is shown in Fig. 11.1 [622]. The selectivity can be further enhanced by optimizing the conditions applied during extraction.

As the final step, a pure metal compound can be obtained from the solution by precipitation under controlled conditions (e.g., pH) using a suitable precipitant. In some cases, only minor modifications to the methods used commercially in hydrometallurgical industry are required before they can be applied for metal reclamation from the spent hydroprocessing catalysts.

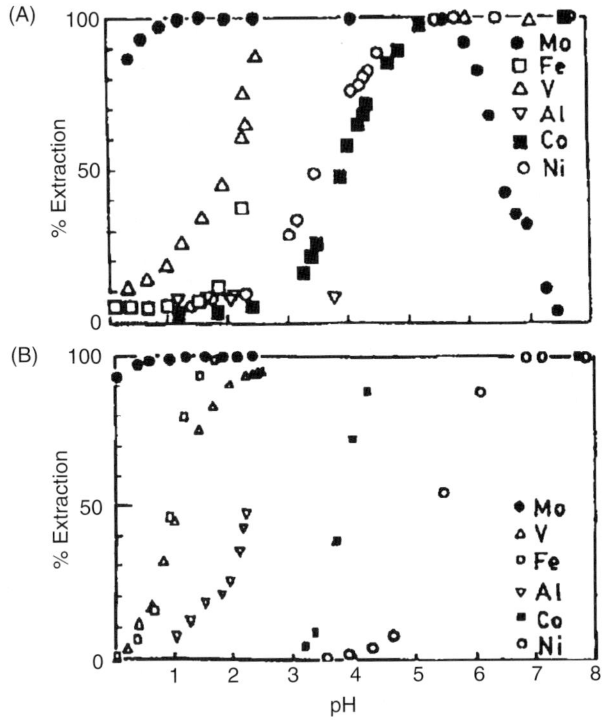

Figure 11.21: Extraction of metals (A) with 20 vol.% LIX 63 in EXXSOL D80 and (B) with 20 vol.% Cyanex 272 in EXXSOL D80 [From ref. 622. Reprinted with permission].

Some typical examples involving spent hydroprocessing catalysts were selected to describe these events in more details.

An important contribution to the development and testing of the organic agents used for the extraction of metal compounds from various solutions obtained by dissolution of spent catalysts was made by Inoue et al. [622,623]. Before the extraction method could be applied, the spent $CoMo/Al_2O_3$ catalyst deposited with V and Ni was roasted at 700 °C, then roast was suspended in 63% H_2SO_4 and evaporated to dryness. After dissolving in water and filtered, the filtrate was diluted to attain pH of 1.2 and extracted. With respect to the purity of isolated metals, pH of the solution may be the most critical parameter. This was demonstrated by the results shown in Fig. 11.21 [622]. In this case, before the extraction, the solution contained 2.7, 0.75, 0.03, 13.5, 1.0, and 0.17 g/L of Mo, V, Fe, Al, Co, and Ni, respectively. An excellent separation of the Mo from the other metals in the scrub solution could be achieved at pH approaching zero when 20 vol.% Cyanex 272 in EXXSOL D 80 as diluent was used as extracting agent (Fig. 11.21b). Subsequently, the aqueous solution of ammonia could be used to strip Mo from the extract. In this case, a good separation was achieved when pH of the extract was maintained between 8.0 and 8.4. After the Mo separation, pH of the scrub solution was increased to 1.5 by the addition of $Ca(OH)_2$ powder. At this pH, V was separated with

Cyanex 272 and subsequently stripped from the extract with the aqueous solution of ammonia. However, if present, Fe would interfere with the V separation. After the separation of Mo and V, the efficient separation of Ni and Co from Al could be achieved using the mixture of LIX 63 and SYNEX DN-052 in the EXXSOL D 80 (Fig. 11.21a).

In another study, Zhang et al. [677] evaluated commercial reagents, such as TR-83, PC-88A, and PIA-8, under similar conditions. Among these reagents, PIA-8 exhibited a similar performance as CYANEX-272. Thus, Mo could be nearly completely extracted from the solution at a pH approaching zero together with V, Fe and small amounts of Al. These co-extracted metals could be scrubbed from the solvent phase using H_2SO_4 before Mo was recovered by stripping using 5 to 7% ammonia. Vanadium could be further recovered from the scrub solution still containing Fe and Al after adjusting pH to about 1.5 using $Ca(OH)_2$. Subsequently, V could be recovered by stripping with 6% ammonia. Figure 11.22 [622] shows that for the 20 vol.% TR-83 in EXXSOL D80 and PC-88A in EXXSOL D80, the separation of Mo was interfered with Fe.

Figure 11.21 [622] shows that the separation of Ni and Co from Al is possible, but it may require large adjustment of pH. This problem was elevated using LIX 63 and/or the mixture of

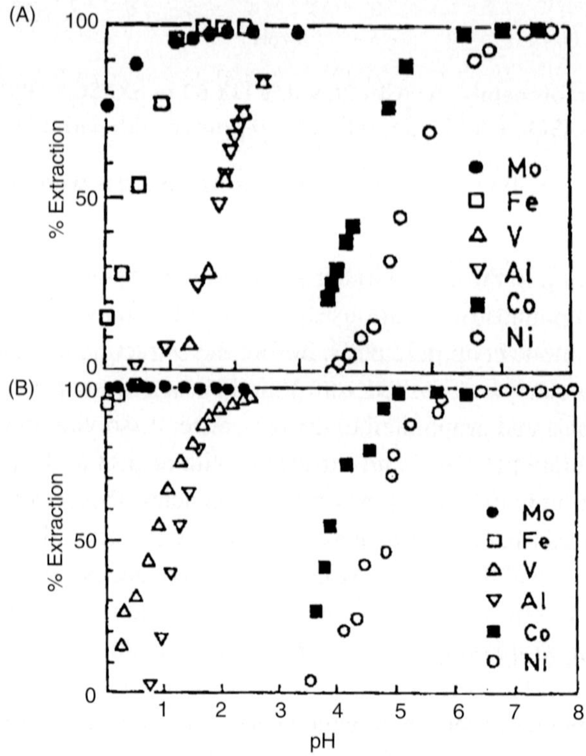

Figure 11.22: Extraction of metals (A) with TR-85 in EXXSOL D80 and (B) with PC-88A in EXXSOL D80 [From ref. 622. Reprinted with permission].

SYNEX DN-052 in the EXXSOL D80 [678]. In this case, Ni and Co can be readily separated from Al after separation of Mo by a slight adjustment of pH. The separation of Ni from Co can then be easily achieved using conventional methods described by Ritcey [679]. In this regard, long chains alkyl amines exhibited a good performance. The Al, as the last metal left in the original solution, can be isolated by precipitation as pure $Al(OH)_3$.

The METREX process can be used as another example illustrating the use of organic agents for the extraction of metals from leachates. The order of metals separation was identical as those observed by Inoue et al. [622,623] and Zhang et al. [677]. The optimization of conditions (e.g., type of agents, pH of extraction, ratio of agent/diluent, etc.) for separation of metals from solutions using organic agents was carried out by Olazabal et al. [502,503]. Another study involving numerous extractants is that of Sato et al. [680].

Phosphoric acid derivatives have received the attention as potential agents for separation of Mo from V in aqueous solutions. For example, Hirai et al. [681,682] used bis(2-ethylhexyl)monothiophosphoric acid and bis(2-ethylhexyl)phosphoric acid. The latter agent was efficient for the metal separation from the sulfuric acid solutions [683], whereas tri-n-butyl phosphate was suitable for that from the hydrochloric acid solutions [684,685].

In the leachates containing Mo and V, the latter could be isolated by the precipitation with $(NH_4)_2SO_4$ at pH of 8.6 [683]. The Mo was separated from the remaining V in 0.05 M sulfite ion medium by extracting with tri-n-octylamine, stripping with aqueous ammonia and precipitating by acidification of the stripped solution. The method based on the use of ion-exchange resins was also suitable for the separation of Mo from V in the same solution [660]. Among several resins, the polystyrene resin cross-linked with bis-(22-hydroxyethyl) amino group was the most efficient for the separation of Mo from V [686]. In this case, the solution containing Mo and V was passed through the column of the water swollen resin and the effluent analysed.

Membrane separation is another potential method which may be applied to metal recovery from spent hydroprocessing catalysts. The separation of active metals could be enhanced by introducing an iminodiacetate group on the membrane [687]. The metal adsorbed from the solution by the agent could be eluted by permeation of an HCl solution. So far, this method was only tested for recovery of noble metals from spent catalysts. It is believed that the membrane separation deserves attention as potential method for metal recovery from spent hydroprocessing catalysts.

11.3 Commercial Processes

There are several companies in the world which specialize in the recovery and reprocessing of spent catalysts. Following are the main companies involved: Gulf Chemical & Metallurgical Corporation (GCMC) (USA), CRI-MET (USA), Taiyo Koko Co. (Japan), EURECAT

(France), Spent Catalyst Recycling (Germany), Taiyo Mining and Industrial Co. (Japan), Aura Metallurgic (Germany), Sadaci (Belgium), Full Yield Industries (Taiwan), Metallurgy Vanadium (USA), Metal-Tech (Isreal), Nippon Catalyst Cycle Co. (Japan), Moxba-Metrex (The Netherlands), and Quanzhou Jing-Tai Industry Co. (China). The process technologies used in these industries and the products recovered from spent hydroprocessing catalysts are reviewed in this section. It is to be noted that for proprietary reasons, only limited information is available in the open literature. This prevents technical/economic evaluations and/or comparison of the commercial processes.

11.3.1 Gulf Chemical & Metallurgical Corporation Process

At GCMC's metals reclamation plant, an integrated process involving soda ash roasting and electric smelting is used to recover all metals and alumina from spent hydroprocessing catalysts [566,567]. A schematic diagram of GCMC's metal recovery process is shown in Fig. 11.17 [567]. Spent catalysts and sodium carbonate are roasted in a multiple hearth furnace to burn off the hydrocarbons and some of the sulfur, and to convert the molybdenum, the vanadium, and the remaining sulfur to water-soluble salts. The roasted material was milled, leached with water to dissolve the molybdenum and vanadium compounds, and then filtered to separate the solid containing alumina, cobalt, and nickel from the molybdenum and vanadium in solution.

The Mo/V laden leach solution obtained from the GCMS process is first treated to remove phosphorous, aluminum, and arsenic, and then, the purified solution is mixed with ammonium sulfate and chloride to precipitate AMV. The AMV is calcined at 400–600 °C to decompose it into ammonia and vanadium pentoxide. The granular V_2O_5 is fused and quenched on a rotating wheel to produce flakes typically containing over 99% V_2O_5. Ammonia is recovered in a series of scrubbers using dilute hydrochloric and sulfuric acid and recycled for the precipitation of AMV.

The filtrate from the AMV precipitation, containing the molybdenum, is treated with a reducing agent, heated, and acidified to precipitate molybdic acid. After filtration and washing, the molybdic acid is calcined to produce molybdic trioxide that is over 98% pure. In a separate operation, molybdic acid is converted into high purity ammonium molybdate solution by treatment with ammonia and nitric acid. The ammonium molybdate solution is sold to catalyst manufacturers. Molybdenum and vanadium left in the filtrate from molybdic acid precipitation are recovered by solvent extraction. The filter cake is smelted onsite in GCMC's electric arc furnace. The products are high-grade fused alumina for refractory and abrasive applications and an alloy containing 37 to 43% nickel and 12 to 17% cobalt, which is sold to nickel-cobalt refineries.

11.3.2 CRI-MET Process

CRI-MET's plant uses a two-stage caustic pressure leaching process to recover metals from spent hydroprocessing catalysts [688,689]. In this process, spent catalysts are milled in a solution of sodium aluminate and sodium hydroxide. The resulting slurry is fed into an autoclave and leached under oxidizing conditions at elevated temperature and pressure to convert the sulfur to sulfate, oxidize the organic compounds, and dissolve the molybdenum and vanadium. The autoclaved material is thickened and filtered to separate the liquid from the solids. The molybdenum from the first stage leach is separated from the vanadium as molybdenum trisulfide, by H_2S treatment. The molybdenum trisulfide is then converted to low-grade molybdenum trioxide, which is then purified and calcined to pure molybdenum trioxide. The vanadium solution is the feed for the vanadium hydroxide precipitation step with soda ash. This hydroxide is water washed, dried, and calcined to form a vanadium oxide product.

The solids from the first leach, containing alumina, cobalt, and nickel, are leached a second time at high temperature and pressure with a strong caustic to solubilize aluminium. The nickel-cobalt solids from the second leach are separated, washed, and dried, then shipped to another plant to be calcined. The calcined nickel-cobalt material is then exported to a nickel-cobalt refiner [689,690]. Sodium aluminate solution is treated further to produce an alumina trihydrate product which is sold to producers of aluminium sulfate and sodium aluminate.

11.3.3 EURECAT Process

Initially, EURECAT was using a combination of hydrometallurgical and pyrometallurgical steps in its metal reclamation process [252,307,691]. The spent catalyst was heated in an oxidizing environment to remove coke, part of the sulfur and other hydrocarbon impurities. The decoked catalyst was then subjected to alkaline leaching with caustic soda. Mo, V and P, and part of the Al present in the spent catalyst were converted to soluble sodium salts during this treatment. The Ni, Co, Fe and most of the alumina were not leached out and remain in the residual solid after filtration. The leaching efficiency was controlled by the parameters, such as pH, concentration, liquid/solid ratio, residence time, etc. The process was continuous with a counter current percolation using 12 tanks in series.

The filtrate contained Na salts of molybdate and/or tungstate, vanadate, and impurities, such as arsenate and phosphate. The filtrate was purified to remove arsenate, and small amount of aluminate before the separation of Mo and/or W and V. These impurities were removed by precipitation. The ion exchange was used to separate Mo and V. Vanadyl sulfate and ammonium molybdate were the final products. The filtrate cake containing between 2 and 4% of Ni and/or Co was processed by pyrometallurgy to recover Ni and Co. The pyrometallurgical process was carried out by EURECAT's partner, a large mining company specialized in nickel metallurgy [7]. The process consisted of a fusion of the solid in an arc furnace to separate cobalt and nickel in a matte form from the aluminium and residual impurities, which are eliminated in the slag. The nickel/cobalt matte was then treated conventionally by solvent extraction to separate nickel and cobalt before electrolysis of the metals. The alumina containing slag was completely inert and it could be disposed safely or could be used as a road construction material. The valuable metals extracted from the spent catalysts were electrolytical nickel and cobalt metals, molybdenum oxide, ammonium molybdate (or tungstate), and vanadium sulfate. The yield recovery for molybdenum, vanadium, and tungsten was over 90%, and for cobalt and nickel as high as 97%. This process is now abandoned and a complete pyrometallurgical route is used [7]. The pyrometallurgical route starts with dry catalysts that are melted in a furnace at temperatures around 1200–1500 °C. Heavy metals sink to the bottom as alloys and are separated from slag, containing the alumina or silica support. It is claimed that the pyrometallurgical route has the advantage of a complete recycling scheme: metals for manufacturing in special alloys, the inert slag for manufacturing insulation material, such as rock wool.

11.3.4 Taiyo Koko Company Process

In Taiyo Koko Company's metal recovery plant, soda ash roasting process is used to recover Mo and V from spent desulfurization catalysts [692]. A simplified schematic diagram of the process is shown in Fig. 11.23. The spent catalyst is mixed with soda ash and roasted in a kiln at 850 °C for about 2–5 h in an oxidizing atmosphere to convert the Mo and V to their sodium salts. The roasted material is wet ground in a ball mill, leached with water and filtered. The filtered leach solution containing sodium molybdate and sodium vanadate is treated with magnesium chloride and ammonium chloride for the removal of aluminum and phosphorous. The purified solution is mixed with ammonium chloride to precipitate AMV. The AMV is calcined to remove ammonia and the resulting V_2O_5 is fused and flaked. The vanadium product contains over 98% V_2O_5 with iron and molybdenum as minor impurities. After the removal of vanadium, the solution containing Mo is acidified with HCl to precipitate molybdic acid which is calcined to produce MoO_3.

11.3.5 Full Yield Industry Process

Full Yield Industry in Taiwan is reprocessing spent hydrotreating catalysts to produce molybdenum salts and oxide, vanadium salts and oxide, cobalt salts, nickel salts, and ferrovanadium alloy. The process [569] involves soda ash roasting followed by leaching and separation of individual metals by precipitation, solvent extraction/ion exchange methods.

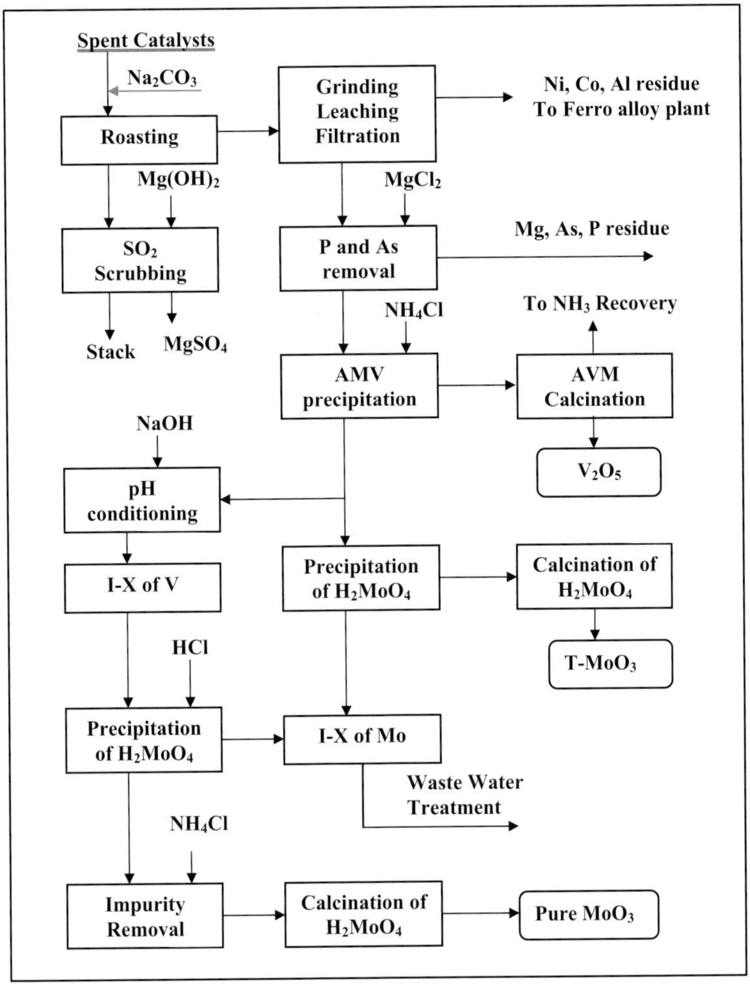

Figure 11.23: TAIYO KOKO process [From ref. 692. Reprinted with permission].

11.3.6 Moxba-Metrex Process

Moxba-Metrex, which is a Dutch Company specialized in the processing of spent catalysts and metal containing residues, recovers all major metals present in spent hydroprocessing catalyst by sulfuric acid leaching process [311,693]. Prior to the leaching, the spent catalyst is decoked in a furnace. The metal oxides are then dissolved in the acid leaving the alumina support as residue which is separated by means of decanting, washing and filtration from the liquor containing metals. The metals are isolated from the liquor by solvent extraction carried out in several stages, i.e. the first Mo, followed by V, Co, and Ni. MoO_3 is sold to steel industry whereas ammonium molybdate and vanadate are sold as intermediate chemicals. Co and Ni sulfates are used for production of Ni and Co powders. The alumina residue is used for production of the refractories. Recovery efficiencies of more than 90% for Mo and Co are achieved in this process.

11.3.7 Quanzhou Jing-Tai Industry Process

In China, Quanzhou Jing-Tai Industry Co. has developed a simple process for the recovery of metals from spent hydrotreating catalysts using H_2O_2 for leaching [694]. The process first subjects the catalysts to a low-temperature calcination step, followed by grinding of the calcined material. The ground material is then leached with 10–20% v/v hydrogen peroxide to recover metals, such as Mo, V, and Ni or Co. This step has a selectivity that is greater than 95% for the metals. The undissolved residue, which contains mainly Al_2O_3 and SiO_2, can be dewatered and used in the production of masonry bricks. The pH of the leach liquor is then adjusted to between 0.5 and 2.5 by adding dilute sulfuric acid to co-precipitate Mo and V with a recovery rate of greater than 99%. The lean liquor contains mostly dissolved Ni or Co, which can be recovered through precipitation at a pH of between 8.5 and 9.5 by caustic additions. The final liquor is then sent to an ion exchange step for final recovery and cleanup of residual metals. The effluent is then discharged after a simple waste water treatment process, which has been found to meet all environmental requirements. The entire process has been found to be simple, cost-effective with minimal environmental impacts while achieving the goal of complete resource recovery from spent catalysts commercially.

11.3.8 Metallurg Vanadium Process

Metallurg Vanadium (Ohio, USA) uses a proprietary pyrometallurgical process to convert metals (V, Ni, Mo, Co, W) present in spent hydroprocessing catalysts to alloys, such as ferrovanadium, (FeNiMo), and other ferroalloys, which are sold to steel companies [695].

11.3.9 German Processes

In Germany, there are three metal reclamation companies, namely GfE Metalle and Materialien GMBH, AURA Metallurgie GMBH, and spent catalyst recycling (SCR) GMBH. Both pyrometallurgical and hydrometallurgical processes are applied at GfE's metal recovery plant [696]. Hydrometallurgical process is used to produce molybdenum chemicals, Co/Ni solution, and aluminium oxides. The pyrometallurgical process serves for the generation of cast vanadium concentrate (CVC) from which high purity vanadium chemicals are extracted. At SCR, spent metal catalysts and metal residues containing Ni, Co, Mo, V, and W are processed to reclaim the metals. Process details are not available for this plant [697]. Aura's metal recovery facility at Helba, Germany, has been in operation since 2001 with a capacity to treat up to 10 000 tons of spent HDS catalyst ($NiMo/Al_2O_3$ and $CoMo/Al_2O_3$) per year.

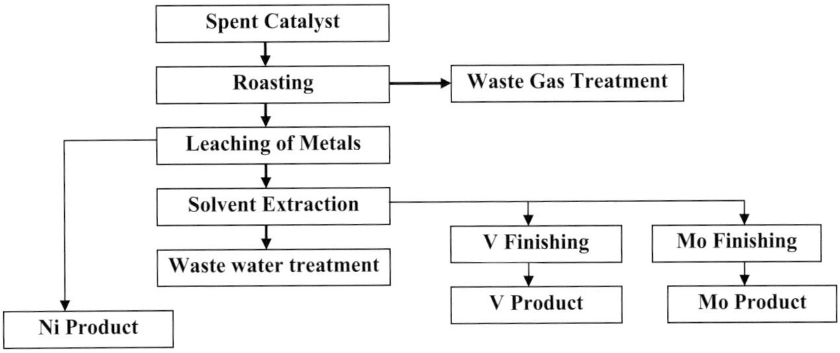

Figure 11.24: Recovery of metals in Nippon Catalyst Cycle Co. [From ref. 699. Reprinted with permission].

Molybdenum oxide hydrate, molybdenum concentrate, ammonium molybdate, cobalt-nickel sulfate, gypsum and aluminium oxide are produced in this plant [698].

11.3.10 Nippon Catalyst Cycle Co. Process

Nippon Catalyst Cycle Co. Ltd (NCC), a subsidiary company of Sumitomo Metal Mining Co., Japan, recovers high purity Mo and V salts from spent HDS catalysts [699]. Roasting, leaching, and solvent extraction techniques are used in their process (Fig. 11.24). Full process details are not available. The recovered metals are reused in the production of new catalysts, steel, metallic salts or metals within the Sumitomo group.

CHAPTER 12

Markets and Price Trends for Metals in Spent Hydroprocessing Catalysts

It is again emphasized that the focus is on the conventional hydroprocessing catalysts. Thus, the primary interest is in the metals such as Mo, W, Co and Ni [700]. In recent years, interest in the reclamation of alumina from spent hydroprocessing catalysts have been growing. It is noted that the bulk of alumina disposed from other refinery operations, e.g., alkylation, Claus, chloride guard, etc., is generally disposed in landfills rather than being recovered. The availability of spent hydroprocessing catalysts either for metal recovery or other utilization routes is dictated by the overall catalyst consumption by the petroleum refining industry. According to Silvy [701], the efficiency of refining capacity should grow by about 2.4% annually. There is a difference between the refining efficiency and the overall refining capacity. For the former, the type and the amount of catalysts utilized by refining industry play a key role. Nevertheless, the demand for hydroprocessing catalysts will continuously grow, i.e., by about 4 and almost 8%, for hydrotreating and hydrocracking catalysts, respectively. Taking into consideration these facts, the prices of hydrotreating and hydrocracking catalysts may exceed \$ 10 and \$ 30 per kilogram. It is to be noted that positive payment to refinery depends on the metals concentration in spent catalysts and fluctuations in the market prices of metals. Apparently, the metal reclaiming companies are taking necessary measures in anticipation of these developments.

12.1 Molybdenum

Steel industry has been the largest consumer of Mo, accounting for almost 75% of demand in 2002 mostly used as an alloy in stainless steels used in a wide range of applications as well as superalloys for use in gas turbine, aircraft parts, missiles, chemical processing equipment, etc. Since 2000, stainless steel production increased at about 4–5% annually. Non-metallurgical applications of Mo are likely to increase by 3–5% annually, mainly due to the increased demand for hydroprocessing catalysts. Thus, more stringent emissions regulations and growing emphasis on fuel efficiency will require higher efficiency of petroleum refining. Consequently, the consumption of hydroprocessing catalysts will increase. A part of Mo is used in the form of Mo sulfide as lubricant for high loads and high temperatures applications, e.g., in space vehicles. Because of high melting point (2610 °C), Mo metal is used as filament supports for light bulbs, metal-working dies, furnace parts and as Mo cathodes in special electrical applications. In fact, the list of products available on the market, which contain various amounts of Mo, is rather long. Detailed accounts of Mo recovery and recycling in US prior 1998 were given by Blossom [702].

A boom in Mo demand may result from the international maritime regulations, which ask for decommissioning of all tankers built before 1987 by 2010. Moreover, a worldwide boom in construction of crude oil and natural gas pipelines may cause shortages in Mo supply. It is not easy to make an estimate of these markets. For example, some constructions of pipelines are being announced without any anticipation of public.

Molybdenum prices increased sharply in the middle of 2002 before being sustained at around $ US 12 per kilogram through 2003. Subsequently, temporary closure of mines in China pushed prices slowly above $ US 16 per kilogram. According to the Price-Waterhouse-Coopers survey, the Mo net revenues showed the most significant annual growth in 2004, i.e., increase by 209%. During this period, the average Mo price rose from about $ US 14 to almost 40 per kilogram. In 2005, price was in the range of $ US 70 per kilogram. It was predicted that in the following years price would increase by about 20%. In early 2008, the price (molyoxide) was about $ US 70 per kilogram.

12.2 Tungsten

Compared with Mo, the information on W recovery from spent hydroprocessing catalysts is rather limited although the role of the latter during catalysis may be similar as that of Mo. Apparently, the methods, which are used for the recovery of Mo, can be applied to that of W. Historically, the W prices fluctuated widely as the market alternated between periods of scarcity and oversupply [703]. Thus, since 1990 to 2000 the price fluctuated between $ US 4 and 13 per kilogram, whereas in 2000 the price fluctuated between $ US 6 and 13 per kilogram. It is unlikely that the price of W will approach that of Mo in a near future.

12.3 Nickel

Nickel is used in a wide range of applications as part of the stainless steels, non-ferrous alloys, and superalloys and in electroplating. In the US (2001), stainless steels and related alloys applications approached 40% of the total use of Ni. A similar amount of Ni was used in non-ferrous alloys and superalloys. As alloying component, Ni plays an important role in chemical, petroleum and aerospace industries. Also, Ni is an active constituent of various catalysts, i.e., hydroprocessing, hydrogenation, dehydrogenation, hydroisomerization, dewaxing, etc. As far as the end-use sectors are concerned, commercial, industrial, transportation and construction materials account for about 29, 27, 24 and 20% of Ni utilization, respectively.

For nickel, the fluctuation in metal prices was evident similarly as for the other metals. For example, in 2005, the prices fluctuated between about 12 and 17 $/kg. Between February and July 2006, the price of nickel increased from ~30 $/kg to more than 50 $/kg before declining to ~32 $/kg. In overall, the demand for nickel has been growing at a rate of 4% annually during the last decade with 7% growth during the last two years.

12.4 Cobalt

Cobalt is used in superalloys for production of components of gas turbines and aircraft engines. Cobalt is also used to make corrosion and wear resistant alloys, high speed steels, magnets, cemented carbides and diamond tools. As component of various catalysts, Co plays a key role in petroleum industry. Apparently, battery sector is the fastest growing application of Co. For example, in 2003, this sector consumed about 11% of the Co available on the market [704].

Cobalt prices were relatively stable until the late 1970s. Then, the price increased rapidly to more than $ US 80 per kilogram, because of the uncertainties in the Co supply. After stabilizing at previous levels ($ US 20), the price increased to about $ US 60 per kilogram in early 1990s because of the unstable situation in the Co supplying countries. The years 1996 to 1998 can be used to illustrate the Co price fluctuation on the market. Thus, in 1996, the price decreased to about $ US 45 per kilogram. In 1997, the price fluctuated between about $ US 40 and 50 per kilogram. In June 1998, the price declined from about $ US 50 to 25 per kilogram, while the average 1998 price was about $ US 45 per kilogram. In 2004 and 2005, the price averaged at about $ US 50 and 40 per kilogram, respectively.

12.5 Vanadium

The sources of vanadium include metallurgical slugs (60%), magnetite ore (35%), and remaining includes spent catalysts and refinery residues. In 2004, the capacity utilization approached 80%. Used in metal alloys such as steel alloys (carbon steel, low and fool alloy steel and tool steel) for structural applications, oil and gas pipelines, buildings and cars, in aerospace titanium alloys for aircraft components, rocket motors, gas turbines, and for production of gas and ceramics. As catalyst, V is used for production of H_2SO_4, maleic anhydride, etc. However, metallurgical uses account for about 90% of the vanadium consumption.

As the result of demand, the price trends of vanadium may reveal wide fluctuations within relatively short period of time. Compared with 2003, the consumption in 2004 increased by about 13%, whereas during the previous years, the consumption was increasing only by about 2% annually. Consequently, in 2004 the price of V increased to $ US 27 from about 15 $ US per kilogram due to the increase in steel production. The rapid increase in demand in 2005 resulted in the price rise to 128 $ US. Subsequently, the price decreased to 38 $ US because of the China entering the market. Then, in 2005, the average price was 70 $ US compared with 27 $ US in 2004. In 2006, the price continued to fall because of the strong supply from China. There are all indications that in a long term the demand for vanadium will be strong [705].

In US, the price of V_2O_5 increased from $ US 1.34 per pound in 2002 to $ US 16.28 in 2005 and then decreased to $ US 8.08 in 2006. In Europe, V_2O_5 prices ranged from $ 6.125 to $ 6.388 per pound in February, as compared with $ 6.333 to $ 6.722 in January 2007. According to Magyar [706], in US domestic consumption of vanadium in February 2007 was about 2% less than that of the previous month, and was about 7% more than that in February 2006. The US ferrovanadium (FeV) prices ranged from $ 15.325 to $ 15.675 per pound of vanadium content in February, as compared with $ 15.722 to $ 16.222 in January 2007. European FeV prices ranged from $ 32.625 to $ 34.125 per kilogram in February, as compared with $ 30.778 to $ 31.778 in January 2007.

12.6 Alumina

Alumina is the main component of hydroprocessing catalysts, accounting for more than 70 wt.% of the fresh catalysts. Depending on the amount of contaminants, the alumina content in spent hydroprocessing catalysts may fall below 60 wt.%. It should be noted that for the metal reclaiming companies, the recovery of Al metal from spent catalysts might not be an attractive option. For this purpose, shipping alumina to Al smelters makes more sense. However, the recovery of a pure Al_2O_3 deserves attention because of its wide applications in various parts of industry, i.e., for reuse as catalyst support, for preparation of refractories, hot gas clean up sorbents, etc. Until few years ago (e.g., in 2004), there was an insufficient worldwide alumina capacity. This resulted in a sharp increase in the spot prices for alumina. This was mostly attributed to a remarkable growth of China's economy. After the alumina production capacity is expanded (e.g., in China), the gap between supply and demand will narrow. However, wild swings in alumina prices may not be avoided. Thus, the response of the alumina producers to fluctuation in the alumina may take several months. Therefore, during the periods of alumina oversupply may push price of alumina below $ 300 per ton down from current spot price of about $ 600 per ton.

CHAPTER 13

Future Perspectives

The consumption of hydroprocessing catalysts and the associated generation of spent catalysts will be influenced by the future developments in petroleum refining. It is anticipated that during the next few years the consumption of crude will increase until it may level off before it will begin to decline under the pressure of environmental lobby. In view of the political turmoil and other problems in various parts of the world, it is not easy to make an accurate prediction. However, a gradual decline in the supply of light and medium crudes, offset by an increased supply of heavy and extra heavy crudes, seems to be inevitable. Then, even during the period of an unchanged crude consumption, generation of spent catalysts will increase because heavy and extra heavy crudes will account for a greater portion of the overall crude envelope.

Transportation sector accounts for a large portion of emissions, such as CO_2, SO_X, NO_X, and particulates. In the case of diesel fuel, sulfur in the fuel affects the performance of catalysts in catalytic converters. Consequently, emissions of NO_X and particulate matter are increased. These problems may be alleviated by using an ultra low sulfur fuels. In petroleum refineries, this translates into an increased consumption of catalyst and hydrogen. The development of advanced catalysts, which can out perform currently used catalysts, may be another option to deal with these problems. Nevertheless, it is unlikely that all challenges, which petroleum industry will be facing, can be met without increased catalyst consumption.

The companies who received certification from regulatory authorities, which allows them to carry out environmental services to petroleum refineries, have to be ready to deal with the increased volume of spent catalysts. Also, novel catalysts introduced on the market may require modifications of the established procedures. In this regard, some exploratory research may be necessary to ensure safety and environmental compliance.

The stringent environmental regulations that are in effect today have been continuously evolving, i.e., becoming even more stringent. Therefore, all activities and developments in this area have to be carefully monitored. In this regard, a team of experts and/or expert companies, who obtained certifications from regulatory authorities, can play a key role in assisting petroleum refineries in making decision on the final fate of spent hydroprocessing catalysts. The companies with the expertise dedicated to TCM can be called up to provide necessary technical support and economic evaluations.

It has been observed that the efficiency of petroleum refining improves with the increasing size of refinery. These trends are favorable for an efficient utilization of by-products, such as refinery gases and residues. Moreover, large quantities of spent catalysts, which may be generated in a large size refinery, could make the integration of spent catalyst recycling (e.g., regeneration, rejuvenation, and reprocessing) with refining operation economically attractive. Thus, rather than to deal with complex regulatory issues concerning packaging, transportation, disposal, etc. spent catalysts handling will be limited to refinery site. At present time, there is no information suggesting an interest in such a venue, although very close cooperation between some refineries and certified companies has been noted. However, a potential integration of catalyst recycling with petroleum refineries in the future cannot be ruled out.

Until now, recycling of spent hydroprocessing catalysts has been dominated by oxidative regeneration. At the same time, rejuvenation has been making little contribution in spite of significant advances in developing a commercial process have been made. Thus, the database required for the design and construction of a commercial rejuvenation plant has been established. It is believed that time for introducing the first commercial rejuvenation plant is approaching. In this regard, organic acids alone or in the mixture with various oxidation agents appear to be the best alternative. Apparently, the potential of reprocessing spent catalysts has not yet been fully realized, although very encouraging results have been obtained. It is anticipated that both rejuvenation and reprocessing will be contributing to the overall recycling of spent hydroprocessing catalysts.

It has been indicated that for extra heavy feeds (200–300 ppm of V + Ni) and particularly ultra heavy feeds (>300 ppm of V + Ni), carbon rejection route (e.g., coking) may be the best option for primary upgrading. In this case, the composition of primary liquids may differ from that of the liquids derived either from conventional crudes or produced during catalytic upgrading, such as hydroprocessing. Because of more severe conditions employed, aromatic and heteroaromatic compounds in coking liquids are more refractory, requiring more catalyst and hydrogen, compared with a similar boiling point liquids obtained under less severe conditions.

Numerous attempts have been made to find new outlets for spent catalysts with the aim to avoid disposal in landfills as the least attractive option. Various degrees of success have been indicated in the studies on incorporation of spent catalysts in a wide range of materials.

Growing concerns about the fate of spent hydroprocessing catalysts prompted research activities in metal reclamation. The published results show that the efficiency of metal recovery using different methods is similar. In this regard, it is not easy to identify a method of choice. However, it is essential that other relevant factors are also taken into consideration. For example, it is necessary that all waste by-products of metal reclamation, i.e., liquid streams containing strong acids and bases, toxic gases, etc. are neutralized before their disposal. In a wide range of organic and inorganic agents, water-soluble organic agents offer the efficiency of metal recovery, which is comparable or better than that using other methods. At the same time, metal recovery with organic agents is conducted under rather mild conditions, which are free of corrosive environment. Therefore, a much higher level of safety precautions has to be taken in the case of the methods that are based on the use of strong acids, bases, and chlorine than that of the organic agent. Also, material requirements on the design and construction of equipment used for the recovery are rather different. The fluctuation in metal prices on the world market introduces some uncertainties in metal reclamation. Therefore, a more comprehensive approach to metal reclamation from spent catalysts may be necessary. Ultimately, this should benefit both refiner and metal reclaiming company.

One may anticipate that landfills will continue to be scrutinized by regulatory authorities. It is believed that new and more stringent regulations will be implemented. Advancement in chemical analysis may lead to discovery of new contaminants, which previously were either overlooked or were undetected. This will require modification of regulatory acts, e.g., comprehensive liability act. Therefore, avoiding landfilling, as much as possible, may be the best recommendation to petroleum refiners. Thus, even after a costly pretreatment of spent catalyst before landfilling, their long-term stability in landfills has not yet been clearly determined. It is anticipated that recycling of spent catalysts will gain increasing momentum in the future. Continuous research in this area is needed for improving the technologies currently available and for developing new technologies. This redissolution phenomenon was, however, limited by the low solubility of this Anderson HPA. In the case of the CoMoP dried catalyst (P/Mo ratio greater than 0.4), characterization of the additive containing catalyst evidenced $PCoMo_{11}O_{407-}$ formation. Redissolution and redispersion of the active components of the catalyst by chelating agents, thus, appeared to be responsible for the activity enhancement.

References

[1] J.G. Speight, The Chemistry and Technology of Petroleum, Marcel Dekker, New York (1999).
[2] J.J. McKetta, Petroleum Processing Handbook, Marcel Dekker, New York (1991).
[3] R.E. Boening, N.K. McDaniel, R.D. Petersen, R.P. Van Driesen, Hydrocarbon Process 66 (1987) 59.
[4] F.E. Biasca, F.E. Dickenson, R.L. Chang, H. Johnson, H.E. Bailey, D.R. Simbeck, Upgrading heavy crude oils and residue to transportation fuel: technology economics, and outlook, Phase 7. (R. Garland Report Coordinator), SAF Pacific Inc. Eng. & Economic Consultant, May 2003.
[5] K.W. Robschlager, W.J. Deelen, J.E. Naber, Caracas, Venezuela, vol. 4, 5th UNITAR/UNDP Int. Conf. Heavy Crude and Tar Sands (Aug. 1991) 249.
[6] R.P. Silvy, Appl. Catal. 261 (2004) 247.
[7] P. Dufresne, Appl. Catal. A 322 (2007) 67.
[8] M. Marafi, A. Stanislaus, Appl. Catal. 47 (1989) 96.
[9] M. Marafi, S. Omani, H. Al-Sheeha, A. Al-Barood, A. Stanislaus, Ind. Eng. Chem. Res. 46 (2007) 1968.
[10] E. Furimsky, F.E. Massoth, Catal. Today 52 (1999) 381.
[11] M. Absi-Halabi, A. Stanislaus, H. Qabazard, Hydrocarbon Process. 76 (1997) 45.
[12] K. Al-Dalama, A. Stanislaus, Chem. Eng. J. 120 (2006) 33.
[13] E. Furimsky, F.E. Massoth, Catal. Today 17 (1993) 537.
[14] D.L. Trimm, Stud. Surf. Sci. Catal. 53 (1990) 41.
[15] E. Furimsky, Catal. Today 30 (1996) 223.
[16] Fed. Regist. August 6 (1998).
[17] Fed. Regist. 68 (202) (2003) 59935–59940.
[18] D. Rapaport, Hydrocarbon Process. 79 (2000) 11.
[19] Y. Ohtsuka, E. Furimsky, Energy Fuels 9 (1995) 439.
[20] E. Furimsky, A. Siukola, A. Turenne, Ind. Eng. Chem. Res. 35 (1996) 4406.
[21] R.K. Clifford, Pet. Technol. Q. (Spring) (1997) 33.
[22] M. Marafi, A. Stanislaus, Resour. Conserv. Recycl. J. 52 (2008) 859.
[23] M. Marafi, A. Stanislaus, Resources, Conserv. Recyc. J. 53 (2008) 1.
[24] C. Cameron, March 14–16, Paris, France, 6th European Fuels Conf. (2005).
[25] H. Bodewes, Hydrocarbon Asia (Jan/Feb 2006).
[26] E. Furimsky, Oil Gas Sci. Technol. Rev. IFP 54 (1999) 597.
[27] E. Furimsky, Stud. Surf. Sci. Catal. 169 (2007) 12.
[28] H. Toshima, Albermarle Techn. Seminar, Aug. 4, 2005, Banff, Canada.
[29] A. Sharpe, B. Jones, G. Baumgartner, R. Anderson, R. Adarme, M. Boer, Oil Gas J. (Jan. 21 2008).
[30] S.L. Lee, Courrier 59 (2008) 4.
[31] UltraMix Hydroprocessing Reactors, UOP Honeywell Co. Report, 2008.
[32] R.H. Patel, E. Bingham, P. Christensen and M. Muller, 1st Indian Refining Round Table, Dec. 1–2, New Delhi, India.
[33] Ballard et al., US Patent 3,218,249, 1965.
[34] Dreyer et al., US Patent 3,235,344, 1965.
[35] Fouad et al., US Patent 4,836,989, 1989.

[36] Grosboll et al., US Patent 4,126,540, 1978.
[37] Riopelle et al., US Patent 3,353,924, 1967.
[38] J. Swain, M. Zonnevylle, D. Pohl, F. S Geerdes, Rome, Italy, European Refing. Conf. Nov. 15 (2000).
[39] A. Gragnani, 6th European Fuels Conf., March 14–16, Paris, France.
[40] H. Topsoe, B.S. Clausen, N.Y. Topsoe, P. Zeuthen, Stud. Surf. Sci. Catal. 53 (1999) 77.
[41] M.S. Rana, J. Ancheyta, S.K. Maity, G. Marroquin, Int. J. Oil Gas Technol. 1 (2008) 250.
[42] P. Zuideveld, J. Wolf, Hydrocarbon Process. 85 (2006) 73.
[43] E. Furimsky, Ind. Eng. Chem. Res. 48 (2009) 2752.
[44] A. Prakash, Refining Processes Handbook, Elsevier, Amsterdam (2002).
[45] L. Reyes, C. Zarpa, J.H. Krasuk, in: B. Delmon, G.F. Froment (Eds.), Catalyst Deactivation, vol. 88, Elsevier Sci, Amsterdam (1994).
[46] D. Alvarez, R. Galiasso, P. Andreu, J. Jpn. Pet. Inst. 22 (1979) 234.
[47] T.H. Lindstrom, F. Levesque, M. Cathcart, Hydrocarbon Process. 82 (2003) 49.
[48] F.E. Massoth, Adv. Catal. 27 (1978) 265.
[49] E. Furimsky, F.E. Massoth, Catal. Rev. Sci. Eng. 47 (2005) 297.
[50] R. Prins, Adv. Catal. 46 (2001) 399.
[51] E. Furimsky, Appl. Catal. 199 (2000) 147.
[52] A. Stanislaus, B.H. Cooper, Catal. Rev. Sci. Eng. 36 (1994) 75.
[53] H. Topsoe, B.S. Clausen, F.E. Massoth, in: J. Anderson, M. Boudart (Eds.), Catalysis, Science and Technology, vol. 11, Springer (1996) 1.
[54] A.C. Jacobsen, B.H. Cooper, P.N. Hannerup, Proc 12th Petrol. World Congr, vol. 4 (1987) 97.
[55] M. Breysse, E. Furimsky, S. Kasztelan, M. Lacroix, G. Perot, Catal. Rev. Sci. Eng. 44 (2002) 651.
[56] S. Kasztelan, B. MacGarvey, J. Catal. 147 (1994) 476.
[57] B. MacGarvey, S. Kasztelan, J. Catal. 148 (1994) 149.
[58] M. Breysse, B.A. Bennett, D. Chadwick, M. Vrinat, Bull. Soc. Chim. Belg. 90 (1981) 1271.
[59] S.M.A.M. Bouwens, F.B.M. van Zon, M.P. van Dijk, A.M. van der Kraan, V.H.J. de Beer, J.A.R. van Veen, D.C. Koningsberger, J. Phys. Chem. 146 (1994) 375.
[60] S.P.A. Louwers, R. Prins, J. Catal. 133 (1992) 94.
[61] A.I. Dugulan, M.W.J. Craje, A.R. Overweg, G.J. Kearley, J. Catal. 229 (2005) 276.
[62] M. Brorson, A. Carlson, H. Topsoe, Catal. Today 123 (2007) 31.
[63] H. Topsoe, Appl. Catal. 322 (2007) 3.
[64] F. Besenbacher, M. Brorson, B.S. Clausen, S. Helveg, B. Hinnemann, J. Kibsgard, J.V. Lauristen, P.G. Moses, J.K. Norskov, H. Topsoe, Catal. Today 130 (2008) 86.
[65] C. Moreau, P. Genest, Theoretical Aspects of Heterogeneous Catalysis, Kluwer, New York, pp. 256
[66] A. Logadottir, P.G. Moses, B. Hinnemann, N.Y. Topsoe, K.G. Knudsen, H. Topsoe, J.K. Norskov, Catal. Today 111 (2006) 44.
[67] X.D. Wen, Z. Cao, Y.W. Li, J. Wang, H. Jiao, J. Phys. Chem B 110 (2006) 23860.
[68] R.R. Chianelli, G. Berhault, Catal. Today 53 (1999) 357.
[69] G. Berhault, A. Mehta, A.C. Pavel, J. Yang, L. Rendon, M.J. Yacaman, L.C. Araiza, A.D. Moller, R.R. Chianelli, J. Catal. 198 (2001) 9.
[70] S. Kasztelan, C. R. Acad. Sci. Paris, Ser. II 307 (1988) 727.
[71] J. Kibsgaard, J.V. Laurentsen, E. Loegsgaard, B.S. Clausen, H. Topsoe, F. Besenbacher, J. Am. Chem. Soc. 128 (2006) 13950.
[72] M.S. Rana, J. Ancheyta, P. Rayo, S.K. Maity, Catal. Today 98 (2004) 151.
[73] M. Breysse, P. Afanasiev, C. Geantet, M. Vrinat, Catal. Today 86 (2003) 5.
[74] S.K. Maity, M.S. Rana, S.K. Bej, J. Ancheyta-Juarez, G. Murali Dhar, T.S.R. Prasada Rao, Appl. Catal. 205 (2001) 215.
[75] E. Furimsky, Carbons and Carbon Supported Catalysts in Hydroprocessing, Royal Soc. Chem, Cambridge, UK (2008).

[76] J. Ancheyta, G. Betancourt, G. Centeno, G. Marroquin, F. Alonso, E. Garciafigueroa, Energy Fuels 16 (2002) 1438.
[77] B.M. Moyse, B.H. Cooper, A. Albjerg, NPRA Annual Meeting Paper AM- 84-59, San Antonio, TX, March 1984.
[78] A. de Bruijn, I. Naka, J.W.M. Sonnemans, Ind. Eng. Chem. Process. Des. Dev. 20 (1981) 40.
[79] S. Kobayashi, S. Kushiyama, R. Aizawa, Y. Koinuma, K. Inoue, Y. Shimizu, K. Egi, Ind. Eng. Chem. Res. 26 (1987) 2245.
[80] J. Bartholdy, B.H. Cooper, Prepr. -Am. Chem. Soc., Div. Pet. Chem. 38 (1993) 386.
[81] G.M.K. Abotsi, A.W. Scaroni, Fuel Process. Technol. 22 (1989) 107.
[82] R. Hubaut, J. Altafulla, A. Rives, C. Scott, Fuel 86 (2007) 743.
[83] J. Ancheyta, M.S. Rana, E. Furimsky, Catal. Today 109 (2005) 3.
[84] H. Topsoe, B. Hinnemann, J.K. Norskov, J.V. Lauristen, F. Besenbacher, P.L. Hansen, G. Hytoft, R.G. Egeberg, K.G. Knudsen, Catal. Today 107/108 (2005) 12.
[85] J. Ramirez, S. Fuentes, G. Diaz, M. Vrinat, M. Breysse, M. Lacroix, Appl. Catal. 52 (1989) 211.
[86] S. Eijsbouts, J.J.L. Hinerman, H.J.W. Elzerman, Appl. Catal. 105 (1993) 53.
[87] M. Absi-Halabi, A. Stanislaus, T. Al-Mughni, S. Khan, A. Qamra, Fuel 74 (1995) 1211.
[88] H. Beuther, B.K. Schmid, Proc. World. Petr. Congr. Section III, pap. 20, 1963.
[89] A. Hauser, A. Marafi, A. Almutairi, A. Stanislaus, Energy Fuels 22 (2008).
[90] J.M. Oelderik, S.T. Sie, D. Bode, Appl. Catal. 47 (1981) 1.
[91] B.H. Cooper, B.B.L. Donnis, B. Moyse, Oil Gas J. (Dec. 8 1986) 39.
[92] A. Stanislaus nd, M. Absi-Halabi, Rio de Janeiro Brazil, Proc. 17th World Petrol. Congr. 1–5 Sept (2002).
[93] T. Fujikawa, H. Kimura, K. Kiriyama, K. Hagiwara, Catal. Today 111 (2006) 188–193.
[94] S. Eijsbouts, F. Plantenga, B. Leiiveld, Y. Inoue, Y, K. Fujita. Presented at the 226th ACS National Meeting, New York 7–11 Sept., 2003.
[95] T. S Remans, Prague, Feb, Eur. Catal. Technol. Conf. (2003).
[96] H. Topsoe, B.S. Clausen, N.Y. Topsoe, P. Zeuthen, Stud. Surf. Sci. Catal. 53 (1990) 77.
[97] B. Delmon, Catal. Lett. 22 (1993) 1.
[98] B. Delmon, Stud. Surf. Sci. Catal. 53 (1990) 1.
[99] E.J.M. Hensen, P.J. Kooyman, Y. Van der Meer, A.M. Van der Kraan, V.H.J. De Beer, J.A.R. van Veen, R.A. Van Santen, J. Catal. 199 (2001) 224.
[100] E.J.M. Hensen, V.H.J. De Beer, F.A.R. Van veen, R.A. Van Santen, Catal. Lett. 84 (2002) 1.
[101] M. Breysse, G.D. Mariadassou, G. Geantet, M. Vincient, M. Lemarie, Catal. Today 84 (2003) 129.
[102] C. Thomazeau, C. Geantet, M. Lacroix, M. Danot, V. Harle, P. Raybaud, Appl. Catal. A 322 (2007) 92.
[103] H. Topsoe, Appl. Catal. A 322 (2007) 3.
[104] P. Raybaud, Appl. Catal. A 322 (2007) 76.
[105] S. Eijsbouts, S.W. Mayo, K. Fujita, Appl. Catal. 322 (2007) 58.
[106] Y. Okamoto, S. Ishihara, M. Kawano, M. Satoh, T. Kubota, J. Catal. 217 (2003) 12.
[107] P. Blanchard, C. Lamonier, A. Griboval, E. Payen, Appl. Catal. 322 (2007) 33.
[108] R.R. Chianelli, G. Berhault, P. Raybaud, S. Kasztelan, Appl. Catal. 227 (2002) 83.
[109] H. Topsoe, K.G. Knudsen, L. Skyum anf B.H. Copper, NPRA Ann. Meet. 2005, Technical Papers. p16.
[110] L. Skyum, P. Zeuthen, Hydrocarbon Eng. 11 (2006) 59.
[111] J. Reinhardt, U. Balfanz, E. Brevoord, S. Mayo, F.L. Plantenga, Noordwijk, The Netherlands June 10–13, Akzonobel Catal. Symp. (2001).
[112] S.L. Lee, Albemarle Catal. Courier 63 (2006) 8.
[113] E. Brevoord, Catal. Courier 60 (2005) 4.
[114] F.L. Plantenga, Catalyst Courier No. 47, 2002. Akzo Nobel.
[115] L. Kane, S. Romannow, Hydrocarbon Process. 83 (2004) 26.
[116] S. Fukase, S. Akashah, Hydrocarbon Asia 14 (2004) 24.
[117] S. Torrisi, AIChE Spring Nat. Meet. New Orleans, March 2002.
[118] T. Remans, Eur. Catal. Technol. Conf. (March 2004).

[119] D. Krenzke, M. Zehender, Hydrocarbon Asia 12 (2002) 44.
[120] F. Morel, S. Kressman, V. Harle, S. Kasztelan, Stud. Surf. Sci. Catal. 106 (1997) 1.
[121] H. Qabazard, R. Adarme, in: D.L. Trimm (Ed.), Catalysts in Petroleum Refining 1989, Elsevier, Amsterdam (1990) 61.
[122] S. Togawa, E.K.T. Kam, M. Al Shamali, Prepr. -Am. Chem. Soc., Div. Pet. Chem. 46 (2001) 3.
[123] A. Kundu, K.D.P. Nigam, A.M. Duquenne, H. Delmas, Rev. Chem. Eng. 19 (2003) 531.
[124] F.M. Nooy, Oil Gas J. (12 November 1984) 152.
[125] C.J. Pereira, J.W. Beeckman, Ind. Eng. Chem. Res. 28 (1989) 422.
[126] C.J. Pereira, W.-C. Cheng, J.W. Beeckman, W. Suarez, Appl. Catal. 42 (1988) 47.
[127] G.L. Hamilton, M. Fitzgerald, V. D'Amico, Proc. 1985 AIChE Nat. Meet. Nov. 2–7, 1986, Miami, FL, p. 1.
[128] H.R. Jackson, K.E. Whitehead, Oil Gas J. (Feb. 7 1977) 95.
[129] A.M. Henke, Oil Gas J. (April 6 1970) 68.
[130] H.R. Jackson, K.E. Whitehead, Oil Gas J. (Feb. 7 1977) 66.
[131] A.M. Henke, Oil Gas J. 97 (April 6 1970) 97.
[132] S.G. Paradis, G.D. Gould, D.A. Bea, E.M. Reed, Chem. Eng. Prog. 67 (1971) 57.
[133] B.J. Young, R.L. Richardson, Hydrocarbon Process. (Sept. 1977) 103.
[134] A. Billon, F. Morel, M.E. Morrison, J.P. Peries, Rev. Inst. Fr. Pet. 49 (1994).
[135] A. Al-Nasser, S.R. Chaudhuri, S. Battacharya, Stud. Surf. Sci. Catal. 100 (1996) 171.
[136] M. Skripek, E. De la Fuente, W.H. Smith, Fifth UNITAR/UNDP Int. Conf. On Heavy Oils and Tar Sands, August 1991, Caracas, Venezuela, vol. 4, p. 37.
[137] W.C. Van Zijl Langhout, C. Ouwerkek, K.M. Pronk, Oil Gas J. 1 (1980) 120.
[138] S. Kressmann, F. Morel, V. Harle, S. Kasztelan, Catal. Today 43 (1998) 203.
[139] B. Scheffer, M.A. van Koten, K.W. Robschlager, F.C. de Bocks, Catal. Today 43. (1998).
[140] R.P. van Driesen, J. Caspers, A.R. Campbell, G. Lunin, Hydrocarbon Process. 58 (1979) 107.
[141] R.M. Eccles, Fuel Process. Technol. 35 (1993) 21.
[142] E. Furimsky, Appl. Catal. 171 (1998) 177.
[143] M. Absi-Halabi and A. Stanislaus, in Deactivation and Testing of Hydroprocessing Catalysts, P, O'Connor, T. Takatsuka and G.L. Woolery (Eds.), ACS Symp. Series 634, 1996, p. 229.
[144] E. Furimsky, Erdol Kohle 35 (1982) 455.
[145] G. Gualda, S. Kasztelan, J. Catal. 161 (1996) 319.
[146] M.T. Martinez, M.A. Callejas, E. Carbo, A. Hernandez, Stud. Surf. Sci. Catal. (1997) 565.
[147] J. Bartholdy, B.H. Cooper, Stud. Surf. Sci. Catal. 100 (1996) 117.
[148] M. De Wind, Y. Miyauchi, K. Fujita, Stud. Surf. Sci. Catal. 199 (1996) 157.
[149] A. Al-Nasser, S.R. Chaudhari, S. Bhattacharya, Stud. Surf. Sci. Catal. 100 (1996) 171.
[150] M.S. Rana, J. Ancheyta, Catal. Today 109 (2005) 24.
[151] Y. Yokoyama, N. Ishikawa, K. Nakanishi, K. Satoh, A. Nishijima, H. Shimada, N. Matsubayashi, M. Nomura, Catal. Today 29 (1996) 261.
[152] B.M. Vogelaar, P. Steiner, T.F. van der Zijden, A.D. van Langeveld, S. Eijsbouts, J.A. Moulijn, Appl. Catal. 318 (2007) 28.
[153] Y. Tanaka, H. Shimada, N. Matsubayashi, A. Nishijima, M. Nomura, Catal. Today 45 (1998) 319.
[154] S. Gamez, R. Bacaus, M. Vrinat, Prepr. - Am. Chem. Soc., Div. Pet. Chem. 50 (2005) 375.
[155] S. Eijsbouts, L.C.A. van den Oetelaar, R.R. van Puijenbroek, J. Catal. 229 (2005) 352.
[156] S. Eijsbouts, Stud. Surf. Sci. Catal. 127 (1999) 21.
[157] S. Eijsbouts, Appl. Catal. 158 (1997) 53.
[158] U. Usman, T. Kubota, Y. Okamoto, Ind. Eng. Chem. Res. 45 (2006) 3537.
[159] B. Guichard, M.R. Auberger, E. Devers, C. Legens, P. Raybaud, Catal. Today 130 (2008) 97.
[160] H. Toulhoat, J.C. Plumail, G. Martino, Y. Jacquin, Prepr. - Am. Chem. Soc., Div. Pet. Chem. 30 (1985) 85.
[161] J. Devanneaux, J.P. Gallez, L. Mariette, P.A. Engelhard, Prepr. - Am. Chem. Soc., Div. Pet. Chem. 32 (1987) 84.
[162] S. Dejonghe, R. Hubaut, J. Grimblot, J.P. Bonnelle, T. Des Courieres, D. Faure, Catal. Today 7 (1990) 569.

[163] C. Guillard, M. Lacroix, M. Vrinat, M. Breysse, B. Mocaer, J. Grimblot, T. Des Courieres, D. Faure, Catal. Today 7 (1990) 587.
[164] C.E. Scott, B.P. Embaid, F. Gonzalez-Himenez, R. Hubaut, J. Grimblot, J. Catal. 166 (1997) 333.
[165] L. Trytten, M.R. Gray, E.C. Sanford, Ind. Eng. Chem. Res. 29 (1990) 725.
[166] W. Khanda, I. Siu, J. Adjaye, A.E. Nelson, M.R. Gray, Energy Fuels 18 (2004) 539.
[167] W. Kearnbach, W. Kisielow, L. Warzecha, K. Miga, R. Klecan, Fuel 69 (1990) 221.
[168] M. Sau, K. Basak, U. Manna, M. Santra, R.P. Verma, Catal. Today 109 (2005) 112.
[169] D. Dong, S. Jeong, F.E. Massoth, Catal. Today 37 (1999) 267.
[170] B.D. Muegge, F.E. Massoth, Stud. Surf. Sci. Catal. 68 (1991) 297.
[171] M. Anemia, Y. Korai, I. Mochida, J. Jpn. Pet. Inst. 46 (2003) 99.
[172] N. Koizumi, Y. Urabe, K. Inamura, T. Itoh, M. Yamada, Catal. Today 106 (2005) 211.
[173] K.P. de Jong, Ind. Eng. Chem. Res. 33 (1994) 821.
[174] K.P. de Jong, D. Reinalda, C.A. Emeis, in: B. Delmon, F.G. Froment (Eds.), Catalyst Deactivation, vol. 88, Elsevier, Amsterdam (1994) 155.
[175] K.P. de Jong, H.P.C.E. Kuipers, J.A.R. van Veen, in: C.H. Bartholomew, J.B. Dutt (Eds.), Catalyst Deactivation 1991, Elsevier Publ, Amsterdam (1991) 28.
[176] J. Van Dorn, J.A. Moulijn, G. Djega-Mariadassou, Appl. Catal. 60 (1990) 77.
[177] J. Van Dorn, J.A. Moulijn, Fuel. Process. Technol. 35 (1993) 275.
[178] A.D. van Langeveld, E.L.J. Ariens, J.A. Moulijn, R. De Jong, Prepr. - Am. Chem. Soc., Div. Pet. Chem. (1998) 55.
[179] P. Wiwel, P. Zeuthen, A.C. Jacobsen, Stud. Surf. Sci. Catal. 68 (1991) 257.
[180] P. Zeuthen, J. Bartholdy, P. Wiwel, B.H. Cooper, Stud. Surf. Sci. Catal. 88 (1994) 1998.
[181] M.A. Callejas, M.T. Martinez, T. Blasco, E. Sastre, Appl. Catal. 218 (2001) 181.
[182] K. Matsushita, R. Koide, S. Fukase, A. Stanislaus, A. Al-Barood, F. Al-Jasem, M. Absi-Halabi, Prepr. - Am. Chem. Soc., Div. Pet. Chem.
[183] C.H. Yang, S.F. Chen, H. Chen, F. Du, Prepr.- Am. Chem. Soc., Div. Pet. Chem. 47 (2002) 318.
[184] R.R. Galiasso Tailleur, Energy Fuels 22 (2008) 1509.
[185] M. Millan, C. Adell, C. Hinojosa, A.A. Herod, R. Kandiyoti, Oil Gas Sci. Technol. Rev. IFP 63 (2008) 69.
[186] B.M. Vogelaar, A.D. van Langeveld, S. Eijsbouts, J.A. Moulijn, Fuel 86 (2007) 1122.
[187] H. Hauser, A. Marafi, A. Stanislaus, A. Al-Abduani, Energy Fuels 19 (2005) 544.
[188] T.E. Myers, B.L. Myers, T.H. Fleisch, G.W. Zajac, AIChE Symp. Ser. 85(273) (1989) 21.
[189] M. Nunez Isaza, A. Pachon, V. Kafarov, D.E. Resasco, Appl. Catal. 199 (2000) 263.
[190] M. Nunez, M. Villamizar, Appl. Catal. 252 (2003) 51.
[191] C. Yang, F. Du, H. Zheng, K.H. Chung, Fuel (2004).
[192] A. Morales, R.B. Solari, Stud. Surf. Sci. Catal. 100 (1996) 125.
[193] A. Marafi, H.A. Stanislaus, M. Absi-Halabi, A. Hauser, K. Matsushita, Pet. Sci. Technol. 23 (2005) 385.
[194] P.R. Fozard, J.W. McMillan, P. Zeuthen, J. Catal. 129 (1995) 43.
[195] P. Zeuthen, J. Bartholdy, F.E. Massoth, Appl. Catal. 152 (1995) 103.
[196] R. Galiasso Tailleur, L. Caprioli, Catal. Today 109 (2005) 185.
[197] P.W. Tamm, H.F. Harnsberger abd, A.G. Bridge, Ind. Eng. Chem. Process. Des. Dev. 20 (1981) 262.
[198] H. Nagaishi, E.W. Chan, E.C. Sanford, M.R. Gray, Energy Fuels 11 (1997) 402.
[199] J.C. Plumail, H. Toulhoat, Y. Jacquin, G. Martino, Prod. 8th World Congr. Catal. vol. 11, p. 317, 1984 (Berlin).
[200] F.M. Dautzemberg, J.C. de Deken, Metal Complexes in Fossil Fuels, Am. Chem. Soc. Symp. Series, vol. 344, 1987, pp. 233.
[201] M. Yumoto, S.G. Kukes, M.T. Klein, B.C. Gates, Ind. Eng. Chem. Res. 40 (2001) 131.
[202] M. Yumoto, S.G. Kukes, M.T. Klein, B.C. Gates, Ind. Eng. Chem. Res. 35 (1996) 3203.
[203] H. Koyama, E. Nagai, H. Kumagai, Prepr. - Am. Chem. Soc., Div. Pet. Chem. (1995) 446.
[204] P. Gripka, S. Torrisi, Hydrocarbon Eng. (Dec. 1999) 39.
[205] M. Ternan, E. Furimsky, B.I. Parsons, Fuel Process. Technol.

[206] G. Gualda and S. Kasztelan, Catalyst Deactivation, B. Delmon, G.F. Froment (Eds.), vol. 88, 1994, pp. 145.
[207] S. Kobayashi, S. Kushiyama, R. Aizawa, Y. Koinuma, K. Inoue, Y Shimizu, K. Egi, Ind. Eng. Chem. Res. 26 (1987) 2241.
[208] M. Seki, M. Yoshimoto, J. Jpn. Pet. Inst. 44 (2001) 259.
[209] H. Seki, S. Sato, M. Yoshimoto, J. Jpn. Pet. Inst. 44 (2001) 154.
[210] T. Takahashi, H. Higashi, T. Kai, Catal. Today 104 (2005) 76.
[211] A. Marafi, A. Hauser, A. Stanislaus, Energy Fuels 20 (2006) 445.
[212] S.M. Richardson, H. Nagaishi, M.R. Gray, Ind. Eng. Chem. Res. 35 (1996) 3940.
[213] H. Higashi, T. Takahashi, T. Kai, Stud. Surf. Sci. Catal. 145 (2003) 339.
[214] F. Kumata, H. Seki, T. Saito, M. Yoshimoto, J. Jpn. Pet. Inst. 44 (2001) 252.
[215] I.C. Lewis, L.S. Singer, Prepr.- Am. Chem. Soc., Div. Pet. Chem. 30 (1985) 834.
[216] I.C. Lewis, Carbon 20 (1981) 519.
[217] J.W. Bunger, Prepr.- Am. Chem. Soc., Div. Pet. Chem. 30 (1985) 549.
[218] D. Nohara, T. Sakai, Ind. Eng. Chem. Res. 31 (1992) 14.
[219] M. Absi-Halabi, A. Stanislaus, D.L. Trimm, Appl. Catal. 72 (1991) 193.
[220] J. Kubo, H. Higashi, Y. Ohmoto, H. Arao, Energy Fuels 10 (1996) 478.
[221] K.P. de Jong, Ind. Eng. Chem. Res. 33 (1994) 3141.
[222] J.C. Fetzer, Stud. Surf. Sci. Catal. 100 (1996) 263.
[223] I. Nakamura, K. Aimoto, K. Fujimoto, AIChE Symp. Ser. 85(273) (1989) 15.
[224] E. Furimsky, Fuel Process. Technol. 6 (1981) 1.
[225] P. Kovacic, J. Oziomek, J. Org. Chem. 29 (1964) 100.
[226] M. Meot-Ner, J. Phys. Chem. 88 (1980) 2724.
[227] B.D. Flockart, I.M. Sesay, R.C. Pink, J. Chem. Soc. Chem. Commun. 10 (1980) 439.
[228] B.C. Gates, J.R. Katzer, G.C.A. Schuit, Chemistry of Catalytic Processes, McGraw Hill, New York (1979).
[229] A.W. Scaroni, R.G. Jenkins, Prepr. - Am. Chem. Soc., Div. Pet. Chem. 30 (1985) 544.
[230] P. Magnoux, A. Rabeharitsara, H.S. Cerqueira, Appl. Catal. 304 (2006) 142.
[231] E. Furimsky, Carbons and Carbon Supported Catalysts in Hydroprocessing, Royal Society of Chemistry, Cambridge, UK (2008).
[232] S.K. Sahoo, S.S. Ray, I.D. Singh, Appl. Catal. 278 (2004) 83.
[233] H. Seki, F. Kumata, Stud. Surf. Sci. Catal. 126 (1999) 357.
[234] H. Seki, F. Kumata, Energy Fuels 14 (2000) 980.
[235] A. Fonseca, P. Zeuthen, J.B. Nagy, Fuel 74 (1995) 1267.
[236] A. Fonseca, P. Zeuthen, J.B. Nagy, Fuel 75 (1996) 1363.
[237] A. Fonseca, P. Zeuthen, J.B. Nagy, Fuel 75 (1996) 1413.
[238] F. Ali, N. Ghaloum, A. Hauser, Energy Fuels 20 (2006) 231.
[239] A. Hauser, A. Stanislaus, A. Marafi, A. Al-Adwani, Fuel 84 (2005) 259.
[240] K. Matsushita, A. Hauser, A. Marafi, R. Koide, A. Stanislaus, Fuel 83 (2004) 1031.
[241] H. Beuther, O.A. Larson, A.J. Perrotta, in: B. Delmon, G.F. Froment (Eds.), Catalyst Deactivation, Elsevier, Amsterdam (1980) 271.
[242] V.A. Munoz, S.V. Ghorpadkas, M.R. Gray, Energy Fuels 8 (1994) 426.
[243] M.R. Gray, Y. Zhao, C.M. McKnight, Fuel 79 (2000) 285.
[244] M.M. Boduszynski, Prepr.- Am. Chem. Soc., Div. Pet. Chem. 20 (1983) 1376.
[245] R.F. Sullivan, M.M. Boduszinski, J. Fetzer, Energy Fuels 3 (1989) 603.
[246] M.A. Bannayan, H.L. Lemke, W.K. Stephenson, Stud. Surf. Sci. Catal. 100 (1996) 273.
[247] K. Matsushita, A. Marafi, A. Hauser, A. Stanislaus, Fuel 83 (2004) 1669.
[248] J. Bartholdy, S.I. Andersen, Energy Fuels 14 (2000) 52.
[249] F. Ali, G. Narjes, A. Hauser, Energy Fuels 20 (2006) 231.
[250] I. Mochida, X.Z. Zhao, K. Sakanishi, Ind. Eng. Chem. Res. 29 (1990) 334.
[251] I. Mochida, X.Z. Zhao, K. Sakanishi, Ind. Eng. Chem. Res. 29 (1990) 2324.
[252] G. Berrebi, P. Dufresne, Y. Jacquier, Resour. Conserv. Recycl. 10 (1994) 1.

[253] J. Guitian, A. Souto, L. Ramirez, R. Marzin, B. Solari, Proc. of Fourth UNITAR/UNDP Int. Conf. on Heavy Crudes and Tar Sands, August 1988, Edmonton, Canada, vol. 4, p. 237.
[254] H. Knozinger, D. Cordischi, B. Vielhaber, Catal. Today 7 (1990) 447.
[255] M. Loos, I. Ascone, C. Goulon-Ginet, J. Goulon, C. Guillard, M. Lacroix, M. Breysse, D. Faure, T. des Courieres, Catal. Today 7 (1990) 515.
[256] J.P. Janssen, A.D. Van Langeveld, J.A. Moulijn, Appl. Catal. 179 (1999) 229.
[257] S. Asaoka, S. Nakata, Y. Shiroto and C. Takeuchi, Metal Complexes in Fossil Fuels 344, R.H. Filby, J.F. Branthaver (Eds.), Am. Chem. Soc. Symp. Series, 1987, pp. 275.
[258] C.S. Kim, F.E. Massoth, Fuel Process. Technol. 35 (1993) 289.
[259] F.X. Long, B.S. Gevert, J. Catal. 222 (2004) 1.
[260] F.X. Long, B.S. Gevert, P. Abrahamsson, J. Catal. 222 (2004) 6.
[261] M.A. Callejas, M.T. Martinez, Energy Fuels 13 (1999) 629.
[262] C. Botchwey, A.K. Dalai, J. Adjaye, Can. J. Chem. Eng. 82 (2004) 478.
[263] B.P. Embaid, F. Gonzales-Jimenez, C.E. Scott, Appl. Catal. 287 (2005) 75.
[264] Y. Yamamoto, Y. Mizutani, Y. Shibata, Y. Kitou, H. Yamazaki, Stud. Surf. Sci. Catal. 100 (1996) 181.
[265] C. Takeuchi, Y. Fukui, Y. Shiroto, M. Nakamura, Proc. 11th World Petr. Congr. vol. 4, 187.
[266] C. Takeuchi, S. Asaoka, S. Nakata, Y. Shiroto, Prepr. - Am. Chem. Soc., Div. Pet. Chem. 30 (1985) 96.
[267] M. Sato, N. Takayama, S. Kurita, T. Kwan, Nippon Kagaku Zasshi 92 (1971) 834.
[268] E. Newson, Ind. Eng. Chem. Process. Des. Dev. 14 (1975) 27.
[269] J. Wei, in: C.H. Bartholomew, J.B. Butt (Eds.), Catalyst Deactivation, Elsevier, Amsterdam (1991) 333.
[270] R.R. Melkote, K.F. Jensen, Chem. Eng. Sci. 44 (1989) 649.
[271] L.O. Oyekunle, O.B. Ikpekri, Ind. Eng. Chem. Res. 43 (2004) 66.
[272] L.O. Oyekunle, O.B. Ikpekri, A. Jaiyola, Catal. Today 109 (2005) 128.
[273] J. Ancheyta, G. Betancourt, G. Centeno, G. Marroquin, Energy Fuels 17 (2003) 462.
[274] J. Ancheyta, G. Betancourt, G. Marroquin, G. Centeno, L.C. Castaneda, F. Alonso, J.F. Munoz, M.T. Gomez, P. Rayo, Appl. Catal. 233 (2002) 159.
[275] C.J. Pereira, Ind. Eng. Chem. Res. 29 (1990) 512.
[276] C.J. Pereira, J.W. Beeckman, W.C. Cheng, W. Suarez, Ind. Eng. Chem. Res. 29 (1990) 520.
[277] P. Bourseau, G. Muratet, C. Saillоour, H. Toulhoat, Rev. Inst. Fr. Pet. 40 (1985) 595.
[278] H. Toulhoat, R. Szymanski, J.C. Plumail, Catal. Today 7 (1990) 531.
[279] F.M. Dautzemberg, J. van Klinken, K.M.A. Pronk, S.T. Sie, J.B. Wijffels, ACS Symp. Ser. 65 (1978) 254.
[280] P.N. Hannerup, A.C. Jacobsen, ACS Symp. Ser. 28 (1983) 576.
[281] L.O. Oyekunle, R. Hughes, Ind. Eng. Chem. Res. 26 (1987) 1945.
[282] S.J. Khang, J.F. Mosby, Ind. Eng. Chem. Process. Des. Dev. 25 (1986) 437.
[283] E.K.T. Kam, M. Al-Shamali, M. Al-Jeraidan, Studies on deactivation mechanism of hydroprocessing catalysts using Kuwaiti feedstocks Report # KISR 7037, Dec. 2003.
[284] E.K.T. Kam, M. Al-Shamali, M. Juraidan, H. Qabazard, Energy Fuels 19 (2005) 753.
[285] M. Juraidan, M. Al-Shamali, H. Qabazard, E.K.T. Kam, Energy Fuels 20 (2006) 1354.
[286] T. Takatsuka, Y. Wada and S. Inoue, Deactivation and Testing of Hydrocarbon-Processing Catalysts, Am. Chem. Soc., Chap. 30, 1996, pp. 414.
[287] J.P. Janssens, B.J. Bezemer, A.D. van Langeveld, S.T. Siie, J.A. Moulijn, in: B. Delmon, G.F. Froment (Eds.), Catalyst Deactivation, Elsevier, Amsterdam (1994) 335.
[288] J.P. Janssens, A.D. van Langeveld, S.T. Sie, J.A. Moulijn, ACS Series, Chap. 18, 1997, pp. 238.
[289] B.M. Vogelaar, R.J. Berger, B. Bezemer, J.-P. Janssen, A.D. van Langeveld, S. Eisbouts, J.A. Moulijn, Chem. Eng. Sci. (2006).
[290] H. Toulhoat, D. Hudebine, P. Raybaud, D. Guillame, S. Kressmann, Catal. Today (2005).
[291] S.H. Kodama, H. Nitta, T. Takatsuka, T. Yokoyama, J. Jpn. Pet. Inst. 23 (1980) 310.
[292] H.A.H. Al-Adwani, H.M.S. Lababidi, I.M. Alatigi, F.S. Al-Dafferi, Can. J. Chem. Eng. 83 (2005) 281.
[293] E.K.T. Kam, M.M. Al-Mashan, H. Dashti, Stud. Surf. Sci. Catal. 100 (1996) 283.
[294] V.T. Sinha, M.S. Butensky, D. Hyman, Ind. Eng. Chem. Process. Des. Dev. 25 (1986) 321.

[295] API Publication 303, June 1992.

[296] D.M. Nielsen, Practical Handbook of Environmental Site Characterization and Ground Water Monitoring, Taylor & Francis, New York (2006).

[297] L.K. Wang, N.C. Pereira, Y.-T. Hung, Handbook of Environmental Engineering, vol. 1 Air Pollution Control, Humana Press, Totowa, NJ (2004).

[298] D.A. Vallero, Fundamentals of Air Pollution, 4th Ed., Elsevier, Amsterdam (2006).

[299] Fed. Regist. 67 (89) May 8 (2002).

[300] D. Rappoport, Hydrocarbon Process. 79 (2000) 7.

[301] EPA Environmental Fact Sheet, EPA530—01-013, July 2001.

[302] H.J. Feldman, API Comments to 2004 Draft Report to Congress on Cost and Benefits of Federal Regulations (69 FR 7987, Feb. 2004).

[303] C. Cutchey, Malaysian Oil Sci. Technol. 11 (2002) 31.

[304] D. Hanson, Chem. Eng. News (9 October 1995) 31.

[305] P. Dufresne, N. Brahma, F. Girardier, Rev. Inst. Fr. Pet. 50 (1995) 283.

[306] Export and Import of Hazardous Waste Regulations; User's Guide to Hazardous Waste Classification, Environment Canada, 1993.

[307] G. Berebi, P. Dufresne, Y. Jacquier, Environ. Prog. 12 (1993) 97.

[308] Albermarle Catalysts Concept for Total Catalyst Management, Report TCM-HPC 2007.

[309] F.D. Brent Jr., R.G. Nelson, W.A. Ruff, US Patent Doc. 4,900,713/A, 1990.

[310] W. Lavers, Oil Fats Int. (Sept. 1 2003).

[311] J.S.M. Jocker, Prepr. - Am. Chem. Soc., Div. Pet. Chem. 38 (1993) 74.

[312] H.J. Pietrzak, Oil Gas J. (28 July 1986) 93.

[313] Guidelines for Management of Spent Catalyst, Report of Eur. Catal. Manufact. Assoc., CEFIC, 2001.

[314] Environmental and Safety Guidelines for Petroleum Industry, Int. Finance Corp., April 30, 2007.

[315] A.K. Rhodes, Oil Gas J. (29 Nov. 1993) 39.

[316] G.L. Sexton, Oil Gas J. (28 July 1993) 94.

[317] R.C. Kunzelman, R.W. Goerlich, J.G. Welch, C.A. Vuitel, Prepr. Pap. - Am. Chem. Soc., Div. Fuel Chem. (1987).

[318] F.A. Ali, A. Hauser, H.A. Abdullah, A. Al-Adwani, Energy Fuels 20 (2006) 45.

[319] H.A. Abdullah, A. Hauser, F.A. Ali, A. Al-Adwami, Energy Fuels 20 (2006) 320.

[320] T. Kajima, US Patent, 4,912,071, 1990.

[321] Kawakami, Japan Patent, 44,755, 1977.

[322] E. Furimsky, A. Siukola, A. Turrene, Ind. Eng. Chem. Res. 35 (1996) 440.

[323] Anon, Oil Gas J. (Oct. 12 1992) 56.

[324] J.D. Seamans, J.A. Partin, E.R. Samonte, J.R. Lockemeyer, US Patent, WO 9,425,157, A1, 1994.

[325] J.A. Lassner, L.B. Lasher, R.L. Koppel, J.N. Hamilton, Chem. Eng. Prog. (1994) 95.

[326] J. Stone, Oil Gas J. (May 5 1986) 160.

[327] Alder, K. 1995. New approach ends catalyst headaches. NPRA Q& A. Oct. 3, 1995.

[328] Criterian Affiliated Companies. 2007. Total integrated catalyst services management program. http://www.criterioncatalysts.com/english/index.htm.

[329] S. Eijsbouts, A.A. Battiston, G.C. Van Leerdam, Catal. Today 130 (2008) 361.

[330] J. McKenna, E. Pickering, Chem. Eng. Nov. (1995) 175.

[331] E. Furimsky, Fuel Process. Technol. 27 (1991) 131.

[332] E. Furimsky, D.D. Duguay, J. Houle, Fuel 67 (1988) 182.

[333] J.B. Krider, Oil Gas J. (July 28 1986) 94.

[334] J.C. Afonso, T.S. De Lima, P.C. Campos, Stud. Surf. Sci. Catal. 130 (2000) 2849.

[335] R.D. Erdman, Oil Gas J. (May 25 1985).

[336] C. Dekker, Oil Gas J. 1 (1999) 44.

[337] E. Furimsky, L.C.G. Janke, R. Dureau, Fuel Process. Technol. 33 (1993) 77.

[338] J. Workman, P. Dellavalle, US Patent Appl. 9,413,392.
[339] W.M.A. Kox, E. Van den Vlist, Conserv. Recycl. 4 (1981) 29.
[340] C. Shi, J. Fernandez-Jimenez, J. Hazard. Mater. B 137 (2006) 1566.
[341] D. Sun, X.Z. Li, M. Brungs, D. Trimm, Water Sci. Technol. 38 (1998) 211.
[342] G.M. Valles, G. Avila, S. Martinez, R. Terradas, J.M. Nogues, Chemosphere 68 (2007) 1946.
[343] M. Cervinkova, M. Vondruzka, V. Bednarik, A. Pazdera, J. Hazard. Mater. 142 (2007) 222.
[344] P.A. Birgham, R.J. Hand, Adv. Appl. Ceram. 105 (2006) 21.
[345] D.L. Trimm, in: D.L. Trimm, S. Akashah, M. AbsiHalabi, A. Bishara (Eds.), Catalysis in Petroleum Refining, Elsevier, Amsterdam (1988).
[346] W.E. Lee, M.I. Ojovan, M.C. Stennet, N.C. Hyatt, Adv. Appl. Ceram. 105 (2006) 3.
[347] I.A. Sobolev, S.A. Dmitriev, F.A. Lifanov, A.P. Kobelev, S.V. Stefanovsky, M.I. Ojovan, Glass Technol. 46 (2005) 28.
[348] R. Randall, S. Chattopadhyay, J. Hazard. Mater. 114 (2004) 211.
[349] S. Shelly, Environ. Eng. World (March–April 1995) 40.
[350] J.T. Schofield, Toxic Hazard. Disp. Process. Stabiliz/Solidif. 1 (1979) 297.
[351] T.S. Singh, K.K. Pant, J. Hazard. Mater. 131 (2006) 29.
[352] R. Malviya, R. Chaudhary, J. Hazard. Mater. 137 (2006) 207.
[353] R. Malviya, R. Chaudhary, J. Hazard. Mater. 137 (2006) 267.
[354] U. Rattanasak, C. Jaturapitakkul, T. Sudaprasert, Waste Manage. Res. 19 (2001) 456.
[355] D.D. Sun, Energy Sources 25 (2003) 607.
[356] D.D. Sun, J.H. Tay, J.D. Navratil, C. Easton, Water Sci. Technol. 44 (2001) 285.
[357] D.G. Sun, L. Chang, J.H. Tay, J.D. Navratil, C. Easton, J. Environ. Eng. 127 (2001) 916.
[358] K. Shih, T. White, J.O. Locke, Environ. Sci. Technol. 40 (2006) 5520.
[359] R.A. Shawabkeh, J. Hazard. Mater. 125 (2005) 237.
[360] A. Stanislaus, G.R. Gouda, S. Al-Fulaij, Prepr. - Am. Chem. Soc., Div. Pet. Chem. 43 (1998) 491.
[361] D.D. Sun, J.H. Tay, C. Easton, Water Sci. Technol. 42 (2000) 71.
[362] D.D. Sun, J.H. Tay, H.K. Cheong, D.L.K. Leung, G. Qian, J. Hazard. Mater. B 87 (2001) 213.
[363] HydroMet Corporation Limited. (The Hydromet Story). Presented at Waste Management Association Australia, Annual Conference 4th June 2003.
[364] M. Smart, Oil Gas J. (Apr. 16 1984) 67.
[365] R.W. King, Oil Gas J. (March 25 1985) 114.
[366] P. Dufresne, Prepr. - Am. Chem. Soc., Div. Pet. Chem. 38 (1993) 53.
[367] P. Dufresne, F. Valeri, S. Abotteen, Stud. Surf. Sci. Catal. 100 (1996) 253.
[368] T. Suzuki, P. Dufresne, Stud. Surf. Sci. Catal. 92 (1995) 215.
[369] Y. Yoshimura, E. Furimsky, Fuel 65 (1986) 1385.
[370] P. Zeuthen, P. Bloom, M. Muegge, F.E. Massoth, Appl. Catal. 68 (1991) 117.
[371] P. Zeuthen, P. Bloom, M. Muegge, F.E. Massoth, Appl. Catal. 78 (1991) 265.
[372] J. Van Dorn, J.L. Bosch, R.J. Bakkum, J.A. Moulijn, in: B. Delmon, G.F. Froment (Eds.), Catalyst Deactivation, Elsevier, Amsterdam (1987) 391.
[373] J. Van Dorn, H.A.A. Barbolina, J.A. Moulijn, Ind. Eng. Chem. Res. 31 (1992) 101.
[374] E. Furimsky, M. Nielsen, P. Jurasek, Energy Fuels 9 (1995) 439.
[375] G. De Soete, Rev. Inst. Fr. Pet. 4 (1993) 113.
[376] J.P. Hamalainen, M.J. Aho, J.L. Tummavuori, Fuel 12 (1994) 1894.
[377] E. Furimsky, Catal. Today 46 (1998) 3.
[378] E. Furimsky, Appl. Catal. 204 (2003) 1.
[379] Y. Noguchi, T. Itoh, Y. Obayashi, K. Komine, Prepr. - Am. Chem. Soc., Div. Pet. Chem. 38 (1993) 50.
[380] Y. Ohtsuka, E. Furimsky, Energy Fuels 11 (1997) 1073.
[381] Y. Yoshimura, E. Furimsky, T. Sato, H. Shimada, N. Matsubayashi, A. Nishijima, Calgary Canada, Proc 9th Int. Congr. Catal. (1988).

[382] R. Iwamoto, N. Kagami, J. Jpn. Inst. Pet. 50 (2007) 147.
[383] Y. Yoshimura, E. Furimsky, T. Sato, H. Shimada, N. Matsubayashi, A. S Nishijima, Calgary Canada, Proc. 9th Int. Congr. Catal (1988).
[384] H.A. Abdou, Oil Gas J. (Apr. 16 1984) 66.
[385] Y. Yoshimura, N. Matsubayashi, H. Yokokawa, T. Sato, H. Shimada, A. Nishijima, Ind. Eng. Chem. Res. 30 (1991) 1092.
[386] D. Mahadjev, E. Zhecheva, B.R. Aleksic, B.D. Aleksic, B. Markovic, S. Bogdanov, React. Kinet. Catal. Lett. 1 (1991) 7.
[387] Y. Yoshimura, E. Furimsky, Appl. Catal. 23 (1986) 157.
[388] E. Furimsky, P. Sears, T. Suzuki, Energy Fuels 2 (1988) 634.
[389] E. Furimsky, Appl. Catal. 196 (2000) 55.
[390] Y. Yoshimura, T. Sato, H. Shimada, N. Matsubayashi, M. Imamura, A. Nishijima, S. Yoshitomo, T. Kameoka, H. Yanase, Energy Fuels 8 (1994) 435.
[391] E. Furimsky, Y. Yoshimura, Ind. Eng. Chem. Res. 26 (1987) 657.
[392] J.M. Bogdanor, H.F. Rase, Ind. Eng. Chem. Prod. Res. Dev. 25 (1986) 220.
[393] J.S. Brinen, W.D. Amstrong, J. Catal. 54 (1978) 57.
[394] A.I. Kharlamov, React. Kinet. Catal. Lett. 15 (1980) 339.
[395] T.G. Parham, R.P. Merril, J. Catal. 85 (1984) 295.
[396] R.V. Nalitham, A.R. Tarrer, J.A. Guin, C.W. Curtis, Ind. Eng. Chem. Res. 24 (1985) 160.
[397] P.B. Weisz, R.D. Goodwin, J. Catal. 2 (1963) 397.
[398] P.B. Weisz, R.D. Goodwin, J. Catal. 6 (1966) 227.
[399] Y. Chang, Thesis, Univ. of Pensylvania, 1987.
[400] K. Klusacek, H. Davidova, P. Fott, P. Schneider, Chem. Eng. Sci. 9 (1985) 1717.
[401] A. Stanislaus, M. Marafi, M. Absi Halabi, Arab. J. Sci. Eng. 21 (1996) 273.
[402] J.P. Ammus, G.P. Androutsopoulos, A.H. Tsetsekou, Ind. Eng. Chem. Res. 26 (1987) 1323.
[403] H.P. Stephens, F.V. Stohl, Prepr. Pap. - Am. Chem. Soc., Div. Fuel Chem. 29 (1984) 79.
[404] F.V. Stephens, Q.A. Qader, F.E. Massoth, D.S. Thakur, Ind. Eng. Chem. Res. 26 (1987) 846.
[405] B.G. Johnson, F.E. Massoth, J. Bartholdy, AIChE J. 32 (1986) 1980.
[406] T.L. Cable, F.E. Massoth, M.G. Thomas, Fuel Process. Technol. 10 (1985) 105.
[407] M. Inoguchi, S. Sakurada, Y. Satomi, K. Inaba, H. Kagaya, K. Tate, T. Mizutori, R. Nishiyama, T. Nagai, S. Onishi, Bull. Jpn. Pet. Inst. 14 (1977) 153.
[408] T. Sakabe, T. Yagi, Hydrocarbon Process. 12 (1979) 103.
[409] P.A. Hertan, F.P. Larkin, W.R. Jackson, Fuel Process. Technol. 10 (1985) 121.
[410] A. Artega, J.L.G. Fierro, P. Grange, B. Delmon, Stud. Surf. Sci. Catal. 34 (1987) 59.
[411] C.S. Kim, F.E. Massoth, E. Furimsky, Fuel Process. Technol. 32 (1992) 39.
[412] A. Billon, J. Bousquet, J. Rossarie, Proc. 1988 NPRAAnn. Meet., March 1988, San Antonio, TX.
[413] A. Billon, F. Morel, M.E. Morrison, J.P. Peries, Rev. Inst. Fr. Pet. 49 (1994) 495.
[414] P.T. Clark, A.L. Hensley, J.Z. Shyu, J.A. Kaduk, G.J. Ray, Stud. Surf. Sci. Catal. 62 (1991) 417.
[415] L. Guibard, S. Kressmann, F. Morel, V. Harle, P. Dufresne, Hydrocarbon Eng. 30 (2000) 54.
[416] V.L.S. Teixeira da Silva, R. Frety, M. Schmal, Ind. Eng. Chem. Res. 33 (1994) 1692.
[417] D. Madeley, S.E. Wanke, Appl. Catal. 39 (1988) 295.
[418] M.M. Gellerman, R.R. Aliev, V.G. Sidelkovskaya, Chem. Technol. Fuel Oils 29 (1993) 32.
[419] E. Furimsky, Appl. Catal. 44 (1985) 189.
[420] R.G. Copperthwaite, G.J. Hutchings, P. Johnson, S.W. Orchard, J. Chem. Soc., Faraday Trans. 82 (1986) 1007.
[421] L. Carlton, R.G. Copperthwaite, G.J. Hutchings, E. Reynhardt, J. Chem. Soc. Chem. Commun. (1987) 1008.
[422] Y.I. Solovetskii, V.V. Lunin, P.V. Ryabchenko, React. Kinet. Catal. Lett. 55 (1995) 463.
[423] I. Barin, O. Knacke, Thermochemical Properties of Inorganic Substances, Springer-Verlag, Berlin (1973).
[424] E. Furimsky, Energy Fuels 22 (2008) 237.
[425] E. Furimsky, J. Houle, Y. Yoshimura, Appl. Catal. 33 (1987) 97.

[426] E. Furimsky, Fuel Process. Technol. 56 (1998) 263.
[427] D.E. Sherwood, Jr., J.R. Hardee, Jr., US Patent 5,445,728 Aug. 29, 1995.
[428] G.J. Hutchings, H. Comninos, R.G. Copperthwaite, L.J. van Rensburg, R. Hunter, T. Temistocleous, J. Chem. Soc., Faraday Trans. 85 (1989) 633.
[429] G.J. Hutchings, R. Hunter, L.J. van Rensburg, Appl. Catal. 41 (1988) 11.
[430] REACT Technology, ALBERMARLE, www.albermarle.com.
[431] J.M.-R. Ginestra, K.C. Lee, J.D. Seamans, Int. Appl. Nu., PCT/US2005/001825., Nov. 2005.
[432] S. Eijsbouts, F.W. Houtert, M.A. Jansen, T. Kamo, F.M. Platenga, Int. Appl. No. PCT/2000/006050, Jan. 2001.
[433] ReFRESH Topsoe's Proprietary Technology, Haldor Topsoe, www.topsoe.com.
[434] P. Mazoyer, C. Geantet, F. Diehl, S. Loridant, M. Lacroix, Catal. Today 130 (2008) 75.
[435] V. Costa, K. Marchand, M. Digne, C. Geantet, Catal. Today 130 (2008) 69.
[436] B. Scheffer, N.J.J. Dekker, P.J. Magnus, J.A. Moulijn, J. Catal. 121 (1990) 31.
[437] P. Marecot, S. Peyrovi, D. Bahloul, J. Barbier, Appl. Catal. 66 (1990) 181.
[438] P. Marecot, S. Peyrovi, D. Blahoul, J. Barbier, Appl. Catal. 66 (1990) 191.
[439] Y. Zhao, M.R. Gray, F. Wei, Catal. Lett. (2008).
[440] S. Eijsbouts, WO200176740, 2001.
[441] P. Dufresne, S. Eijsbouts, F. Labruyere, F.L. Plantenga, L.A. Gerrisen, WO20010176741, 2001.
[442] F. Valeri, F. Girardier, Y. Haquet, E. Aillard, Sub-Sahara Afr. Catal. Symp., 7–10 Feb. 1999, pap. 18.
[443] N. Frizi, P. Blanchard, E. Payen, P. Baranek, C. Lancelot, M. Rebeilleau, C. Dupuy, J.P. Dath, Catal. Today 130 (2008) 32.
[444] P. Mazoyer, C. Geantet, F. Diehl, C. Pichon, T.S. Nguyen, M. Lacroix, Oil Gas Sci. Technol. Rev. IFP 60 (2005) 791.
[445] A.I. Dugulan, E.J.M. Hensen, J.A.R. van Veen, Catal. Today 130 (2008) 126.
[446] C. Bruyninckx, D.J. Neuman, M. van Vliet, G.K. Semper, NPRA Ann. Meeting, 15-17 March, San Francisco, 1998.
[447] E. Gobina, Global Market for Catalyst Regeneration, BCC Research Report CHM0456.
[448] J. Wilson, AIChE Meeting, Aug. 19–22, San Diego, CA, 1990.
[449] L.N. Osipov, A.A. Gimbutas, A.N. Chagovets, N.Y. Vinogradova, G. Karbauskas, B.V. Bednov, Chem. Technol. Fuel Oils 31 (1995) 5.
[450] M. Tamayama, Kogaku Kagaku 50 (1986) 624.
[451] R.E. Ellingham, J. Garret, Appl. Ind. Catal. 3 (1984) 25.
[452] T. Chang, Oil Gas J. 79 (19 October 1989).
[453] G. Berebi, Amsterdam The Netherlands, Proc. Ketjen Catal. Symp. May 27–30 (1984).
[454] S.R. Blashka, Hydrocarbon Eng. (Jan. 1999) 21.
[455] D.J. Neuman, Energy Citations Database 9503162, 1995.
[456] J.M.R. Ginestra, J.D. Seamans, K.S. Lee, USPAP, 2005, 0159295.
[457] P.J. Shukis, J.D. Carruthers, V.J. Lostaglio, US Patent 6239054, 2001.
[458] B.B. Lejlveld, Albemarle Catal. Courier 68 (2007) 11.
[459] T. Remans, Proc. Europ. Catal. Technol. Conf., Feb. 2003, Prague.
[460] T. Remans, Proc. Europ. Catal. Technol. Conf., March 2001.
[461] S. Murff, Oil Gas J., Int. Catal. Conf. & Exhib. Feb. 1–2, Houston, TX, 1996.
[462] M. Marafi, A. Stanislaus, E. Kam, J. Environ. Manag. 86 (2008) 665.
[463] S.R. Blashka, J.G. Welch, K. Nite, A.P. Furfaro, Oil Gas J. (Sept. 18 1995) 62.
[464] G.R. Dyrkacz, L. Ruscic, L. Marshall, Energy Fuels 10 (1996) 849.
[465] J.D. Seamans, J.G. Welch, C.A. Vuitel, Prepr. Pap. - Am. Chem. Soc., Div. Fuel Chem. 33 (1988) 549.
[466] R.D. Erdman, Oil Gas J. (March 25 1985) 115.
[467] J.R. Moses, L.E. Auger, Energy Prog. (March 1983).
[468] C. Vuitel, NPRA Ann. Meeting, Oct. 8–10, 1997.
[469] C. Vuitel, Hydrocarbon Eng. (Dec. /Jan. 1998/1999).

[470] D.L. Holbrook, B.J. Arena, T. Verachtert, J.C. Bricker, NPRA Ann. Meet. San Antonio, TX, 1993, pap. AM-93-61.
[471] D.J. Neuman, NPRA Ann. Meeting, March 19–21, San Francisco, CA, 1995, pap. AM-95-41.
[472] Anon, Hydrocarbon Eng. 12 (2007) 62.
[473] D.J. Neuman, NPRA Ann. Meeting, March 15-17, San Francisco, CA, 1999, pap. AM-98-59.
[474] P. Zimmerman, NAPRA Ann. Meet., Arizona, 2006, p. 39.
[475] A. Stanislaus, M. Marafi, M. Absi-Halabi, Appl. Catal. 105 (1993) 195.
[476] A. Stanislaus, M. Marafi, M. Absi-Halabi, in: M.L. Ocelli, R.R. Chianelli (Eds.), Hydrotreating Technology for Pollution Control, Marcel Dekker (1996) 327.
[477] M. Marafi, A. Stanislaus, J. Mol. Catal. A 202 (2003) 417.
[478] M. Marafi, A. Stanislaus, J. Hazard. Mater. B 101 (2003) 123.
[479] H. Beuther, R.A. Flinn, Ind. Eng. Chem. Prod. Res. Dev. 2 (1963) 53.
[480] M. Marafi, A. Stanislaus, C.J. Mumford, M. Fahim, Appl. Catal. 47 (1989) 85.
[481] M. Marafi, A. Stanislaus, C.J. Mumford, F. Fahim, in: D.L. Trimm (Ed.), Catalysis in Petroleum Refining, Elsevier, Amsterdam (1989) 213.
[482] M. Marafi, The Regeneration of Deactivated Hydrotreating Catalysts, M.Sc. Thesis, Univ. of Aston, Birmingham, 1988.
[483] A. Stanislaus, M. Marafi, M. Absi-Halabi, Prepr. - Am. Chem. Soc., Div. Pet. Chem. 39 (1994) 627.
[484] A. Stanislaus, M. Marafi, M. Absi-Halabi, Prepr. - Am. Chem. Soc., Div. Pet. Chem. 38 (1993) 62.
[485] M. Marafi, A. Stanislaus, F. Jasem, Prepr. - Am. Chem. Soc., Div. Pet. Chem. 42 (1998) 486.
[486] G.B. McVicker, J.L. Carter, L.L. Murell, J.J.Remiak, US Patent, 4,522,928, June 11, 1985.
[487] D.R. Farell, J.W. Ward, US Patent 4,122,000, Oct. 24, 1978.
[488] M. Marafi, A. Stanislaus, M. Absi-Halabi, Appl. Catal. B Environ. 4 (1994) 19.
[489] M.R. Reda, Ind. Eng. Chem. Res. 30 (1991) 2148.
[490] M. Marafi, E.K.T. Kam, A. Stanislaus, M. Absi-Halabi, Appl. Catal. 147 (1996) 35.
[491] O. Levenspiel, Chemical Reaction Engineering, 2nd Edn., Wiley, New York (1972).
[492] R.C. Reid, J.M. Prasnitz, B.E. Poling, The Properties of Gases and Liquids, 4th Edn., McGraw-Hill, New York (1987).
[493] T.N. Angelidis, E. Tourasanidis, E. Marinou, G.A. Stalidis, Resour. Conserv. Recycl. 13 (1995) 269.
[494] D.E. Sherwood, Jr., J.R. Hardie Jr., US Patent 5.254.513, Oct. 19, 1993.
[495] D.E. Sherwood, Jr., US Patent 5,230,791, July 27, 1993.
[496] D.E. Sherwood, Jr., J.R. Hardie, Jr. US Patent 5,445,728, Aug. 29, 1995.506.
[497] J.E. Duddy, L.M. Abrams, S.J. Hildebrandt, US Patent 5,906,953, May 25, 1999.
[498] K.G. Tesker, J.D. Milligen, US Patent, 4,863,884, Sept. 5, 1989.
[499] W.R. Ernst, L.H. Hiltzik, A.R. Garcia, M.D. Franke, A.S. Mayerson, J.D. Carruthers, Miner. Metall. Process 4 (1987) 78.
[500] K.W. Babcock, L.H. Hiltzik, W.R. Ernstabnd, J.D. Carruthers, Appl. Catal. 47 (1989) 295.
[501] S.J. Hildebrandt, R.O. Koseoglu, J.F. Duddy, D.E. Sherwood, Prepr. - Am. Chem. Soc., Div. Pet. Chem. 38 (1993) 40.
[502] M.A. Olazabal, L.A. Fernandez, J.A. Madariaga, J. Met. Solvent Extr. Ion Exch. 9 (1991) 735.
[503] M.A. Olazabal, M.M. Orive, L.A. Fernandez, J.A. Madariaga, J. Met. Solvent Extr. Ion Exch. 10 (1992) 623.
[504] M.M. Orive, M.A. Olazabal, L.A. Fernandez, J.A. Madariaga, J. Met. Solvent Extr. Ion Exch. 10 (1992) 787.
[505] S. Gryglewicz, M. Rutkowski, Erdol and Kohle Erdgas 47 (1994) 334.
[506] N.K. Nadirov, S.A. Petrsov, Chem. Technol. Fuels 25 (1990) 424.
[507] M. Seapan, Z. Guohui, Am. Chem. Soc. Symp. Ser. 411 (1989) 9.
[508] J.S. MacPartland, R.G. Bautista, Sep. Sci. Technol. 25 (1990) 919.
[509] E. Furimsky, Erdol and Kohle Erdgas 32 (1979) 383.
[510] W.P. Hettinger Jr., Catal. Today 13 (1992) 157.
[511] D.J. Neuman, W. Roller, Hart's Fuel Technol & Manag. Special Report, May 1997, 39.
[512] Proc. 46th Ann. NPRA Symp., Oct. 1993, Dallas, TX, p. 110.

[513] J. Martin, Regeneration of Hydrotreating Catalysts, Ketjen Catal. Symp. (1981).
[514] D.E. Sherwood, Jr., J.R. Hardee, Jr., J.A. Lemen, US Patent 5,209,840, May 11, 1991.
[515] K.G. Mittal, S.C. Gupta, R.K. Agarwal, J.R. Rai, J. Chem. Technol. Biotechnol. 39 (1987) 59.
[516] C. Sudhakar, G.G. Sandford, A.K. Bhattacharya, US Patent 5,423,975, 1995.
[517] A.J. Suchanek, Oil Gas J. (March 25 1985) 115.
[518] J. Myers, Oil Gas J. (March 25 1985) 114.
[519] J.L. Rockwell, US Patent 7,033,488, April 25, 2007.
[520] R.A. Persak, Oil Gas J. (March 25 1985) 112.
[521] J.E. Duddy, L.M. Abrams, S.J. Hildebrand, US Patent 5,925,238, 1999.
[522] M. Marafi, H. Al-Sheeha, S. Al-Omani, A. Al-Barrod, Fuel Process. Technol. 90 (2009) 264.
[523] M. Marafi, A. Stanislaus, Appl. Catal. B 71 (2007) 199.
[524] M. Marafi, A. Stanislaus, Catal. Today 134 (2008) 421.
[525] D. Santhiya, Y.P. Ting, J. Biotechnol. 116 (2005) 171.
[526] D. Berman, PTQCatalysis 2003, www.eptq.com.
[527] E.L. Gardner, K.R. Dennis, US Patent 4,888,316, 1989.
[528] M. de Boer, US Patent 6,030,915, 2000.
[529] A. Stanislaus, K. Al-Dalama, M. Absi-Halabi, J. Mol. Catal. A 181 (2002) 33.
[530] D.L. Trimm, A. Stanislaus, Appl. Catal. 21 (1986) 215.
[531] F.M. Lee, R.D. Knudsen, D.R. Kidd, Ind Eng. Chem. Res. 31 (1992) 487.
[532] C.D. Cronauer, B.R. Bjorklund, US Patent 4,350,609, 1982.
[533] V.V. Lunin, I.I. Miroshnichenko, D.M. Panteleev, Y.I. Solovetskii, Kinet. Catal. 3 (1993) 646.
[534] V.V. Lunin, I.I. Miroshnichenko, Y.I. Solovetskii, Kinet. Catal. 34 (1993) 658.
[535] Y.I. Solovetskii, V.V. Lunin, I.I. Miroshnichenko, Kinet. Catal. 34 (1993) 655.
[536] V.V. Lunin, I.I. Miroshnichenko, S.A. Sergienko, Y.I. Solovetskii, G.D. Chukin, Kinet. Catal. 34 (1993) 658.
[537] F. Silva, G. Guitian, R. Galiasso, J. Krasuk, 11th World Petr. Congr., 1983, London, UK, vol. 4, p. 199.
[538] K. Niemann, F. Wenzel, Fuel Process. Technol. 35 (1993) 1.
[539] S. Zhang, D. Liu, W. Deng, G. Que, Energy Fuels 21 (2007) 3057.
[540] S. Beret, S.G. Yu, C.R. Hsie, US Patent 4,761,220, Aug. 1, 1988.
[541] T. Itoh and H. Sue 5th UNITAR/UNDP Int. Conf. on Heavy Crudes and Tar Sands, August 1991, Caracas, Venezuela, vol. 4, p. 43.
[542] A. Matsumura, T. Kondo, Y. Sasaki, K. Unegawa, Erdol and Kohle Erdgas 47 (1994) 280.
[543] B.C. Ku, Y.W. Kim, Y.T. Choi, US Patent 6,171,566, 2001.
[544] S.H. Lee, C.W. Shin, J. Seong, B.J. Kim, K.I. Choi, US Patent 6,800,585, 2004.
[545] K.H. Choi, N. Kunisada, Y. Korai, I. Mochida, K. Nakano, Catal. Today 86 (2003) 277.
[546] K.H. Choi, H. Sang, K.R. Shin, J. Seong, J.H. Kim, B.J. Kim, US Patent 6,673,740, 2004.
[547] Anon. Appl. Catal. 65 (1990) News Brief.
[548] Z.G. Zulfuganov, E.M. Dzhafarova, S.A. Dadasheva, S.A. Dzhafarova, Y.G. Nagdalieva, Azerb. Neft. Khoz. 10 (1989) 56.
[549] E. Furimsky, Appl. Catal. 156 (1997) 207.
[550] E. Furimsky, M. Yumura, Erdol and Kohle Erdgas 39 (1986) 163.
[551] E. Furimsky, M. Biagini, Fuel Process. Technol. 46 (1996) 17.
[552] N. Su, H.Y. Fang, Z.H. Chen, F.S. Liu, Cem. Concr. Res. 30 (2000) 1773.
[553] M. Bukowska, B. Pacewska, I. Wilinska, J. Therm. Anal. Calorim. 74 (2003) 931.
[554] M. Bukowska, B. Pacewska, I. Wilinska, Cem. Concr. Res. 34 (2004) 759.
[555] B. Pacewska, I. Wilinska, M. Bukowska, W. Nocuri Wczelik, Cem. Concr. Res. 32 (2002) 1823.
[556] Y.S. Tseng, C.L. Huang, K.C. Hsu, Cem. Concr. Res. 35 (2005) 782.
[557] N. Su, Z.H. Chen, H.Y. Fang, Cem. Concr. Compos. 23 (2001) 111.
[558] H.L. Chen, Y.S. Tseng, K.C. Hsu, Cem. Concr. Compos. 26 (2004) 657.
[559] U. Rattanasak, C. Jaturapitakkul, T. Sudapraset, Waste Manage. Res. 19 (2001) 56.
[560] S. Sanga, Y. Nishimura, US Patent 3,960,760, 1976.

[561] A.W. Liles, R.D. Schwartz, US Patent 3,968,036, 1976.
[562] A. Escardina, A. Barda, E. Sanchezand, V. Cantavella, Br. Ceram. Trans. 98 (1999) 172.
[563] E.I. Basaldella, J.C. Paladino, M. Solari, G.M. Valle, Appl. Catal. B 66 (2006) 186.
[564] H. Zeiringer, US Patent 4,142,871, 1979.
[565] P. Cichy, US Patent 4,349,381, 1982.
[566] Z.R. Lianos, W.G. Deering, Proc. TMS Ann. Meeting 3 (1995) 425.
[567] Z.R. Llanos, W.G. Derering, Proc. TMS Fall Extrac. & Proc. Conf. (2000) 759.
[568] A. Paschenko, Binding Mater. 13 (1975).
[569] V.M. Sokolov, L.G. Litvin, V.V. Martynenko, A.V. Suvorin, A.N. Mekhed, Waste Treat. Clean Technol. (2005) 381.
[570] S. Al-Fulaij, G.R. Gouda, A. Stanislaus, S. Dieb, H. Dashti, H. Ibrahim, Spent catalyst treatment and utilization, PET 39 Final Report, KISR 1614, 1985, Kuwait Inst.for Scientific Research, Kuwait.
[571] W. Acchar, B.M. Rulf, A.M. Segadaes, Appl. Clay Sci. (2008).
[572] T.R. Farrel, J.U.A. Zakarian, Oil Gas J. (May 19 1986) 49.
[573] R. Bertram, F. Danzinger, NPRA Ann. Meeting, San Antonio, TX, March 1994, M-94-50.
[574] J. Weitkamp, Erdol and Kohle Erdgas, Petrochem. 31 (1978) 13.
[575] H.F. Schultz, J. Weitkamp, Ind. Eng. Chem. Prod. Res. Dev. 11 (1972) 46.
[576] H. Pines, Chemistry of Catalytic Hydrocarbon Conversion, Academic Press, New York (1981).
[577] M.J. Girgis, Y.P. Tsao, Ind. Eng. Chem. Res. 35 (1996) 386.
[578] C.L. O'Young, W.Q. Xu, M. Simon, S.L. Suib, Stud. Surf. Sci. Catal. 84 (1994) 1.
[579] J.G. Bendoraitis, A.W. Chester, F.G. Dwyer, W.E. Garwood, Tokyo, Proc. 7th Int. Zeol. Conf (1986) 669.
[580] N.Y. Chen, W.E. Garwood, Ind. Eng. Chem. Res. 25 (1986) 641.
[581] N.Y. Chen, W.E. Garwood, F.G. Dwyer, Shape Selective Catalysis in Industrial Applications, Marcel Dekker, New York (1989).
[582] R.D. Gillespie, US Patent 6,818,589, Nov. 2004.
[583] R.D. Gillespie, US Patent 7,041,866, May 2006.
[584] R.D. Gillespie, US Patent 6,977,322, Dec. 2005.
[585] J.G. Vassilakis, R.R. Rosin, S.A. Bradley, R.D. Gillespie, M.J. Cohn, F. Xu, US Patent 7,015,175, May 2006.
[586] A. de Klerk, E. Furimsky, Upgrading Primary Products from Fischer-Tropsch Synthesis.
[587] M. Cowley, Energy Fuels 20 (2006) 1771.
[588] J.M. Campelo, F. Lafont, J.M. Marinas, Appl. Catal. 170 (1998) 139.
[589] R.T. Jacobsen, Chem. Eng. Prog. 4 (2005) 41.
[590] R.T. Jacobsen, Chem. Eng. Prog. 3 (2005) 20.
[591] R.T. Jacobsen, Chem. Eng. Prog. 2 (2005) 22.
[592] S.C. Fung, Prepr. - Am. Chem. Soc., Div. Pet. Chem. 38 (1993).
[593] J.N. Beltraminin, in: G.J. Antos, A.M. Aitani, J.M. Pertera (Eds.), Catalytic Naphtha Reforming, Marcel Dekker, New York (1995).
[594] N. Pherson, Oil Gas J. (Feb. 22 1988) 55.
[595] S.R. Blashka, J.G. Welch, K. Nite, A.P. Furfaro, Oil Gas J. (18 Sept. 1995) 62.
[596] J.P. Rosso, Chem. Eng. Prog. 12 (1992) 66.
[597] J.P. Rosso, M.I. El Guindy, in: G.J. Antos, A.M. Aitani, J.M. Parera (Eds.), Catalytic Naphtha Reforming, Marcel Dekker, New York (1995) 395.
[598] P. Scholtz, Chem. Technol. 40 (1988) 451.
[599] R.K. Mishra, New Port Rhode Island, Proc 17th Precious Metals Conf. (1993) 449.
[600] D.R. Tyson, R.G. Bautista, Sep. Sci. Technol. 22 (1987) 1149.
[601] L. Mastny, M. Bumbova, E. Kalalova, O. Svajgl, V. Prazak, Chem. Prum. 36 (1986) 243.
[602] P.L. Sibrell, G.B. Atkinson, US Bur. Mines, Rep. PB- 95-18172/XAB, 1995.
[603] F.T. Awadalla, R.E.Molnar, G.M. Ritcey, US Patent 5,304,233, 1994.
[604] G.B.J. de Boer, M. Hausselaar, US Patent 5,302,183, 1994.
[605] N. Ezawa, H. Inoue, S. Takada, H. Masuda, US Patent 5,252,305, 1993.

[606] G.Q. Li, S. Konishi, K. Saito, T. Sugo, J. Membr. Sci. 95 (1994) 63.
[607] R.T. Jacobsen. Proc. 1993 AIChE Spring Nat. Mtg., Houston, TX, Pep. 69c, p. 12.
[608] D.T. Liang, 2nd Asian Petrol. Technol. Symp. Program, 2006, 1/11.
[609] G. Gutnikov, U.S Patent 3567433, 1971.
[610] W. Millsap, N. Reisler, Eng. Min. J. 179 (1978) 107.
[611] G.L. Hubred, G.L., 1984a.. U.S Patent 4514368.
[612] G.L. Hubred, D.A. Van Leirsburg. U.S Patent 4514369, 1985.
[613] J.P. Marcantonio, U.S Patent 5066469, 1991.
[614] L.G. Hubred, US Patent 4514368, 1985.
[615] J.B. Lopez, A.P. Eugene, V.C. Anthony, U.S Patent 4762812, 1988.
[616] J.P. Marcantonio; US Patent 7,485,267, Feb. 3, 2009.
[617] S.M. Villarreal, B.L. Kharisov, L.M. Torres Martinez, V.N. Elizondo, Ind. Eng. Chem. Res. 38 (1999) 4624.
[618] J.M. Yoo, J.C. Lee, B.S. Kim, H.S. Lee, J.K. Jeong, J. Chem. Eng. Jpn. 37 (2004) 1129.
[619] I.M. Valverde Jr., J.F. Paulino, J.C. Afonso, J. Hazard. Mater. 160 (2008) 310.
[620] D.E. Hyatt, U.S Patent 4657745, 1987.
[621] K. Toyabe, K Kirishima, H. Shibayama, H. Hanawa, U.SPatent 5431892 1995.
[622] K. Inoue, P. Zhang, Prepr. - Am. Chem. Soc., Div. Pet. Chem. 38 (1993) 77.
[623] K. Inoue, Z. Pingwei, H. Tsuyama, 3rd Int. Symp. Recycl. Metals & Eng. Materials. (1995) 339.
[624] P. Zhang, K. Inoue, Energy Fuels 9 (1995) 231.
[625] G. Parkinson, Chem. Eng. (Feb. 16, 1987) 25.
[626] R.E. Siemens, B.W. Jong, D.W. Matson, Ind. Eng. Chem. Res. 9 (1986) 189.
[627] J.K. Rastas, K.J. Karpale, H. Tiitimen, US Patent 4,474,735, 1984.
[628] M.A. Rabah, I.F. Hewaidy, F.E. Farghaly, Powder Metall. 40 (1997) 283.
[629] Y.C. Lai, W.J. Lee, K.L. Huang, C.M. Wu, J. Hazard. Mater. 154 (2008) 588.
[630] P.R. Raisoni, S.G. Dixit, Miner. Eng. 1 (1988) 225.
[631] M. Marafi, E. Furimsky, Erdol and Erdgas Kohle 121 (2005) 93.
[632] W. Mulak, A. Szymczycha, A. Lesniewicz, W. Zyrnicki, Physicochem. Problems Miner. Process. 40 (2006) 69.
[633] N. Rokukawa, U.S Patent 4382068, 1983.
[634] K.H. Park, D. Mohapatra, B.R. Reddy, C.W. Nam, Int. J. Miner. Process. 80 (2006) 261.
[635] K.H. Park, D. Mohapatra, B.R. Reddy, J. Hazard. Mater. B 138 (2006) 311.
[636] B.B. Jong, S.C. Rhoads, A.M. Stubbs, T.R. Stoelting, U.S. Bureau of Mines, Report No. 9252, 1989.
[637] T.N. Angelids, E. Tourasanidis, E. Marinaou, G.A. Stalidis, Resour. Conserv. Recycl. 13 (1995) 269.
[638] D.D. Sun, L. Chang, J.H. Tay, J.D. Navratil, C. Easton, J. Environ. Eng. 127 (2001) 916.
[639] J.T. Veal, K.A. Anderson, R.M. Kowaleski, U.S Patent 6180072, 2001.
[640] H. Brandl, Microbial leaching of metals, in: H.J. Rehm, G. Reed (Eds.), Biotechnology, vol. 10, Wiley-VHC, Weinheim (2001) 191–224.
[641] C. Bombarcher, R. Bachofen, H. Brandl, Appl. Environ. Microbiol. 64 (1998) 1237.
[642] K.M.M. Aung, Y.P. Ting, J. Biotechnol. 121 (2006) 62.
[643] B. Blaustein, D. Suhy, E. Spana, Prepr. Pap. - Am Chem. Soc., Div. Fuel Chem. 35 (1990) 1613.
[644] B. Blaustein, J.T. Hauck, G.J. Olson, J.P. Baltrus, Fuel 72 (1993) 1612.
[645] P. Joffe, G.T. Sperll, DOE Rep., PC-92119-T4 1993.
[646] K.A. Sandback, P.M. Joffe, DOE Rep. 1993, PC-92119- T2 & T3.
[647] K.A. Sandback, DOE Rep. 1995, PC/92119-T6.
[648] L. Briand, H. Thomas, A. De la Vega Alonso, E. Donati, Proc. Int. Biohydrometallurgy Symp., Part A, Elsevier, Amsterdam (1999) 263.
[649] H. Strasser, W. Burgstaller, F. Schinner, FEMS Microbiol. Lett. 119 (1994) 365.
[650] D. Santhiya, Y.P. Ting, J. Biotechnol. 121 (2006) 62.
[651] D.J. Kim, D. Mishra, K.H. Park, J.G. Ahn, D.E. Ralph, A.J. Parker, Mater. Trans. 49 (2008) 2383.
[652] D. Mishra, D.J. Kim, D.E. Ralph, J.G. Ann, Y.H. Rhee, Hydrometallurgy 88 (2007) 202.

[653] D. Mishra, D.J. Kim, D.E. Ralph, J.G. Ahn, Y.H. Rhee, J. Hazard. Mater. 152 (2008) 1082.
[654] R.F. Sebenik, R.A. S Ference, Kansas, 12–17 Sept, Symp. Recov. Spent Catal. ACS (1982).
[655] B. Wang, Y. Meng, C.S. Duan, Mod. Chem. Industry 25 (2005) 204.
[656] Y. Chen, Q. Feng, Y. Shao, G. Zhang, L. Qu, Y. Lu, Miner. Eng. 19 (2006) 94.
[657] B.B. Karr, P. Datta, V.N. Misra, Hydrometallurgy 72 (2004) 87.
[658] B.B. Karr, B.V.R. Murthy, V.N. Misra, Int. J. Miner. Process. 76 (2005) 143.
[659] Y. Chen, Q. Feng, Y. Shao, G. Zhang, L. Qu, Y. Lu, Int. J. Miner. Process. 79 (2006) 42.
[660] Z.R. Llanos, F.G. Provoost, G.W. Deering, J.F. Debaene, J.F., U.S Patent 5702500, 1997.
[661] Z.R. Llanos, G.W. Deering, Air and Waste Manag. Assoc. 90th Annual Meeting, Toronto, Canada, 1997, June 8-13.
[662] R.G. Busnardo, N.G. Busnardo, G.N. Salvato, J.C. Afonso, J. Hazard. Mater. B 139 (2007) 391.
[663] G. Gravey, J. Le Goff, C. Gonin, US Patent 4182747, 1980.
[664] T. Yoshida, M. Ushiyama, T. Yokohama, US Patent 4216118, 1980.
[665] J.Y. Welsh, P.C. Picquet, P.D. Schyns, US Patent 4,292,282, 1981.
[666] I. Gaballah, M. Djona, J.C. Mugica, R. Solozobal, Resour. Conserv. Recycl. 10 (1994) 87.
[667] I. Gaballah, M. Djona, Metall. Mater. Trans. B 26 (1995) 41.
[668] M.W. Ojeda, M. Ruiz, J.B. Rivarola, TMS Fall Extract. Process. Conf. 1 (1999) 113.
[669] M.W. Ojeda, J.B. Rivarola, O.D. Quiroga, Miner. Eng. 15 (2002) 585.
[670] M.W. Ojeda, J.B. Rivarola, O.D. Quiroga, Latin American Appl. Res. 33 (2003) 195.
[671] M.W. Ojeda, J.B. Rivarola, O.D. Quiroga, Recovery of molybdenum from a hydrodesulfurization exhausted catalyst: Modeling of the carbochlorination reaction. 2005.
[672] B.B. Karr, Int. J. Miner. Process. 75 (2005) 249.
[673] L.J. Silva, L.A. Bray, D.W. Matson, Ind. Eng. Chem. Res. 32 (1993) 2485.
[674] L.J. Silva, L.A. Bray, US Patent, 5,420,088, 1995.
[675] Y.C. Lai, W.J.L. Lee, K.L. Huang, C.M. Wu, J. Hazard. Mater. (2007).
[676] F.F. Wong, K.C. Chiang, C.Y. Chen, K.L. Chen, J.J. Huang, M.Y. Yeh, Plasma Chem. Plasma Process. 28 (2008) 353.
[677] P. Zhang, K. Inoui, H. Tatsumi, Energy Fuels 9 (1995) 231.
[678] Amax Inc. Jap. Patent (Kokai) 53-86612, 1977.
[679] G.M. Ritcey, Handbook of Solvent Extraction, Wiley Interscience, New York (1981).
[680] Y. Sato, K. Kondo, F. Nakashio, J. Chem. Eng. Jpn. 22 (1989) 200.
[681] T. Hirai, T. Hashimoto, I. Tsuboi, A. Hino, I. Komasawa, J. Chem. Eng. Jpn. 28 (1995) 85.
[682] T. Hirai, N. Onoe, I. Komasawa, J. Chem. Eng. Jpn. 26 (1993) 416.
[683] R.K. Biswas, M. Wakihara, M. Taniguchi, Hydrometallurgy 14 (1985) 219.
[684] I. Komasawa, H. Hosoba, N. Kurokowa, T. Otake, J. Chem. Eng. Jpn. 20 (1987) 176.
[685] I. Komasawa, H. Hosoba, N. Kurokawa, T. Otake, J. Chem. Eng. Jpn. 20 (1987) 183.
[686] M. Nishizawa, T. Yokoyama, T. Kimuta, T.M. Suzuki, Bull. Chem. Soc. Jpn. 57 (1994) 2859.
[687] G.Q. Li, S. Konishi, K. Saito, T. Sugo, J. Membr. Sci. 95 (1984) 63.
[688] E. Wiewiorowski, R. Tinnin, R.A. Cronojevich, SME Annual Meeting, Phoenix, January 25, 1988.
[689] A. Case, G. Garretson, E. Wiewiorowski, 3rd Int. Symp. Recycl. Metals & Eng. Mater. Nov. 12–15 (1995) 449.
[690] R. Crnojevich, E. Wiewiorowski, L.R. Tinnin, A.B. Case, Int. Symp. Recycl. of Metals & Eng. Materials (October 28-31 1990).
[691] P. Dufresne, Y. Jacquier, F. Valeri, Manama Bahrain, 1st Specialty Conf. on Environ. Issues in Petrol. and Petrochem. Indust. December 4-6, 1995 (1995).
[692] S. Toda, J. Jpn. Soc. Min. 105 (1989) 265.
[693] MOXBA-Metrex. The catalyst in waste recycling. 2007. http://www.moxba.com/images1/moxba/documenten/ProcessChart(scheme).doc.
[694] M.V. Wang, Recycling of metals and Engineered Materials, R. Stewart, R. Stephens, J.C. Daley (Eds.), TMS, 2000 (The Minerals, Metals & Materials Society).

[695] Metallurg Vanadium. Company Profile. 2007. http://www.metallurgvanadium.com/spentcatalysts2.html.
[696] GfE Metalle und Materialien GmbH. 2007. Vanadium Sourcing in terms of Environmental Technologies and Services. Company Brochure. www.gfe.com.
[697] SCR. Spent Catalyst Recycling GMBH. Company Brochure, 2007, www.spentcat.net/services.htm.
[698] AURAMET. AURA metallurgical Gmbh. 2007 Company Profile. www.auramet.de/sp_process.asp?1g=english.
[699] Euromet. 2006. Catalyst Recycling. Company Brochure. http://www.euromet.co.uk/page_view/catalyst_recycling/2/hydrotreating_catalyst/11/en 6.
[700] T.A. Millensifer, Oil Gas J. (1977) 82.
[701] R.P. Silvy, Appl. Catal. 261 (2004) 247–252.
[702] J.W. Blossom, 1998 Molybdenum Recycling in US. US Geological Survey, L1. 2004.
[703] K.B. Shedd, US Geological Survey, Open-File Report 2005, 1028.
[704] K.B. Shedd, Cobalt Recycling in United States in 1998. US Geological Survey. 2004.
[705] J.H. Roderick, J. Smith, F. Fin, Metal. Bull. (Nov. 2006) 1.
[706] M.J. Magyar, Mineral Industry Survey, US Geological Survey, August 2007.

Index

S

235.00